De Gruyter Graduate

Höpfner · Asymptotic Statistics

Reinhard Höpfner

Asymptotic Statistics

With a View to Stochastic Processes

De Gruyter

Mathematics Subject Classification 2010: 62F12, 62M05, 60J60, 60F17, 60J25

Author
Prof. Dr. Reinhard Höpfner
Johannes Gutenberg University Mainz
Faculty 8: Physics, Mathematics and Computer Science
Institute of Mathematics
Staudingerweg 9
55099 Mainz
Germany
hoepfner@mathematik.uni-mainz.de

ISBN 978-3-11-025024-4
e-ISBN 978-3-11-025028-2

Library of Congress Cataloging-in-Publication Data
A CIP catalog record for this book has been applied for at the Library of Congress.

Bibliographic information published by the Deutsche Nationalbibliothek
The Deutsche Nationalbibliothek lists this publication in the Deutsche Nationalbibliografie; detailed bibliographic data are available in the Internet at http://dnb.dnb.de.

© 2014 Walter de Gruyter GmbH, Berlin/Boston

Typesetting: PTP-Berlin Protago-T$_E$X-Production GmbH, www.ptp-berlin.de
Printing and binding: CPI buch bücher.de GmbH, Birkach
♾ Printed on acid-free paper

Printed in Germany

www.degruyter.com

To the students who like both statistics and stochastic processes

Preface

At Freiburg in 1991 I gave my very first lectures on local asymptotic normality; the first time I explained likelihood ratio processes or minimum distance estimators to students in a course was 1995 at Bonn, during a visiting professorship. Then, during a short time at Paderborn and now almost 15 years at Mainz, I lectured from time to time on related topics, or on parts of these. In cooperation with colleagues in the field of statistics of stochastic processes (where most of the time I was learning) and in discussions with good students (who exposed me to very precise questions on the underlying mathematical notions), the scope of subjects entering the topic increases steadily; initially, my personal preference was on null recurrent Markov processes and on local asymptotic mixed normality. Later, lecturing on such topics, a first tentative version of a script came into life at some point. It went on augmenting and completing itself in successive steps of approximation. It is now my hope that the combination of topics may serve students and interested readers to get acquainted with the purely statistical theory on one hand, developed carefully in a 'probabilistic' style, and on the other hand with some (there are others) typical applications to statistics of stochastic processes.

The present book can be read in different ways, according to possibly different mathematical preferences of a reader. In the author's view, the core of the book are the Chapters 5 (Gaussian shift models), 6 (mixed normal and quadratic models), 7 (local asymptotics where the limit model is a Gaussian shift or a mixed normal or a quadratic experiment, often abbreviated as LAN, LAMN or LAQ), and finally 8 (examples of statistical models in a context of diffusion processes where local asymptotics of type LAN, LAMN or LAQ appear).

A reader who wants to concentrate on the statistical theory alone should skip chapters or subsections marked by an asterisk * : he or she would read only the Sections 5.1 and 6.1, and then all subsections of Chapter 7. This route includes a number of examples formulated in the classical i.i.d. framework, and allows to follow the statistical theory without gaps.

In contrast, chapters or subsections marked by an asterisk * are designed for readers with an interest in both statistics and stochastic processes. This reader is assumed to be acquainted with basic knowledge on continuous-time martingales, semi-martingales, Ito formula and Girsanov theorem, and may go through the entire Chapters 5 to 8 consecutively. In view of the stochastic process examples in Chapter 8, he or she may consult from time to time the Appendix Section 9 for further background and for references (on subjects such as Harris recurrence, positive or null, convergence of martingales, and convergence of additive functionals of a Harris process).

In both cases, a reader may previously have consulted or have read the Sections 1.1 and 1.2 as well as Chapters 3 and 4 for statistical notions such as score and information in classical definition, contiguity or L^2-differentiability, to be prepared for the core of the book. Given Sections 1.1 and 1.2, Chapters 3 and 4 can be read independently of each other. Only few basic notions of classical mathematical statistics (such as sufficiency, Rao–Blackwell theorem, exponential families, ...) are assumed to be known.

Sections 1.3 and 1.4 are of complementary character and may be skipped; they discuss naive belief in maximum likelihood and provide some background in order to appreciate the theorems of Chapter 7.

Chapter 2 stands isolated and can be read separately from all other chapters. In i.i.d. framework we study in detail one particular class of estimators for the unknown parameter which 'works reasonably well' in a large variety of statistical problems under weak assumptions. From a theoretical point of view, this allows to explicitly construct estimator sequences which converge at a certain rate. From a practical point of view, we find it interesting to start – prior to all optimality considerations in later chapters – with estimators that tolerate small deviations from theoretical model assumptions.

Fruitful exchanges and cooperations over a long period of time have contributed to the scope of topics treated in this book, and I would like to thank my colleagues, coauthors and friends for those many long and stimulating discussions around successive projects related to our joint papers. Their influence, well visible in the relevant parts of the book, is acknowledged with deep gratitude. In a similar way, I would like to thank my coauthors and partners up to now in other (formal or quite informal) cooperations. There are some teachers and colleagues in probability and statistics to whom I owe much, either for encouragement and help at decisive moments of my mathematical life, or for mathematical discussions on specific topics, and I would like to take this opportunity to express my gratitude. Furthermore, I have to thank those who from the beginning allowed me to learn and to start to do mathematics, and – beyond mathematics, sharing everyday life – my family.

Concerning a more recent time period, I would like to thank my colleague Eva Löcherbach, my PhD student Michael Diether as well as Tobias Berg and Simon Holbach: they agreed to read longer or shorter parts of this text in close-to-final versions and made critical and helpful comments; remaining errors are my own.

Mainz, June 2013   Reinhard Höpfner

Contents

Preface		vii
1	**Score and Information**	**1**
	1.1 Score, Information, Information Bounds	2
	1.2 Estimator Sequences, Asymptotics of Information Bounds	15
	1.3 Heuristics on Maximum Likelihood Estimator Sequences	23
	1.4 Consistency of ML Estimators via Hellinger Distances	30
2	**Minimum Distance Estimators**	**42**
	2.1 Stochastic Processes with Paths in $L^p(T,\mathcal{T},\mu)$	43
	2.2 Minimum Distance Estimator Sequences	55
	2.3 Some Comments on Gaussian Processes	68
	2.4 Asymptotic Normality for Minimum Distance Estimator Sequences ..	75
3	**Contiguity**	**85**
	3.1 Le Cam's First and Third Lemma	86
	3.2 Proofs for Section 3.1 and some Variants	92
4	**L^2-differentiable Statistical Models**	**108**
	4.1 L^r-differentiable Statistical Models	109
	4.2 Le Cam's Second Lemma for i.i.d. Observations	119
5	**Gaussian Shift Models**	**127**
	5.1 Gaussian Shift Experiments	127
	5.2 *Brownian Motion with Unknown Drift as a Gaussian Shift Experiment	141
6	**Quadratic Experiments and Mixed Normal Experiments**	**148**
	6.1 Quadratic and Mixed Normal Experiments	148
	6.2 *Likelihood Ratio Processes in Diffusion Models	160
	6.3 *Time Changes for Brownian Motion with Unknown Drift	168

7	**Local Asymptotics of Type LAN, LAMN, LAQ**	**178**
	7.1 Local Asymptotics of Type LAN, LAMN, LAQ	179
	7.2 Asymptotic optimality of estimators in the LAN or LAMN setting	191
	7.3 Le Cam's One-step Modification of Estimators	200
	7.4 The Case of i.i.d. Observations	206
8	***Some Stochastic Process Examples for Local Asymptotics of Type LAN, LAMN and LAQ**	**212**
	8.1 *Ornstein–Uhlenbeck Process with Unknown Parameter Observed over a Long Time Interval	213
	8.2 *A Null Recurrent Diffusion Model	227
	8.3 *Some Further Remarks	240
***Appendix**		**243**
	9.1 *Convergence of Martingales	244
	9.2 *Harris Recurrent Markov Processes	247
	9.3 *Checking the Harris Condition	253
	9.4 *One-dimensional Diffusions	258
Bibliography		267
Index		275

Chapter 1

Score and Information

Topics for Chapter 1:

1.1 Score, Information, Information Bounds
Statistical models admitting score and information 1.1–1.2'
Example: one-parametric paths in nonparametric models 1.3
Example: location models 1.4
Score and information in product models 1.5
Cramér–Rao bound 1.6–1.7'
Van Trees bound 1.8

1.2 Estimator Sequences, Asymptotics of Information Bounds
Sequences of experiments, sequences of estimators, consistency 1.9
Asymptotic Cramér–Rao bound in i.i.d. models 1.10
Asymptotic van Trees bound in i.i.d. models 1.11
Example: an asymptotic minimax property of the empirical distribution function 1.11'

1.3 Heuristics on Maximum Likelihood Estimator Sequences
Heuristics I: certain regularity conditions 1.12
Heuristics II: maximum likelihood (ML) estimators 1.13
Heuristics III: asymptotics of ML sequences 1.14
Example: a normal distribution model 1.15
A normal distribution model by Neyman and Scott 1.16
Example: likelihoods which are not differentiable in the parameter 1.16'

1.4 Consistency of ML Estimators via Hellinger Distances
Definition of ML estimators and ML sequences 1.17
An example using Kullback divergence 1.17"
Hellinger distance 1.18
Hellinger distances in i.i.d. models: a set of conditions 1.19
Some lemmata on likelihood ratios 1.20–1.23
Main result: consistency via Hellinger distance 1.24
Example: location model generated from the uniform law $\mathcal{R}(-\frac{1}{2}, \frac{1}{2})$ 1.25

Exercises: 1.4', 1.4", 1.8', 1.8", 1.17', 1.24"

The chapter starts with classically defined notions of 'score', 'information' and 'information bounds' in smoothly parameterised statistical models. These are studied in Sections 1.1 and 1.2; the main results of this part are the asymptotic van Trees bounds for i.i.d. models in Theorem 1.11. Here we encounter in a restricted setting the type of information bounds which will play a key role in later chapters, for more general sequences of statistical models, under weaker assumptions on smoothness of parameterisation, and for a broader family of risks. Section 1.3 then discusses 'classical heuristics' which link limit distributions of maximum likelihood estimator sequences to the notion of 'information', together with examples of statistical models where such heuristics either work or do not work at all; we include this – completely informal – discussion since later sections will show how similar aims can be attained in a rigorous way based on essentially different mathematical techniques. Finally, a different route to consistency of ML estimator sequences in i.i.d. models is presented in Section 1.4, with the main result in Theorem 1.24, based on conditions on Hellinger distances in the statistical model.

1.1 Score, Information, Information Bounds

1.1 Definition. (a) A *statistical model* (or *statistical experiment*) is a triplet

$$(\Omega, \mathcal{A}, \mathcal{P})$$

where $\mathcal{P}$ is a specified family of probability measures on a measurable space $(\Omega, \mathcal{A})$. A statistical model is termed *parametric* if there is a 1-1-mapping between $\mathcal{P}$ and some subset of some $\mathbb{R}^d$, $d \geq 1$,

$$\mathcal{P} = \{P_\vartheta : \vartheta \in \Theta\}, \quad \Theta \subset \mathbb{R}^d,$$

and *dominated* if there is some σ-finite measure μ on $(\Omega, \mathcal{A})$ such that

$$P \ll \mu \quad \text{for every } P \in \mathcal{P}.$$

(b) A *statistic on* $(\Omega, \mathcal{A}, \mathcal{P})$ is a measurable mapping from $(\Omega, \mathcal{A})$ to some measurable space $(G, \mathcal{G})$. A statistic T taking values in $(\mathbb{R}^k, \mathcal{B}(\mathbb{R}^k))$, with components $T_1, \ldots, T_k$, is termed *q-integrable* ($q \geq 1$) if

$$T \in L^q(P) \quad \text{(i.e. } T_j \in L^q(P) \text{ for every } 1 \leq j \leq k) \quad \text{for every } P \in \mathcal{P}.$$

(c) In a parametric experiment $(\Omega, \mathcal{A}, \{P_\vartheta : \vartheta \in \Theta\})$ with parameter space $\Theta \subset \mathbb{R}^d$, an *estimator for the unknown parameter* is any statistic T taking values in $(\mathbb{R}^d, \mathcal{B}(\mathbb{R}^d))$. T is termed *unbiased* if

$$T \in L^1(P) \quad \text{and} \quad E_\vartheta(T) = \vartheta \quad \text{for every } \vartheta \in \Theta.$$

Section 1.1 Score, Information, Information Bounds   3

(d) Consider an arbitrary experiment $(\Omega, \mathcal{A}, \mathcal{P})$ and a mapping $\gamma : \mathcal{P} \to \mathbb{R}^k$. An *estimator for γ* is any statistic T taking values in $(\mathbb{R}^k, \mathcal{B}(\mathbb{R}^k))$. T is termed *unbiased for γ* if
$$T \in L^1(P) \quad \text{and} \quad E_P(T) = \gamma(P) \quad \text{for every } P \in \mathcal{P}.$$

1.2 Definition (Score and information, classical definition). Consider a parametric experiment
$$\mathcal{E} := (\Omega, \mathcal{A}, \{P_\vartheta : \vartheta \in \Theta\}), \quad \Theta \subset \mathbb{R}^d \text{ open}$$
with dominating measure μ and with densities
$$\frac{d P_\vartheta}{d \mu}(\omega) := f(\vartheta, \omega) = f_\vartheta(\omega), \quad \vartheta \in \Theta, \, \omega \in \Omega.$$

Assume that for every $\omega \in \Omega$ fixed, $f(\cdot, \omega)$ is continuous, and let partial derivatives exist on Θ: as pointwise limits of measurable functions $\frac{f(\vartheta+he_i,\cdot)-f(\vartheta,\cdot)}{h}$, $h \to 0$, these are measurable functions.

(a) Let ∇ denote the vector of partial derivatives with respect to ϑ and define
$$M_\vartheta := (\nabla \log f)(\vartheta, \omega)$$
$$:= 1_{\{f_\vartheta > 0\}}(\omega) \begin{pmatrix} \frac{\partial}{\partial \vartheta_1} \log f \\ \cdots \\ \frac{\partial}{\partial \vartheta_d} \log f \end{pmatrix} (\vartheta, \omega), \quad \vartheta \in \Theta, \, \omega \in \Omega.$$

This yields a well-defined random variable on $(\Omega, \mathcal{A})$ taking values in $(\mathbb{R}^d, \mathcal{B}(\mathbb{R}^d))$. Assume further

(∗)   for every $\vartheta \in \Theta$, we have $M_\vartheta \in L^2(P_\vartheta)$ and $E_\vartheta(M_\vartheta) = 0$.

If all these conditions are satisfied, M_ϑ is termed *score in ϑ*, and the covariance matrix under P_ϑ
$$I_\vartheta := Cov_\vartheta(M_\vartheta) = E_\vartheta\left(M_\vartheta M_\vartheta^\top\right)$$
Fisher information in ϑ. We then call $\mathcal{E}$ an *experiment admitting score and Fisher information*.

(b) More generally, we may allow for modification of M_ϑ defined in (a) on sets of P_ϑ-measure zero, and call any family of measurable mappings $\{\widetilde{M}_\vartheta : \vartheta \in \Theta\}$ from $(\Omega, \mathcal{A})$ to $(\mathbb{R}^d, \mathcal{B}(\mathbb{R}^d))$ with the property

for every ϑ in Θ: $\quad \widetilde{M}_\vartheta = M_\vartheta \quad P_\vartheta$-almost surely

a version of the score in the statistical model $\{P_\vartheta : \vartheta \in \Theta\}$.

Even if many classically studied parametric statistical models do admit score and Fisher information, it is indeed an assumption that densities $\vartheta \to f(\vartheta, \omega)$ should satisfy the smoothness conditions in Definition 1.2. Densities $\vartheta \to f(\vartheta, \omega)$ can be continuous and non-differentiable, their smoothness being e.g. the smoothness of the Brownian path; densities $\vartheta \to f(\vartheta, \omega)$ can be discontinuous, their jumps corresponding e.g. to the jumps of a Poisson process. For examples, see (1) and (3) in [17] (which goes back to [60]). We will start from the classical setting.

1.2' Remark. For a statistical model, score and information – if they exist – depend essentially on the choice of the parameterisation $\{P_\vartheta : \vartheta \in \Theta\} = \mathcal{P}$ for the model, but not on the choice of a dominating measure: in a dominated experiment, for different measures μ_1 and μ_2 dominating $\mathcal{P}$, we have with respect to $\mu_1 + \mu_2$

$$\frac{d P_\vartheta}{d \mu_1} \cdot \frac{d \mu_1}{d(\mu_1 + \mu_2)} = \frac{d P_\vartheta}{d(\mu_1 + \mu_2)} = \frac{d P_\vartheta}{d \mu_2} \cdot \frac{d \mu_2}{d(\mu_1 + \mu_2)}$$

where the second factor on the r.h.s. and the second factor on the l.h.s. do not involve ϑ, hence do not contribute to the score $(\nabla \log f)(\vartheta, \cdot)$.

1.3 Example. ($d = 1$, one-parametric paths in nonparametric models) Fix any probability measure F on $(\mathbb{R}, \mathcal{B}(\mathbb{R}))$, fix any function $h : (\mathbb{R}, \mathcal{B}(\mathbb{R})) \to (\mathbb{R}, \mathcal{B}(\mathbb{R}))$ with the properties

$$(\diamond) \qquad \int h \, dF = 0, \quad 0 < \int h^2 \, dF < \infty,$$

and write $\mathcal{F}$ for the class of all probability measures on $(\mathbb{R}, \mathcal{B}(\mathbb{R}))$. We shall show that for suitable $\varepsilon > 0$ and $\Theta := (-\varepsilon, \varepsilon)$, there are parametric models

$$\{F_\vartheta : \vartheta \in \Theta\} \subset \mathcal{F} \quad \text{with} \quad F_0 = F$$

of mutually equivalent probability measures on $(\mathbb{R}, \mathcal{B}(\mathbb{R}))$ where score M_ϑ and Fisher information I_ϑ exist at every point $\vartheta \in \Theta$, and where we have at $\vartheta = 0$

$$(\diamond\diamond) \qquad M_0(\omega) = h(\omega), \quad I_0 = \int h^2 \, dF.$$

Such models are termed *one-parametric paths through F in direction h*.

(a) We consider first the case of bounded functions h in $(\diamond)$. Let $\sup_{x \in \mathbb{R}} |h(x)| \leq M < \infty$ and put

$$(*) \quad \mathcal{E} = \left(\mathbb{R}, \mathcal{B}(\mathbb{R}), \{F_\vartheta : |\vartheta| < M^{-1}\}\right), \quad F_\vartheta(d\omega) := (1 + \vartheta \, h(\omega)) \, F(d\omega).$$

Then the family $(*)$ is dominated by $\mu := F$, with strictly positive densities

$$f(\vartheta, \omega) = \frac{d F_\vartheta}{d F}(\omega) = 1 + \vartheta \, h(\omega), \quad \vartheta \in \Theta, \; \omega \in \mathbb{R},$$

Section 1.1 Score, Information, Information Bounds 5

hence all probability laws in (∗) are equivalent. For all $|\vartheta| < M^{-1}$, score M_ϑ and Fisher information I_ϑ according to Definition 1.2 exist and have the form

(1.3′)
$$M_\vartheta(\omega) = \left(\frac{h}{1+\vartheta h}\right)(\omega),$$

$$I_\vartheta = \int \left(\frac{h}{1+\vartheta h}\right)^2 dF_\vartheta = \int \frac{h^2}{1+\vartheta h} dF < \infty$$

which at $\vartheta = 0$ gives (⋄⋄) as indicated above.

(b) Now we consider general functions h in (⋄). Select some truncation function $\psi \in \mathcal{C}_0^1(\mathbb{R})$, the class of continuously differentiable functions $\mathbb{R} \to \mathbb{R}$ with compact support, such that the properties

$$\psi(x) = x \quad \text{on } \left\{|x| < \frac{1}{3}\right\}, \quad \psi(x) = 0 \quad \text{on } \{|x| > 1\}, \quad \max_{x \in \mathbb{R}} |\psi| < \frac{1}{2}$$

are satisfied. Put

(∗∗)
$$\mathcal{E} = (\mathbb{R}, \mathcal{B}(\mathbb{R}), \{F_\vartheta : |\vartheta| < 1\}),$$

$$F_\vartheta(d\omega) := \left(1 + [\psi(\vartheta h(\omega)) - \int \psi(\vartheta h) dF]\right) F(d\omega)$$

and note that in the special case of bounded h as considered in (a) above, paths (∗∗) and (∗) coincide when ϑ ranges over some small neighbourhood of 0. By choice of ψ, the densities

$$f(\vartheta, \omega) = 1 + [\psi(\vartheta h(\omega)) - \int \psi(\vartheta h) dF]$$

are strictly positive, hence all probability measures in (∗∗) are equivalent on $(\mathbb{R}, \mathcal{B}(\mathbb{R}))$. Since $\psi(\cdot)$ is in particular Lipschitz, dominated convergence shows

$$\frac{d}{d\vartheta} \int \psi(\vartheta h) dF = \int h \psi'(\vartheta h) dF.$$

This gives scores M_ϑ at $|\vartheta| < 1$

$$M_\vartheta(\omega) = \left(\frac{d}{d\vartheta} \log f\right)(\vartheta, \omega) = \frac{h(\omega)\psi'(\vartheta h(\omega)) - \int h \psi'(\vartheta h) dF}{1 + [\psi(\vartheta h(\omega)) - \int \psi(\vartheta h) dF]}$$

which by (⋄) belong to $L^2(F_\vartheta)$, and are centred under F_ϑ since

$$\int \frac{h(\omega)\psi'(\vartheta h(\omega)) - \int h \psi'(\vartheta h) dF}{1 + [\psi(\vartheta h(\omega)) - \int \psi(\vartheta h) dF]} F_\vartheta(d\omega)$$

$$= \int \left[h(\omega)\psi'(\vartheta h(\omega)) - \int h \psi'(\vartheta h) dF\right] F(d\omega) = 0.$$

Thus with M_ϑ as specified here, pairs

$$(M_\vartheta, I_\vartheta),$$

(1.3") $$I_\vartheta = \int \frac{[h(\omega)\psi'(\vartheta h(\omega)) - \int h\,\psi'(\vartheta h)\,dF]^2}{1 + [\psi(\vartheta h(\omega)) - \int \psi(\vartheta h)dF]} F(d\omega) < \infty$$

satisfy all assumptions of Definition 1.2 for all $|\vartheta| < 1$. At $\vartheta = 0$, expressions (1.3") reduce to $(\diamond\diamond)$. □

1.4 Example. (location models) Fix a probability measure F on $(\mathbb{R}, \mathcal{B}(\mathbb{R}))$ having density f with respect to Lebesgue measure λ such that

f is differentiable on $\mathbb{R}$ with derivative f'
f is strictly positive on (a,b), and $\equiv 0$ outside

for some open interval (a,b) in $\mathbb{R}$ ($-\infty \leq a, b \leq +\infty$), and assume

$$\int_{(a,b)} \left(\frac{f'}{f}\right)^2 dF < \infty.$$

Then the *location model generated by F*

$$\mathcal{E} := (\mathbb{R}, \mathcal{B}(\mathbb{R}), \{F_\vartheta : \vartheta \in \mathbb{R}\}), \quad dF_\vartheta := f(\cdot - \vartheta)\,d\lambda$$

satisfies all assumptions made in Definition 1.2, with score at ϑ

$$M_\vartheta(\omega) = 1_{(a+\vartheta,b+\vartheta)}(\omega)\left(-\frac{f'}{f}\right)(\omega - \vartheta), \quad \omega \in \mathbb{R}$$

since F_ϑ is concentrated on the interval $(a+\vartheta, b+\vartheta)$, and Fisher information

$$I_\vartheta = E_\vartheta(M_\vartheta^2) = \int_{(a,b)} \left(\frac{f'}{f}\right)^2 dF, \quad \vartheta \in \mathbb{R}.$$

In particular, in the location model, the Fisher information does not depend on the parameter. □

1.4' Exercise. In Example 1.4, we may consider $(a,b) := (0,1)$ and $f(x) := c(\alpha)1_{(0,1)}(x)[x(1-x)]^\alpha$ for parameter value $\alpha > 1$ (we write $c(\alpha)$ for the norming constant of the Beta $B(\alpha+1, \alpha+1)$ density). □

1.4" Exercise. (Location-scale models) For laws F on $(\mathbb{R}, \mathcal{B}(\mathbb{R}))$ with density f differentiable on $\mathbb{R}$ and supported by some open interval (a,b) as in Example 1.4, consider the *location-scale model generated by F*

$$\mathcal{E} := (\mathbb{R}, \mathcal{B}(\mathbb{R}), \{F_{(\vartheta_1,\vartheta_2)} : \vartheta_1 \in \mathbb{R}, \vartheta_2 > 0\}), \quad dF_{(\vartheta_1,\vartheta_2)} := \frac{1}{\vartheta_2} f\left(\frac{\cdot - \vartheta_1}{\vartheta_2}\right) d\lambda.$$

Section 1.1 Score, Information, Information Bounds

This model is parameterised by $\Theta := \{\vartheta = (\vartheta_1, \vartheta_2) : \vartheta_1 \in \mathbb{R}, \vartheta_2 > 0\}$. Show that the condition

$$\int_{(a,b)} [1+x^2] \left(\frac{f'}{f}\right)^2 (x) \, F(dx) < \infty$$

guarantees that the location-scale model admits score and Fisher information at every point in Θ. Write

$$G(x) := -\frac{f'(x)}{f(x)}, \quad H(x) := -[1 + \frac{xf'(x)}{f(x)}], \quad a < x < b$$

and show that the score M_ϑ at ϑ has the form

$$M_\vartheta(\omega) = 1_{(\vartheta_1+\vartheta_2 a, \vartheta_1+\vartheta_2 b)}(\omega) \begin{pmatrix} \frac{1}{\vartheta_2} G\left(\frac{\omega - \vartheta_1}{\vartheta_2}\right) \\ \frac{1}{\vartheta_2} H\left(\frac{\omega - \vartheta_1}{\vartheta_2}\right) \end{pmatrix}.$$

Write down the Fisher information I_ϑ at ϑ. Check that I_ϑ is invertible for all $\vartheta \in \Theta$, and that I_ϑ depends on the parameter only through the scaling component ϑ_2 (cf. [127, pp. 181–182]). $\square$

We consider score and information in product models:

1.5 Lemma. Consider as in Definition 1.2 an experiment

$$\mathcal{E} := (\Omega, \mathcal{A}, \{P_\vartheta : \vartheta \in \Theta\}), \quad \Theta \subset \mathbb{R}^d \text{ open}$$

admitting score $\{M_\vartheta : \vartheta \in \Theta\}$ and Fisher information $\{I_\vartheta : \vartheta \in \Theta\}$.

(a) Finite products satisfy again all assumptions of Definition 1.2: for every $n \geq 1$, the product experiment

$$\mathcal{E}_n := \left(\underset{i=1}{\overset{n}{\times}} \Omega, \underset{i=1}{\overset{n}{\otimes}} \mathcal{A}, \{P_{n,\vartheta} := \underset{i=1}{\overset{n}{\otimes}} P_\vartheta : \vartheta \in \Theta\} \right)$$

has score $\{M_{n,\vartheta} : \vartheta \in \Theta\}$ given by

$$M_{n,\vartheta}(\omega_1, \ldots, \omega_n) = \sum_{i=1}^n M_\vartheta(\omega_i) \quad P_{n,\vartheta}\text{-almost surely on } \underset{i=1}{\overset{n}{\times}} \Omega$$

and Fisher information $\{I_{n,\vartheta} : \vartheta \in \Theta\}$ satisfying

$$I_{n,\vartheta} = n \cdot I_\vartheta, \quad \vartheta \in \Theta.$$

(b) Alternatively, it may be convenient to work with infinite product experiments

$$\mathcal{E}_\infty := \left(\underset{i=1}{\overset{\infty}{\times}} \Omega, \underset{i=1}{\overset{\infty}{\otimes}} \mathcal{A}, \{Q_\vartheta := \underset{i=1}{\overset{\infty}{\otimes}} P_\vartheta : \vartheta \in \Theta\} \right),$$

with coordinate projections $X_i : (\omega_1, \omega_2, \ldots) \to \omega_i$, $i \in \mathbb{N}$, and sub-σ-fields

$$\mathcal{F}_n := \sigma(X_1, \ldots, X_n) \subset \bigotimes_{i=1}^{\infty} \mathcal{A}, \quad n \geq 1,$$

and to express n-fold independent replication of the experiment $\mathcal{E}$ as

$$\widetilde{\mathcal{E}}_n := \left(\underset{i=1}{\overset{\infty}{\times}} \Omega, \mathcal{F}_n, \{ P_{n,\vartheta} := Q_\vartheta | \mathcal{F}_n : \vartheta \in \Theta \} \right).$$

Here $Q_\vartheta | \mathcal{F}_n$ denotes the restriction of Q_ϑ to $\mathcal{F}_n$. Again all assumptions of Definition 1.2 hold for $\widetilde{\mathcal{E}}_n$. On $\widetilde{\mathcal{E}}_n$, the score is

$$M_{n,\vartheta}(\omega_1, \omega_2, \ldots) = \sum_{i=1}^{n} M_\vartheta(\omega_i) \quad P_{n,\vartheta}\text{-almost surely on } \underset{i=1}{\overset{\infty}{\times}} \Omega$$

for $\vartheta \in \Theta$, and the Fisher information is

$$I_{n,\vartheta} = n \cdot I_\vartheta, \quad \vartheta \in \Theta.$$

Proof. Select a dominating measure μ for the experiment $\mathcal{E}$, and versions f_ϑ of the densities $\frac{dP_\vartheta}{d\mu}$ which satisfy the assumptions of Definition 1.2.

(1) We prove (a). The product experiment $\mathcal{E}_n$ is dominated by $\mu_n := \bigotimes_{i=1}^{n} \mu$, with densities

$$f_{n,\vartheta}(\omega_1, \ldots, \omega_n) = \frac{dP_{n,\vartheta}}{d\mu_n}(\omega_1, \ldots, \omega_n) = \prod_{i=1}^{n} \frac{dP_\vartheta}{d\mu}(\omega_i) = \prod_{i=1}^{n} f_\vartheta(\omega_i).$$

Then $P_{n,\vartheta}$ is supported by the rectangle

$$A_{n,\vartheta} := \{(\omega_1, \ldots, \omega_n) : f_{n,\vartheta}(\omega_1, \ldots, \omega_n) > 0\} = \underset{i=1}{\overset{n}{\times}} \{\omega_i : f_\vartheta(\omega_i) > 0\}$$

in $\bigotimes_{i=1}^{n} \mathcal{A}$. On the product space $(\underset{i=1}{\overset{n}{\times}} \Omega, \bigotimes_{i=1}^{n} \mathcal{A})$ we have with the conventions of 1.2 (a)

$$M_{n,\vartheta}((\omega_1, \ldots, \omega_n)) = 1_{A_{n,\vartheta}}((\omega_1, \ldots, \omega_n)) \sum_{i=1}^{n} M_\vartheta(\omega_i).$$

Since $P_{n,\vartheta}(A_{n,\vartheta}) = 1$, the measurable mappings

$$(\omega_1, \ldots, \omega_n) \to M_{n,\vartheta}((\omega_1, \ldots, \omega_n)), \quad (\omega_1, \ldots, \omega_n) \to \sum_{i=1}^{n} M_\vartheta(\omega_i)$$

coincide $P_{n,\vartheta}$-almost surely on $(\times_{i=1}^{n} \Omega, \otimes_{i=1}^{n} \mathcal{A})$, and are identified under $P_{n,\vartheta}$ according to Definition 1.2(b). We write $M_{\vartheta,j}$ for the components of M_ϑ, $1 \leq j \leq d$, and $(I_\vartheta)_{j,l} = E_\vartheta(M_{\vartheta,j} M_{\vartheta,l})$.

Section 1.1 Score, Information, Information Bounds

(2) Under $P_{n,\vartheta}$, successive observations $\omega_1, \ldots, \omega_n$ are independent, hence
$$(\omega_1, \ldots, \omega_n) \to M_\vartheta(\omega_i), \quad 1 \le i \le n$$
are $\mathbb{R}^d$-valued i.i.d. random variables on $(\mathsf{X}_{i=1}^n \Omega, \otimes_{i=1}^n \mathcal{A}, P_{n,\vartheta})$. On this space, we have for the components $M_{n,\vartheta,j} = 1_{\{f_{n,\vartheta}>0\}}(\frac{\partial}{\partial \vartheta_j} \log f_{n,\vartheta})$ of $M_{n,\vartheta}$

$$E_{P_{n,\vartheta}}(M_{n,\vartheta,j} M_{n,\vartheta,l}) =$$
$$\int \Big(\sum_{r=1}^n M_{\vartheta,j}(\omega_r) \Big) \Big(\sum_{k=1}^n M_{\vartheta,l}(\omega_k) \Big) P_{n,\vartheta}(d\omega_1, \ldots, d\omega_n) = n \cdot (I_\vartheta)_{j,l}$$

using (∗) in Definition 1.2. This is (a).

(3) The proof of assertion (b) is analogous since on the infinite product space
$$\Big(\mathsf{X}_{i=1}^\infty \Omega, \otimes_{i=1}^\infty \mathcal{A} \Big),$$
in restriction to sub-σ-fields $\mathcal{F}_n = \sigma(X_1, \ldots, X_n)$ where $n < \infty$, the laws $Q_\vartheta | \mathcal{F}_n$ are dominated by $[\otimes_{i=1}^\infty \mu] | \mathcal{F}_n$ with density $(\omega_1, \omega_2, \ldots) \to \prod_{i=1}^n f_\vartheta(\omega_i)$ as above. $\square$

The Fisher information yields bounds for the quality of estimators. We present two types of bounds.

1.6 Proposition (Cramér–Rao bound). Consider as in Definition 1.2 an experiment $\mathcal{E} = (\Omega, \mathcal{A}, \{P_\vartheta : \vartheta \in \Theta\})$, $\Theta \in \mathbb{R}^d$ open, with score $\{M_\vartheta : \vartheta \in \Theta\}$ and Fisher information $\{I_\vartheta : \vartheta \in \Theta\}$. Consider a mapping $\gamma : \Theta \to \mathbb{R}^k$ which is partially differentiable, and let
$$Y : (\Omega, \mathcal{A}) \to (\mathbb{R}^k, \mathcal{B}(\mathbb{R}^k))$$
denote a square-integrable unbiased estimator for γ.

(a) At points $\vartheta \in \Theta$ where the two conditions

(+) \qquad the Fisher information matrix I_ϑ is invertible

(++) $\quad \begin{pmatrix} \frac{\partial}{\partial \vartheta_1}\gamma_1 & \cdots & \frac{\partial}{\partial \vartheta_d}\gamma_1 \\ \cdots & \cdots & \cdots \\ \frac{\partial}{\partial \vartheta_1}\gamma_k & \cdots & \frac{\partial}{\partial \vartheta_d}\gamma_k \end{pmatrix}(\vartheta) = E_\vartheta\left(Y M_\vartheta^\top\right)$

are satisfied, we have a lower bound

(◇) $\qquad \qquad Cov_\vartheta(Y) \ge V_\vartheta I_\vartheta^{-1} V_\vartheta^\top$

at ϑ, where V_ϑ denotes the Jacobi matrix of γ on the l.h.s. of condition (++).

(b) In the setting of (a), Y attains the bound (◇) at ϑ (i.e. achieves $Cov_\vartheta(Y) = V_\vartheta I_\vartheta^{-1} V_\vartheta^\top$) if and only if its estimation error at ϑ admits a representation
$$Y - \gamma(\vartheta) = V_\vartheta I_\vartheta^{-1} M_\vartheta \quad P_\vartheta\text{-almost surely.}$$

(c) In the special case $\gamma = id$ in (a) and (b) above, the bound ($\diamond$) at ϑ reads

$$Cov_\vartheta(Y) \geq I_\vartheta^{-1}$$

and Y attains the bound ($\diamond$) at ϑ if and only if

$$Y - \vartheta = I_\vartheta^{-1} M_\vartheta \quad P_\vartheta\text{-almost surely.}$$

Proof. According to Definition 1.2, we have $M_\vartheta \in L^2(P_\vartheta)$ and $E_\vartheta(M_\vartheta) = 0$ for all $\vartheta \in \Theta$. Necessarily $I_\vartheta = E_\vartheta(M_\vartheta M_\vartheta^\top)$ is symmetric and non-negative definite for all $\vartheta \in \Theta$. We consider a point $\vartheta \in \Theta$ such that (+) holds, together with a mapping $\gamma : \Theta \to \mathbb{R}^k$ and a random variable $Y \in L^2(P_\vartheta)$ satisfying $E_\vartheta(Y) = \gamma(\vartheta)$ and (++).

(1) We start by defining V_ϑ by the right-hand side of (++):

$$V_\vartheta := E_\vartheta\left(Y M_\vartheta^\top\right).$$

ϑ being fixed, introduce a random variable

$$W := (Y - E_\vartheta(Y)) - V_\vartheta I_\vartheta^{-1} M_\vartheta$$

taking values in $\mathbb{R}^k$. Then we have $W \in L^2(P_\vartheta)$ with $E_\vartheta(W) = 0$ and

$$
\begin{aligned}
Cov_\vartheta(W) &= E_\vartheta(W W^\top) \\
&= Cov_\vartheta(Y) - E_\vartheta\left((Y - \gamma(\vartheta)) M_\vartheta^\top I_\vartheta^{-1} V_\vartheta^\top\right) \\
&\quad - E_\vartheta\left(V_\vartheta I_\vartheta^{-1} M_\vartheta (Y - \gamma(\vartheta))^\top\right) + E_\vartheta\left(V_\vartheta I_\vartheta^{-1} M_\vartheta M_\vartheta^\top I_\vartheta^{-1} V_\vartheta^\top\right).
\end{aligned}
$$

On the r.h.s. of this equation, M_ϑ being centred under P_ϑ and $E_\vartheta(Y M_\vartheta^\top) = V_\vartheta$ by definition, both the second and the third summand reduce to $-V_\vartheta I_\vartheta^{-1} V_\vartheta^\top$. By definition of the Fisher information, the fourth summand on the r.h.s. is $+V_\vartheta I_\vartheta^{-1} V_\vartheta^\top$. In the sense of half-ordering of symmetric and non-negative definite matrices, we thus arrive at

($\diamond$) $$0 \leq Cov_\vartheta(W) = Cov_\vartheta(Y) - V_\vartheta I_\vartheta^{-1} V_\vartheta^\top$$

which is (a). Equality in ($\diamond$) is possible only in the case where $W = E_\vartheta(W) = 0$ P_ϑ-almost surely. This is the 'only if' part of assertion (b), the 'if' part is obvious. So far, we did not make use of assumption (++).

(2) We consider the special case $k = d$ and $\gamma = id$: here assumption (++) is needed to identify V_ϑ with the $d \times d$ identity matrix. Then (c) is an immediate consequence of (a) and (b). □

1.7 Remarks. (a) A purely heuristic argument for (++): one should be allowed to differentiate

$$\gamma(\vartheta) = E_\vartheta(Y) = \int \mu(d\omega) f(\vartheta, \omega) Y(\omega)$$

with respect to ϑ under the integral sign, with partial derivatives

$$\frac{\partial}{\partial \vartheta_j} E_\vartheta(Y_i) = \int \mu(d\omega) \frac{\partial}{\partial \vartheta_j} f(\vartheta, \omega) Y_i(\omega)$$

$$= \int P_\vartheta(d\omega) \frac{\partial}{\partial \vartheta_j} \log f(\vartheta, \omega) Y_i(\omega) = E_\vartheta\left((Y M_\vartheta^\top)_{i,j}\right).$$

(b) Condition (++) in 1.6 holds in naturally parameterised d-parametric exponential families when the parameter space is open (see Barra [4, Chap. X and formula (2) on p. 178], Witting [127, pp. 152–153], or van der Vaart [126, Lemma 4.5 on p. 38]): in this case, the score at ϑ is given by the canonical statistic of the exponential family centred under ϑ. See also Barndorff–Nielsen [3, Chap. 2.4] or Küchler and Sørensen [77]. □

1.7' Remark. Within the class of unbiased and square integrable estimators Y for γ, the covariance matrix $Cov_\vartheta(Y) = E_\vartheta\left((Y - \gamma(\vartheta))(Y - \gamma(\vartheta))^\top\right)$, quantifying spread/concentration of estimation errors $Y - \gamma(\vartheta)$ at ϑ, allows to compare different estimators at the point ϑ. The lower bound in ($\diamond$) under the assumptions of Proposition 1.6 involves the *inverse of the Fisher information* I_ϑ^{-1} which thus may indicate an 'optimal concentration'; when $\gamma = id$ the lower bound in ($\diamond$) equals I_ϑ^{-1}.

However, there are two serious drawbacks:

(i) these bounds are *attainable bounds* only in a few classical parametric models, see [27, p. 198], or [127, pp. 312–317], and [86, Theorem 7.15 on p. 300];

(ii) unbiasedness is not 'per se' relevant for good estimation: a famous example due to Stein (see [60, p. 26] or [59, p. 93]) shows that even in a normal distribution model

$$\left(\left(\underset{i=1}{\overset{n}{\times}} \mathbb{R}^k, \underset{i=1}{\overset{n}{\otimes}} \mathcal{B}(\mathbb{R}^k)\right), \left\{\underset{i=1}{\overset{n}{\otimes}} \mathcal{N}(\vartheta, I_k) : \vartheta \in \Theta\right\}\right), \quad \Theta := \mathbb{R}^k \text{ where } k \geq 3$$

estimators admitting bias can be constructed which (with respect to squared loss) concentrate better around the true ϑ than the empirical mean, the best unbiased square integrable estimator in this model.

The following bound, of different nature, allows to consider arbitrary estimators for the unknown parameter. We give it in dimension $d = 1$ only (multivariate generalisations exist, see [29]), and assuming strictly positive densities. The underlying idea is 'Bayesian': some probability measure $\pi(d\vartheta)$ playing on the parameter space Θ selects the true parameter $\vartheta \in \Theta$. Our proof follows Gill and Levit [29].

1.8 Proposition (van Trees inequality). Consider an experiment $\mathcal{E} := (\Omega, \mathcal{A}, \{P_\vartheta : \vartheta \in \Theta\})$ where Θ is an open interval in $\mathbb{R}$, with strictly positive densities $f(\vartheta, \cdot) = \frac{dP_\vartheta}{d\mu}$ on Ω with respect to a dominating measure μ, for all $\vartheta \in \Theta$. Assume that $\mathcal{E}$, satisfying all assumptions of Definition 1.2, has score $\{M_\vartheta : \vartheta \in \Theta\}$ and Fisher information $\{I_\vartheta : \vartheta \in \Theta\}$, and consider a differentiable mapping $\gamma : \Theta \to \mathbb{R}$.

Fix any subinterval (a,b) of Θ and any *a priori law* π with Lebesgue density $g := \frac{d\pi}{d\lambda}$ such that

(i) $\quad g$ is differentiable on $\mathbb{R}$, strictly positive on (a,b), and $\equiv 0$ else

(ii) $\quad \displaystyle\int_{(a,b)} \left(\frac{g'}{g}\right)^2 d\pi =: J < \infty$

hold. Whenever a or b coincide with a boundary point of Θ, assume that $\gamma : \Theta \to \mathbb{R}$ admits at this point a finite limit denoted by $\gamma(a)$ or $\gamma(b)$, and $f(\cdot, \omega)$ a finite limit denoted by $f(a, \omega)$ or $f(b, \omega)$ for fixed $\omega \in \Omega$. Then we have the bound

$$\int_{(a,b)} E_\vartheta\left((T - \gamma(\vartheta))^2\right) \pi(d\vartheta) \geq \frac{\left(\int_{(a,b)} \gamma'(\vartheta)\, \pi(d\vartheta)\right)^2}{\int_{(a,b)} I_\vartheta\, \pi(d\vartheta) + J}$$

for arbitrary estimators T for γ in the experiment $\mathcal{E}$.

Proof. (0) Note that J is the Fisher information in the location model generated by π on $(\mathbb{R}, \mathcal{B}(\mathbb{R}))$ which satisfies all assumptions of Example 1.4.

(1) We show that the assumptions on the densities made in Definition 1.2 imply measurability of

(*) $\qquad\qquad\qquad \Theta \times \Omega \ni (\vartheta, \omega) \to f(\vartheta, \omega) \in (0, \infty)$

when $\Theta \times \Omega$ is equipped with the product σ-field $\mathcal{B}(\Theta) \otimes \mathcal{A}$. In Definition 1.2, we have Θ open, and

$$\forall \vartheta \in \Theta : \quad f(\vartheta, \cdot) : (\Omega, \mathcal{A}) \to (\mathbb{R}, \mathcal{B}(\mathbb{R})) \quad \text{is measurable,}$$
$$\forall \omega \in \Omega : \quad f(\cdot, \omega) : \Theta \to \mathbb{R} \quad \text{is continuous.}$$

Write A_k for the set of all $j \in \mathbb{Z}$ such that $]\frac{j}{2^k}, \frac{j+1}{2^k}]$ has a non-void intersection with Θ, and select some point $\vartheta(k, j)$ in $\Theta \cap]\frac{j}{2^k}, \frac{j+1}{2^k}]$ for $j \in A_k$. Then all mappings $f_k(\cdot, \cdot)$

$$f_k(\vartheta, \omega) := \sum_{j \in A_n} 1_{]\frac{j}{2^k}, \frac{j+1}{2^k}] \cap \Theta}(\vartheta)\, f(\vartheta(k,j), \omega)\,, \quad k \geq 1$$

are measurable in the pair (ϑ, ω), hence the same holds for their pointwise limit (*) as $k \to \infty$.

(2) The product measurability established in (*) allows to view

$$(\vartheta, A) \;\longrightarrow\; P_\vartheta(A) = \int 1_A(\omega)\, f(\vartheta, \omega)\, \mu(d\omega)\,, \quad \vartheta \in \Theta\,, A \in \mathcal{A}$$

as a transition probability from $(\Theta, \mathcal{B}(\Theta))$ to $(\Omega, \mathcal{A})$.

Section 1.1 Score, Information, Information Bounds

(3) We consider estimators $T : (\Omega, \mathcal{A}) \to (\mathbb{R}, \mathcal{B}(\mathbb{R}))$ for γ which have the property

$$\int_{(a,b)} E_\vartheta\left((T - \gamma(\vartheta))^2\right) \pi(d\vartheta) < \infty$$

(otherwise the bound in Proposition 1.8 would be trivial). To prove Proposition 1.8, it is sufficient to consider the restriction of the parameter space Θ to its subset (a, b): thus we identify Θ with (a, b) – then, by assumption, $g = \frac{d\pi}{d\lambda}$ will be strictly positive on Θ with limits $g(a) = g(b) = 0$, and γ as well as $f(\cdot, \omega)$ for all ω will have finite limits at the endpoints of Θ – and work on the product space

$$(\overline{\Omega}, \overline{\mathcal{A}}) := (\Theta \times \Omega, \mathcal{B}(\Theta) \otimes \mathcal{A})$$

equipped with the probability measure

$$\overline{P}(d\vartheta, d\omega) := \pi(d\vartheta)\, P_\vartheta(d\omega) = (\lambda \otimes \mu)(d\vartheta, d\omega)\, g(\vartheta) f(\vartheta, \omega),$$
$\vartheta \in \Theta,\ \omega \in \Omega.$

(4) In the following steps, we write $'$ for the derivative with respect to the parameter (from the set of assumptions in Definition 1.2, recall differentiability of $\vartheta \to f(\vartheta, \omega)$ for fixed ω when $d = 1$). Then

(+) $\quad \int_\Theta d\vartheta\, (f(\vartheta, \omega)g(\vartheta))' = f(b, \omega)g(b) - f(a, \omega)g(a) = 0$

for all $\omega \in \Omega$ since $g(a) = 0 = g(b)$ by our assumption, together with

(++) $\quad \int_\Theta d\vartheta\, \gamma(\vartheta)\, (f(\vartheta, \omega)g(\vartheta))' = -\int_\Theta d\vartheta\, \gamma'(\vartheta)\, (f(\vartheta, \omega)g(\vartheta))$

by partial integration. Combining (+) and (++) we get the equation

$$\int_{\Theta \times \Omega} (\lambda \otimes \mu)(d\vartheta, d\omega)\, (f(\vartheta, \omega)g(\vartheta))'\, (T(\omega) - \gamma(\vartheta))$$

$$= 0 + \int_\Omega \mu(d\omega) \int_\Theta d\vartheta\, \gamma'(\vartheta)\, f(\vartheta, \omega)\, g(\vartheta)$$

$$= \int_\Theta \pi(d\vartheta)\, \gamma'(\vartheta).$$

By strict positivity of the densities and strict positivity of g on Θ, the l.h.s. of the first equality sign is

$$\int_{\Theta \times \Omega} (\lambda \otimes \mu)(d\vartheta, d\omega)\, (f(\vartheta, \omega)g(\vartheta))'\, (T(\omega) - \gamma(\vartheta))$$

$$= \int_\Theta \int_\Omega \pi(d\vartheta) P_\vartheta(d\omega)\, \frac{(f(\vartheta, \omega)g(\vartheta))'}{f(\vartheta, \omega)g(\vartheta)}\, (T(\omega) - \gamma(\vartheta))$$

$$= \int_{\Theta \times \Omega} \overline{P}(d\vartheta, d\omega)\, \left(\frac{g'}{g}(\vartheta) + \frac{f'}{f}(\vartheta, \omega)\right) (T(\omega) - \gamma(\vartheta)).$$

In the last integrand, both factors

$$\left(\frac{g'}{g}(\vartheta) + \frac{f'}{f}(\vartheta,\omega)\right), \quad (T(\omega) - \gamma(\vartheta))$$

are in $L^2(\overline{P})$: the second is the estimation error of an estimator T for γ which at the start of step (3) was assumed to be in $L^2(\overline{P})$, the first is the sum of the score in the location experiment generated by π and the score $\{M_\vartheta : \vartheta \in \Theta\}$ in the experiment $\mathcal{E}$, both necessarily orthogonal in $L^2(\overline{P})$:

$(+++)$
$$\int_{\Theta\times\Omega} \overline{P}(d\vartheta,d\omega)\, \frac{g'}{g}(\vartheta)\, \frac{f'}{f}(\vartheta,\omega) =$$
$$\int_\Theta \pi(d\vartheta)\, \frac{g'}{g}(\vartheta) \int_\Omega P_\vartheta(d\omega)\, \frac{f'}{f}(\vartheta,\omega) = 0\,.$$

Putting the last three blocks of arguments together, the Cauchy–Schwarz inequality with (+++) gives

$$\left(\int_\Theta \pi(d\vartheta)\, \gamma'(\vartheta)\right)^2$$
$$= \left(\int_{\Theta\times\Omega} \overline{P}(d\vartheta,d\omega) \left(\frac{g'}{g}(\vartheta) + \frac{f'}{f}(\vartheta,\omega)\right)(T(\omega) - \gamma(\vartheta))\right)^2$$
$$\leq \int_{\Theta\times\Omega} \overline{P}(d\vartheta,d\omega) \left(\frac{g'}{g}(\vartheta) + \frac{f'}{f}(\vartheta,\omega)\right)^2 \cdot \int_{\Theta\times\Omega} \overline{P}(d\vartheta,d\omega)\, (T(\omega) - \gamma(\vartheta))^2$$
$$= \left(J + \int_\Theta \pi(d\vartheta)\, I_\vartheta\right) \cdot \int_\Theta \pi(d\vartheta)\, E_\vartheta\!\left((T - \gamma(\vartheta))^2\right)$$

which is the assertion. □

1.8' Exercise. For $\Theta \subset \mathbb{R}^d$ open, assuming densities which are continuous in the parameter, cover $\mathbb{R}^d$ with half-open cubes of side length 2^{-k} to prove

$$\Theta \times \Omega \ni (\vartheta,\omega) \to f(\vartheta,\omega) \in [0,\infty) \quad \text{is } (\mathcal{B}(\Theta)\otimes\mathcal{A})\text{-}\mathcal{B}([0,\infty))\text{-measurable}\,.$$

In the case where $d = 1$, the assertion holds under right-continuity (or left-continuity) of the densities in the parameter.

1.8" Exercise. For $\Theta \subset \mathbb{R}^d$ open, for product measurable densities $(\vartheta,\omega) \to f(\vartheta,\omega)$ with respect to some dominating measure μ on $(\Omega,\mathcal{A})$, consider as in Proposition 1.8 the product space $(\overline{\Omega},\overline{\mathcal{A}}) = (\Theta\times\Omega, \mathcal{B}(\Theta)\otimes\mathcal{A})$ equipped with

$$\overline{P}(d\vartheta,d\omega) = \pi(d\vartheta)\, P_\vartheta(d\omega) = (\lambda\otimes\mu)(d\vartheta,d\omega)\, g(\vartheta) f(\vartheta,\omega)\,, \quad \vartheta\in\Theta,\ \omega\in\Omega$$

where π is some probability law on $(\Theta,\mathcal{B}(\Theta))$ with Lebesgue density g, and view $(\vartheta,\omega) \to \vartheta$ and $(\vartheta,\omega) \to \omega$ as random variables on $(\overline{\Omega},\overline{\mathcal{A}})$. We wish to estimate some measurable mapping $\gamma : (\Theta, \mathcal{B}(\Theta)) \to (\mathbb{R}^k, \mathcal{B}(\mathbb{R}^k))$: fixing some loss function $\ell : \mathbb{R}^k \to [0, \infty)$, we call any estimator $T^* : (\Omega, \mathcal{A}) \to (\mathbb{R}^k, \mathcal{B}(\mathbb{R}^k))$ with the property

$$\inf_{T \text{ } \mathcal{A}\text{-mb}} \int_\Theta \pi(d\vartheta) \, E_\vartheta \left(\ell(T - \gamma(\vartheta)) \right) = \int_\Theta \pi(d\vartheta) \, E_\vartheta \left(\ell(T^* - \gamma(\vartheta)) \right)$$

ℓ-*Bayesian with respect to the a priori law* π. Here 'inf' is over the class of all measurable mappings $T : (\Omega, \mathcal{A}) \to (\mathbb{R}^k, \mathcal{B}(\mathbb{R}^k))$, i.e. over the class of all possible estimators for γ. So far, we leave open questions of existence (see Section 37 in Strasser [121]).

In the case where $\ell(y) = |y|^2$ and $\gamma \in L^2(\pi)$, prove that a *squared loss Bayesian* exists and is given by

$$T^*(\omega) = \begin{cases} \int_\Theta \gamma(\xi) \frac{f(\xi,\omega) g(\xi)}{\int_\Theta f(\zeta,\omega) g(\zeta) d\zeta} d\xi & \text{if } \int_\Theta f(\zeta,\omega) g(\zeta) d\zeta > 0 \\ \vartheta_0 & \text{if } \int_\Theta f(\zeta,\omega) g(\zeta) d\zeta = 0 \end{cases}$$

with arbitrary default value $\vartheta_0 \in \Theta$.
Hint: In a squared loss setting, the Bayes property reduces to the L^2-projection property of conditional expectations in $(\overline{\Omega}, \overline{\mathcal{A}}, \overline{P})$. Writing down conditional densities, we have a regular version of the conditional law of $\vartheta : (\vartheta, \omega) \to \vartheta$ given $\omega : (\vartheta, \omega) \to \omega$. The random variable $\gamma : (\vartheta, \omega) \to \gamma(\vartheta)$ belonging to $L^2(\overline{P})$, the conditional expectation of γ given ω under $\overline{P}$ is the integral of $\gamma(\cdot)$ with respect to this conditional law. □

1.2 Estimator Sequences, Asymptotics of Information Bounds

1.9 Definition. Consider a sequence of experiments

$$\mathcal{E}_n = (\Omega_n, \mathcal{A}_n, \{P_{n,\vartheta} : \vartheta \in \Theta\}), \quad n \geq 1$$

parameterised by the same parameter set $\Theta \subset \mathbb{R}^d$ which does not depend on n, and a mapping $\gamma : \Theta \to \mathbb{R}^k$. An *estimator sequence for* γ is a sequence $(Y_n)_n$ of measurable mappings

$$Y_n : (\Omega_n, \mathcal{A}_n) \to (\mathbb{R}^k, \mathcal{B}(\mathbb{R}^k)), \quad n \geq 1 .$$

(a) An estimator sequence $(Y_n)_n$ is called *consistent for* γ if

for every $\vartheta \in \Theta$, every $\varepsilon > 0$: $\quad \lim_{n \to \infty} P_{n,\vartheta} \left(|Y_n - \gamma(\vartheta)| > \varepsilon \right) = 0$

(convergence in $(P_{n,\vartheta})_n$-probability of the sequence $(Y_n)_n$ to $\gamma(\vartheta)$, for every $\vartheta \in \Theta$).

(b) Associate sequences $(\varphi_n(\vartheta))_n$ to parameter values $\vartheta \in \Theta$, either taking values in $(0, \infty)$ and such that $\varphi_n(\vartheta)$ increases to ∞ as $n \to \infty$, or taking values in the space of invertible $d \times d$-matrices such that minimal eigenvalues $\lambda_n^*(\vartheta)$ of $\varphi_n(\vartheta)$ increase to ∞ as $n \to \infty$. Then an estimator sequence $(Y_n)_n$ for γ is called $(\varphi_n)_n$-*consistent* if

for every $\vartheta \in \Theta$, $\{\mathcal{L}(\varphi_n(\vartheta)(Y_n - \gamma(\vartheta)) \mid P_{n,\vartheta}) : n \geq 1\}$ is tight in $\mathbb{R}^k$.

(c) $(\varphi_n)_n$-consistent estimator sequences $(Y_n)_n$ for γ are called *asymptotically normal* if

for every $\vartheta \in \Theta$, $\mathcal{L}(\varphi_n(\vartheta)(Y_n - \gamma(\vartheta)) \mid P_{n,\vartheta}) \longrightarrow \mathcal{N}(0, \Sigma(\vartheta))$ as $n \to \infty$

(weak convergence in $\mathbb{R}^k$), for suitable normal distributions $\mathcal{N}(0, \Sigma(\vartheta))$, $\vartheta \in \Theta$.

For the remaining part of this section we focus on independent replication of an experiment
$$\mathcal{E} := (\Omega, \mathcal{A}, \{P_\vartheta : \vartheta \in \Theta\}), \quad \Theta \subset \mathbb{R}^d \text{ open}$$
which satisfies all assumptions of Definition 1.2, with score $\{M_\vartheta : \vartheta \in \Theta\}$ and Fisher information $\{I_\vartheta : \vartheta \in \Theta\}$. We consider for $n \to \infty$ the sequence of product models

($\diamond$)
$$\mathcal{E}_n := (\Omega_n, \mathcal{A}_n, \{P_{n,\vartheta} : \vartheta \in \Theta\}) = \left(\underset{i=1}{\overset{n}{\times}} \Omega, \underset{i=1}{\overset{n}{\otimes}} \mathcal{A}, \{P_{n,\vartheta} := \underset{i=1}{\overset{n}{\otimes}} P_\vartheta : \vartheta \in \Theta\} \right)$$

as in Lemma 1.5(a), with score $M_{n,\vartheta}$ in ϑ and information $I_{n,\vartheta} = n\, I_\vartheta$. In this setting we present some asymptotic lower bounds for the risk of estimators, in terms of the Fisher information.

1.10 Remark (Asymptotic Cramér–Rao Bound). Consider $(\mathcal{E}_n)_n$ as in ($\diamond$) and assume that I_ϑ is invertible for all $\vartheta \in \Theta$. Let $(T_n)_n$ denote some sequence of unbiased and square integrable estimators for the unknown parameter, $\sqrt{n}$-consistent and asymptotically normal:

(o) for every $\vartheta \in \Theta$: $\mathcal{L}\left(\sqrt{n}(T_n - \vartheta) \mid P_{n,\vartheta}\right) \longrightarrow \mathcal{N}(0, \Sigma(\vartheta))$, $n \to \infty$

(weak convergence in $\mathbb{R}^d$). The Cramér–Rao bound in $\mathcal{E}_n$
$$E_\vartheta \left([\sqrt{n}(T_n - \vartheta)][\sqrt{n}(T_n - \vartheta)]^\top \right) = n\, Cov_\vartheta(T_n) \geq n\, I_{n,\vartheta}^{-1} = I_\vartheta^{-1}$$
makes an 'optimal' limit variance $\Sigma(\vartheta) = I_\vartheta^{-1}$ appear in (o), for every $\vartheta \in \Theta$. Given one estimator sequence $(T_n)_n$ whose rescaled estimation errors at ϑ attain the limit law $\mathcal{N}(0, I_\vartheta^{-1})$, one would like to call this sequence 'optimal'. The problem is that Cramér–Rao bounds do not allow for comparison within a sufficiently broad class of competing estimator sequences. Fix $\vartheta \in \Theta$. Except for unbiasedness of estimators at ϑ, Cramér–Rao needs the assumption (++) in Proposition 1.6 for $\gamma = id$, and needs ($*$) from Definition 1.2: both last assumptions
$$E_\vartheta(M_{n,\vartheta}) = 0, \quad E_\vartheta(T_n M_{n,\vartheta}) = I, \quad n \geq 1$$
(with 0 the zero vector in $\mathbb{R}^d$ and I the identity matrix in $\mathbb{R}^{d \times d}$) combine in particular to

($*$) $\qquad\qquad E_\vartheta([T_n - \vartheta] M_{n,\vartheta}) = I \quad \text{for all } n \geq 1$.

Section 1.2 Estimator Sequences, Asymptotics of Information Bounds

Thus from the very beginning, condition (++) of Proposition 1.6 for $\gamma = id$ establishes a close connection between the *sequence of scores* $(M_{n,\vartheta})_n$ on the one hand and those estimator sequences $(T_n)_n$ to which we may apply Cramér–Rao on the other. Hence the Cramér–Rao setting turns out to be a restricted setting.

1.11 Theorem (Asymptotic van Trees bounds). Consider an experiment $\mathcal{E} := (\Omega, \mathcal{A}, \{P_\vartheta : \vartheta \in \Theta\})$ where Θ is an open interval in $\mathbb{R}$, with strictly positive densities $f(\vartheta, \cdot) = \frac{dP_\vartheta}{d\mu}$ on Ω with respect to a dominating measure μ, for all $\vartheta \in \Theta$. Assume that $\mathcal{E}$, satisfying all assumptions of Definition 1.2, admits score $\{M_\vartheta : \vartheta \in \Theta\}$ and Fisher information $\{I_\vartheta : \vartheta \in \Theta\}$, and assume in addition

(⋄⋄) $\quad\quad\quad\quad \Theta \ni \vartheta \to I_\vartheta \in (0, \infty) \quad$ is continuous.

Then for independent replication (⋄) of the experiment $\mathcal{E}$, for arbitrary choice of estimators T_n for the unknown parameter $\vartheta \in \Theta$ in the product experiments $\mathcal{E}_n$, we have the two bounds (I) and (II)

(I) $\quad\quad \lim_{c \downarrow 0} \liminf_{n \to \infty} \inf_{T_n \, \mathcal{A}_n\text{-mb}} \sup_{|\vartheta - \vartheta_0| < c} E_\vartheta \left([\sqrt{n}(T_n - \vartheta)]^2\right) \geq I_{\vartheta_0}^{-1}$

(II) $\quad\quad \lim_{C \uparrow \infty} \liminf_{n \to \infty} \inf_{T_n \, \mathcal{A}_n\text{-mb}} \sup_{|\vartheta - \vartheta_0| < C/\sqrt{n}} E_\vartheta \left([\sqrt{n}(T_n - \vartheta)]^2\right) \geq I_{\vartheta_0}^{-1}$

at every $\vartheta_0 \in \Theta$.

Proof. Up to easy notational changes it is sufficient to consider the case $\Theta = \mathbb{R}$ and $\vartheta_0 = 0$.

(1) Select a probability measure π on $(\mathbb{R}, \mathcal{B}(\mathbb{R}))$ with Lebesgue density g such that g is differentiable in ϑ, has support $\{g > 0\} = (-1, 1)$, and satisfies

$$J := \int_{(-1,1)} \left(\frac{g'}{g}\right)^2 d\pi < \infty.$$

Then the location model generated by π satisfies the conditions of Example 1.4, and thus conditions (i) and (ii) in Proposition 1.8. The same holds for all probability measures π_r on $(\mathbb{R}, \mathcal{B}(\mathbb{R}))$ obtained from π by scaling

$$\pi_r(d\vartheta) := \frac{1}{r} g\left(\frac{\vartheta}{r}\right) d\vartheta, \quad 0 < r < \infty:$$

π_r is concentrated on $(-r, +r)$, and generates a location model with Fisher information $J_r := \frac{1}{r^2} J$. For arbitrary estimators T_n for $\gamma = id$ in the model $\mathcal{E}_n$ and arbitrary $0 < r < \infty$, the inequality

$$\sup_{|\vartheta - \vartheta_0| < r} E_\vartheta \left([\sqrt{n}(T_n - \vartheta)]^2\right) \geq \int_{-r}^{+r} \pi_r(d\vartheta) \, E_\vartheta \left([\sqrt{n}(T_n - \vartheta)]^2\right)$$

(finite or infinite) is trivial; to its r.h.s. we apply van Trees inequality in Proposition 1.8 and continue

$$(\circ) \qquad \geq n \cdot \frac{1}{\int_{-r}^{+r} I_{n,\vartheta} \, \pi_r(d\vartheta) + J_r} = \frac{1}{\int_{-r}^{+r} I_\vartheta \, \pi_r(d\vartheta) + \frac{1}{r^2 n} J}$$

for arbitrary $0 < r < \infty$.

(2) Consider $c > 0$ small. First, keeping $r := c$ in ($\circ$) fixed, $\frac{1}{c^2 n} J$ vanishes as $n \to \infty$, thus

$$\liminf_{n \to \infty} \inf_{T_n \, \mathcal{A}_n\text{-mb}} \sup_{|\vartheta - 0| < c} E_\vartheta \left([\sqrt{n}(T_n - \vartheta)]^2 \right) \geq \frac{1}{\int_{-c}^{+c} I_\vartheta \, \pi_c(d\vartheta)}.$$

Second, for $c \downarrow 0$ on both sides of the last inequality, continuity ($\circ\circ$) of the Fisher information gives

$$\liminf_{c \downarrow 0} \liminf_{n \to \infty} \inf_{T_n \, \mathcal{A}_n\text{-mb}} \sup_{|\vartheta - 0| < c} E_\vartheta \left([\sqrt{n}(T_n - \vartheta)]^2 \right) \geq I_0^{-1}.$$

On the l.h.s. of the preceding inequality, the term

$$\liminf_{n \to \infty} \inf_{T_n \, \mathcal{A}_n\text{-mb}} \sup_{|\vartheta - 0| < c} R_n(T_n, \vartheta), \qquad R_n(T_n, \vartheta) := E_\vartheta \left([\sqrt{n}(T_n - \vartheta)]^2 \right)$$

is monotone in c since for $c_1 < c_2$

$$\sup_{|\vartheta - 0| < c_2} R_n(T_n, \vartheta) \geq \sup_{|\vartheta - 0| < c_1} R_n(T_n, \vartheta) \geq \inf_{\widetilde{T}_n \, \mathcal{A}_n\text{-mb}} \sup_{|\vartheta - 0| < c_1} R_n(\widetilde{T}_n, \vartheta).$$

Hence '$\liminf_{c \downarrow 0}$' above is in fact '$\lim_{c \downarrow 0}$' which proves the bound (I).

(3) Consider $C < \infty$ large. With $r = C/\sqrt{n}$ in the above chain of inequalities ($\circ$), we exploit again continuity ($\circ\circ$) of the Fisher information and get

$$\liminf_{n \to \infty} \inf_{T_n \, \mathcal{A}_n\text{-mb}} \sup_{|\vartheta - 0| < C/\sqrt{n}} E_\vartheta \left([\sqrt{n}(T_n - \vartheta)]^2 \right) \geq \frac{1}{I_0 + \frac{1}{C^2} J}.$$

Similar to step (2) above, the l.h.s. is monotone in C, thus

$$\lim_{C \uparrow \infty} \liminf_{n \to \infty} \inf_{T_n \, \mathcal{A}_n\text{-mb}} \sup_{|\vartheta - 0| < C/\sqrt{n}} E_\vartheta \left([\sqrt{n}(T_n - \vartheta)]^2 \right) \geq I_0^{-1}$$

which is the bound (II). This concludes the proof of Theorem 1.11. $\square$

Both bounds (I) or (II) in Theorem 1.11 are asymptotic lower bounds of *minimax* type: using 'best possible' estimators for the unknown parameter – where at every stage n of the asymptotics, competition is between *all* estimators which may exist in $\mathcal{E}_n$ – we minimise a maximal risk on small balls around points ϑ_0 in Θ. For independent replication of an experiment $\mathcal{E}$ which satisfies all assumptions of Proposition 1.8, the van Trees inequality thus shows that the maximal risk of estimators on small balls

Section 1.2 Estimator Sequences, Asymptotics of Information Bounds

around ϑ_0 (with respect to squared loss, and with estimation errors rescaled by norming constants $\sqrt{n}$ as $n \to \infty$) will never be better than $I_{\vartheta_0}^{-1}$, the inverse of the Fisher information at ϑ_0.

The two types of bounds (I) and (II) are different in that they imply different notions of 'small neighbourhoods around ϑ_0'. Type (II) neighbourhoods are shrinking balls of radius $O(1/\sqrt{n})$: at first glance this seems to be less natural than the small balls not depending on n which are used in type (I). Let us compare the two types of bounds, writing again $R_n(T_n, \vartheta) := E_\vartheta\left([\sqrt{n}(T_n - \vartheta)]^2\right)$ as in the last proof. For every estimator sequence $(T_n)_n$ and every pair of constants $0 < c < C < \infty$, we have

$$\sup_{|\vartheta - \vartheta_0| < c} R_n(T_n, \vartheta) \geq \sup_{|\vartheta - \vartheta_0| < C/\sqrt{n}} R_n(T_n, \vartheta)$$

$$\geq \inf_{\widetilde{T}_n \; \mathcal{A}_n\text{-mb}} \sup_{|\vartheta - \vartheta_0| < C/\sqrt{n}} R_n(\widetilde{T}_n, \vartheta)$$

for sufficiently large n. It follows that

$$\liminf_{n \to \infty} \inf_{\widetilde{T}_n \; \mathcal{A}_n\text{-mb}} \sup_{|\vartheta - \vartheta_0| < c} R_n(T_n, \vartheta) \geq$$

$$\liminf_{n \to \infty} \inf_{\widetilde{T}_n \; \mathcal{A}_n\text{-mb}} \sup_{|\vartheta - \vartheta_0| < C/\sqrt{n}} R_n(T_n, \vartheta)$$

for arbitrary pairs (c, C), $c > 0$ small and $C < \infty$ large. Using again the monotonicity argument of the last proof, we arrive at a comparison

$$\lim_{c \downarrow 0} \liminf_{n \to \infty} \inf_{\widetilde{T}_n \; \mathcal{A}_n\text{-mb}} \sup_{|\vartheta - \vartheta_0| < c} R_n(T_n, \vartheta) \geq$$

$$\lim_{C \uparrow \infty} \liminf_{n \to \infty} \inf_{\widetilde{T}_n \; \mathcal{A}_n\text{-mb}} \sup_{|\vartheta - \vartheta_0| < C/\sqrt{n}} R_n(T_n, \vartheta)$$

between left-hand sides in the type (I) or type (II) bounds of Theorem 1.11.

The two types of bounds correspond to different traditions. On the one hand, see e.g. Ibragimov and Khasminskii [60, Theorem 12.1, p. 162] and Kutoyants [80, p. 57 or pp. 114–115] who – in more general settings than the present one – work with bounds of type (I). To prove that a given sequence $(\widetilde{T}_n)_n$ attains a bound of type (I):

$$\lim_{c \downarrow 0} \lim_{n \to \infty} \sup_{|\vartheta - \vartheta_0| < c} R_n(\widetilde{T}_n, \vartheta) = I_{\vartheta_0}^{-1}$$

one needs some 'uniformity in the parameter' for weak convergence of rescaled estimation errors which has to be proved separately. On the other hand, Le Cam [81, 82], Hajek [40], Davies [19] or Le Cam and Yang [84] work – in more general settings than the present one – with bounds of type (II). To prove that an estimator sequence $(\widetilde{T}_n)_n$ attains a type (II) bound

$$\lim_{C \uparrow \infty} \lim_{n \to \infty} \sup_{|\vartheta - \vartheta_0| < C/\sqrt{n}} R_n(\widetilde{T}_n, \vartheta) = I_{\vartheta_0}^{-1}$$

no separate proof for 'uniformity in the parameter' for weak convergence of rescaled estimation errors is needed since 'Le Cam's third lemma' settles this problem, see Chapter 3. Our focus in later chapters will be on bounds of type (*II*), see Chapter 7; we will also be interested in loss functions different from squared loss, and in estimator sequences $(\widetilde{T}_n)_n$ which achieve type (*II*) bounds simultaneously with respect to a broad class of loss functions.

We conclude the present section by one example illustrating type (*I*) bounds in the spirit of the references [60] and [80] mentioned above.

1.11' Example. We consider i.i.d. observations $X_i, i \geq 1$, with unknown distribution function F in $\mathscr{F}$, the class of all distribution functions on $(\mathbb{R}, \mathscr{B}(\mathbb{R}))$, and write $\widehat{F}_n$ for the empirical distribution function based on the first n observations $X_1, \ldots, X_n$. For $x \in \mathbb{R}$ fixed, we consider mappings

$$\gamma : \quad \mathscr{F} \ni F \longrightarrow \gamma(F) := F(x) \in [0,1]$$

and wish to prove that the sequence $T_n := \widehat{F}_n(x)$ is asymptotically minimax for γ in a sense which we define below. Assertions of type 'the empirical distribution function provides asymptotically optimal estimators for unknown $F \in \mathscr{F}$' were first proved by Beran [9].

Let $X_i, i \geq 1$, be defined on some $(\Omega, \mathscr{A})$, let $\mathscr{A}_n := \sigma(X_1, \ldots, X_n)$ denote the sub-σ-field of $\mathscr{A}$ generated by the first n observations as in Lemma 1.5(b), and define neighbourhoods of F in $\mathscr{F}$

$$V_\delta(F) := \left\{ \widetilde{F} \in \mathscr{F} : \sup_{y \in \mathbb{R}} |\widetilde{F}(y) - F(y)| < \delta \right\}$$

with radius $\delta > 0$. Then for every $F \in \mathscr{F}$ and every point $x \in \mathbb{R}$ such that $0 < F(x) < 1$, we have a lower bound

(i) $\quad \lim_{\delta \downarrow 0} \liminf_{n \to \infty} \inf_{T_n \ \mathscr{A}_n\text{-mb}} \sup_{\widetilde{F} \in V_\delta(F)} E_{\widetilde{F}} \left(n \, [T_n - \widetilde{F}(x)]^2 \right) \geq F(x)(1 - F(x)),$

and the empirical distribution function satisfies

(ii) $\quad \lim_{\delta \downarrow 0} \lim_{n \to \infty} \sup_{\widetilde{F} \in V_\delta(F)} E_{\widetilde{F}} \left(n \, [\widehat{F}_n(x) - \widetilde{F}(x)]^2 \right) = F(x)(1 - F(x))$

and thus attains the bound specified in (i). The proof is in several steps.

(1) Fix $F \in \mathscr{F}$ and $x \in \mathbb{R}$. It is easy to prove (ii): from

$$\widehat{F}_n(x) - \widetilde{F}(x) = \frac{1}{n} \sum_{i=1}^n \left(1_{(-\infty, x]}(Y_i) - \widetilde{F}(x) \right)$$

for independent random variables $Y_i \sim \widetilde{F}$ we get

$$E_{\widetilde{F}} \left(n \, [\widehat{F}_n(x) - \widetilde{F}(x)]^2 \right) = \widetilde{F}(x)(1 - \widetilde{F}(x)) \quad \text{not depending on } n \geq 1$$

Section 1.2 Estimator Sequences, Asymptotics of Information Bounds

where by definition of $V_\delta(F)$ in terms of the uniform distance $\|f - g\|_\infty = \sup_{y \in \mathbb{R}} |f(y) - g(y)|$

$$\sup_{\widetilde{F} \in V_\delta(F)} \widetilde{F}(x)(1 - \widetilde{F}(x)) \longrightarrow F(x)(1 - F(x)) \quad \text{as } \delta \downarrow 0.$$

Thus the empirical distribution function attains the bound proposed in (i).

(2) It remains to prove the bound (i). Fix $F \in \mathcal{F}$ and $x \in \mathbb{R}$ such that $0 < F(x) < 1$. Let $\mathcal{H}$ denote the system of functions h in $L^2(F)$ which satisfy

$$h \text{ bounded on } \mathbb{R}, \quad \int_{-\infty}^\infty h\,dF = 0, \quad \int_{-\infty}^x h\,dF = 1$$

(in addition to Example 1.3(a), the third condition yields a particular norming of functions $\widetilde{h}$ which satisfy the first two conditions together with $\int_{-\infty}^x \widetilde{h}\,dF \neq 0$). For $h \in \mathcal{H}$ and neighbourhoods $V_\delta(F)$ of F we introduce one-parametric paths $\mathcal{S}^h$ through F by

$$\delta^h := \frac{\delta}{\sup|h|}, \quad \mathcal{S}^h := \{F_\vartheta^h : |\vartheta| < \delta^h\}, \quad dF_\vartheta^h := (1+\vartheta h)\,dF.$$

The choice of δ^h makes sure that $\mathcal{S}^h$ is contained in $V_\delta(F)$. As shown in (1.3'),

$$I_\vartheta^h := \int \frac{h^2}{1+\vartheta h}\,dF, \quad |\vartheta| < \delta^h$$

is the Fisher information in the path $\mathcal{S}^h$; in particular, we have in $\mathcal{S}^h$

$$I_0^h = \int h^2\,dF \quad \text{at } \vartheta = 0.$$

(3) Fix $h \in \mathcal{H}$. Due to the norming factor included in the definition of $\mathcal{H}$ we have in $\mathcal{S}^h$

$$F_\vartheta^h(x) = \int_{-\infty}^x (1+\vartheta h)\,dF = F(x) + \vartheta, \quad |\vartheta| < \delta^h,$$

hence any $\mathcal{A}_n$-measurable estimator T_n for $\gamma : \mathcal{F} \ni \widetilde{F} \longrightarrow \widetilde{F}(x) \in [0,1]$ estimates in restriction to $\mathcal{S}^h$

(+) $\quad\quad\quad\quad\quad\quad \gamma : \mathcal{S}^h \ni F_\vartheta^h \longrightarrow F(x) + \vartheta.$

Note that (+) makes appear a shift by $F(x)$; recall that F and x are fixed. In restriction to $\mathcal{S}^h$, we may associate to γ a new mapping $\overline{\gamma}^h$ and to T_n a new estimator $\overline{T}_n$

$$\overline{\gamma}^h(\vartheta) := \gamma(F_\vartheta^h) - F(x) = \vartheta, \quad F_\vartheta^h \in \mathcal{S}^h, \quad \overline{T}_n := T_n - F(x).$$

Then $\overline{T}_n$ is an $\mathcal{A}_n$-measurable estimator for $\overline{\gamma}^h = id$ on $\{\vartheta : |\vartheta| < \delta^h\}$. An estimator for the unknown parameter ϑ in the model $\{\vartheta : |\vartheta| < \delta^h\}$ corresponds to any $\mathcal{A}_n$-measurable estimator for γ in restriction to $\mathcal{S}^h$. Now we apply the asymptotic van

Trees bound in Theorem 1.11 for estimation of the unknown parameter ϑ in small type (I) neighbourhoods of the point $\vartheta_0 = 0$ in $\{\vartheta : |\vartheta| < \delta^h\}$:

$$\lim_{c \downarrow 0} \liminf_{n \to \infty} \inf_{T_n \mathcal{A}_n\text{-mb}} \sup_{|\vartheta|<c} E_{F_\vartheta^h}\left([\sqrt{n}(\overline{T}_n - \vartheta)]^2\right) \geq \left(I_0^h\right)^{-1}.$$

Transforming this back from $(\overline{T}_n, \overline{\gamma}^h = id)$ to (T_n, γ), we obtain at the point F in $\mathcal{S}^h$ a bound

$(++)$ $$\lim_{c \downarrow 0} \liminf_{n \to \infty} \inf_{T_n \mathcal{A}_n\text{-mb}} \sup_{F_\vartheta^h : |\vartheta|<c} E_{F_\vartheta^h}\left([\sqrt{n}(T_n - \gamma(F_\vartheta^h))]^2\right) \geq \left(I_0^h\right)^{-1}.$$

(4) Now we consider the system of functions $h \in \mathcal{H}$ with associated one-parametric paths $\mathcal{S}^h$ passing through F, and have from $\mathcal{S}^h \subset V_\delta(F)$ and from $(++)$ bounds

$$\lim_{\delta \downarrow 0} \liminf_{n \to \infty} \inf_{T_n \mathcal{A}_n\text{-mb}} \sup_{\widetilde{F} \in V_\delta(F)} E_{\widetilde{F}}\left([\sqrt{n}(T_n - \gamma(\widetilde{F}))]^2\right)$$

$$\geq \lim_{c \downarrow 0} \liminf_{n \to \infty} \inf_{T_n \mathcal{A}_n\text{-mb}} \sup_{F_\vartheta^h : |\vartheta|<c} E_{F_\vartheta^h}\left([\sqrt{n}(T_n - \gamma(F_\vartheta^h))]^2\right) \geq \left(I_0^h\right)^{-1}$$

for arbitrary $h \in \mathcal{H}$, whence

$$\lim_{\delta \downarrow 0} \liminf_{n \to \infty} \inf_{T_n \mathcal{A}_n\text{-mb}} \sup_{\widetilde{F} \in V_\delta(F)} E_{\widetilde{F}}\left([\sqrt{n}(T_n - \widetilde{F}(x))]^2\right) \geq \sup\left\{\left(I_0^h\right)^{-1} : h \in \mathcal{H}\right\}.$$

(5) In order to conclude the proof of (i) on the basis of the last inequality, it is sufficient to prove

$(+++)$ $$\sup\left\{\left(I_0^h\right)^{-1} : h \in \mathcal{H}\right\} = F(x)(1 - F(x)).$$

This is a consequence of the norming conditions imposed on the functions in class $\mathcal{H}$ at the start of step (2), together with Cauchy–Schwarz:

$$1 = \int_{-\infty}^{x} h(y) F(dy) - 0 = \int_{-\infty}^{+\infty} h(y) \left[1_{(-\infty,x]}(y) - F(x)\right] F(dy)$$

$$\leq \left(\int h^2 \, dF\right)^{\frac{1}{2}} \left(\int [1_{(-\infty,x]}(y) - F(x)]^2 F(dy)\right)^{\frac{1}{2}}$$

$$= \left(I_0^h\right)^{\frac{1}{2}} \left(F(x)(1 - F(x))\right)^{\frac{1}{2}}$$

where equality holds if and only if

(o) $\quad$ for some $c \in \mathbb{R}$: $h(y) = c \cdot [1_{(-\infty,x]}(y) - F(x)]$, $y \in \mathbb{R}$

or equivalently (again as a consequence of the norming conditions in class $\mathcal{H}$) if and only if

(oo) $$c = \frac{1}{F(x)(1 - F(x))}.$$

Hence the system $\mathcal{H}$ contains exactly one element $\bar{h}$ determined through (o) and (oo)

$$\bar{h}(y) = \frac{1_{(-\infty,x]}(y) - F(x)}{F(x)(1 - F(x))} = \begin{cases} \frac{1}{F(x)}, & y \le x \\ \frac{-1}{1-F(x)}, & y > x \end{cases}$$

($\bar{h}$ satisfies the norming conditions in $\mathcal{H}$, and is bounded since $0 < F(x) < 1$) with the property

$$\sup\left\{\left(I_0^h\right)^{-1} : h \in \mathcal{H}\right\} = \left(I_0^{\bar{h}}\right)^{-1} = \left(\int \bar{h}^2 dF\right)^{-1} = F(x)(1 - F(x)).$$

This element $\bar{h} \in \mathcal{H}$ is called a *least favourable direction* at F; the associated model $\mathcal{E}^{\bar{h}}$ is a *least favourable one-parametric path* through F. Within the collection of paths $\mathcal{E}^h$, $h \in \mathcal{H}$, the least favourable path $\mathcal{E}^{\bar{h}}$ minimises the Fisher information at F. This is (+++). Hence, by the last inequality in step (4), the asymptotic minimax bound (i) is proved. □

1.3 Heuristics on Maximum Likelihood Estimator Sequences

There is a close connection between score and information and maximum likelihood (ML) estimation. In this section, we discuss – in heuristic terms only – a tradition which seemed convinced that under independent replication of experiments, maximum likelihood should lead in all relevant statistical problems to estimator sequences which are $\sqrt{n}$-consistent, asymptotically normal with covariance matrix of the limit distribution at ϑ equal to the inverse I_ϑ^{-1} of the Fisher information, and such that estimator sequences with better concentrated limit distribution do not exist. These heuristics aim at a representation of rescaled ML estimation errors in terms of score and information, stated under (h6) and (h7) in Heuristics III below. Later (see Chapter 7), based on a mathematically different approach and on assumptions of a different type, a representation of rescaled estimation errors of type (h6) and (h7) will indeed show up again, and will characterise in broad generality estimator sequences which are asymptotically optimal in a rigorous sense. We illustrate by examples where traditional heuristics may work – or do not work at all. Equalities and formulae based on heuristics will be indicated through a notation $\stackrel{(!!)}{=}$ and a numerotation (h1)–(h7), and sometimes by a (!!) in the text.

1.12 Heuristics I (Interchange conditions). Consider an experiment

$$\mathcal{E} := (\Omega, \mathcal{A}, \{P_\vartheta : \vartheta \in \Theta\}), \quad \Theta \subset \mathbb{R}^d \text{ open}$$

admitting score $\{M_\vartheta : \vartheta \in \Theta\}$ and Fisher information $\{I_\vartheta : \vartheta \in \Theta\}$, under all conditions of Definition 1.2. The densities $f_\vartheta = \frac{dP_\vartheta}{d\mu}$ are assumed strictly positive and the parameterisation $\vartheta \to f(\vartheta, \cdot)$ sufficiently smooth on Θ. In such experiments $\mathcal{E}$, we expect the following *interchange conditions* (h1) and (h2) to hold. First, as in Remark 1.7(a) we should have

$$\frac{\partial}{\partial \vartheta_i} E_\vartheta Y = \frac{\partial}{\partial \vartheta_i} \int Y f_\vartheta \, d\mu \stackrel{(!!)}{=} \int Y \frac{\partial}{\partial \vartheta_i} f_\vartheta \, d\mu = \int Y \left(\frac{\partial}{\partial \vartheta_i} \log f_\vartheta \right) f_\vartheta \, d\mu$$

in our model for suitable $Y : (\Omega, \mathcal{A}) \to (\mathbb{R}, \mathcal{B}(\mathbb{R}))$, and thus

(h1) $$\nabla^\top (E_\vartheta Y) \stackrel{(!!)}{=} E_\vartheta (Y \cdot M_\vartheta^\top)$$

which corresponds to the special case of dimension $k = 1$ in condition (++) in Proposition 1.6. Second, accepting (h1) we should also have

$$\frac{\partial}{\partial \vartheta_j} \frac{\partial}{\partial \vartheta_i} E_\vartheta Y \stackrel{(!!)}{=} \int Y \frac{\partial}{\partial \vartheta_j} \left(\left(\frac{\partial}{\partial \vartheta_i} \log f_\vartheta \right) \cdot f_\vartheta \right) d\mu$$

$$= \int Y \left(\left(\frac{\partial}{\partial \vartheta_j} \frac{\partial}{\partial \vartheta_i} \log f_\vartheta \right) \cdot f_\vartheta \right.$$

$$\left. + \left(\frac{\partial}{\partial \vartheta_i} \log f_\vartheta \right) \left(\frac{\partial}{\partial \vartheta_j} \log f_\vartheta \right) \cdot f_\vartheta \right) d\mu$$

$$= E_\vartheta \left(Y \left(\frac{\partial}{\partial \vartheta_j} \frac{\partial}{\partial \vartheta_i} \log f_\vartheta + M_{\vartheta,i} M_{\vartheta,j} \right) \right)$$

where $M_{\vartheta,1}, \ldots, M_{\vartheta,d}$ are the components of M_ϑ. In the particular case $Y \equiv 1$, the l.h.s. of this chain of equations equals 0, hence by definition of score and Fisher information, assuming tacitely P_ϑ-integrability of $(\nabla \nabla^\top \log f)(\vartheta, \cdot)$,

(h2) $$I_\vartheta = E_\vartheta \left(M_\vartheta M_\vartheta^\top \right) \stackrel{(!!)}{=} -E_\vartheta \left((\nabla \nabla^\top \log f)(\vartheta, \cdot) \right).$$

Again for general Y we can write the first line of the above chain of equalities in the alternative form

$$\frac{\partial}{\partial \vartheta_j} \frac{\partial}{\partial \vartheta_i} E_\vartheta Y \stackrel{(!!)}{=} \int Y \frac{\partial}{\partial \vartheta_j} \left(\frac{\partial}{\partial \vartheta_i} f(\vartheta, \cdot) \right) d\mu.$$

Comparing right-hand sides in their original and their alternative form when $Y \equiv 1$, the condition

(h2') $$E_\mu \left((\nabla \nabla^\top f)(\vartheta, \cdot) \right) \stackrel{(!!)}{=} 0$$

should be equivalent to the interchange condition (h2). □

1.13 Heuristics II (ML method). Consider a model $\mathcal{E}$ with score $\{M_\vartheta : \vartheta \in \Theta\}$ and Fisher information $\{I_\vartheta : \vartheta \in \Theta\}$ as in Definition 1.2, with densities $f_\vartheta = \frac{dP_\vartheta}{d\mu}$ strictly

Section 1.3 Heuristics on Maximum Likelihood Estimator Sequences

positive, and such that the parameterisation $\vartheta \to f(\vartheta,\cdot)$ is sufficiently smooth. A maximum likelihood estimator $\widehat{\vartheta}$ for the unknown parameter in $\mathcal{E}$ is a statistic selecting for $\omega \in \Omega$ a point $\widehat{\vartheta}(\omega)$ where the likelihood function

$$\Theta \ni \vartheta \longrightarrow f(\vartheta,\omega) \in (0,\infty)$$

attains its maximum.

1.14 Heuristics III (Asymptotics of ML estimators). Consider independent replication of an experiment $\mathcal{E}$ as in Heuristics 1.12 and 1.13

$$\mathcal{E}_n = (\Omega_n, \mathcal{A}_n, \{P_{n,\vartheta} : \vartheta \in \Theta\}) = \left(\overset{n}{\underset{i=1}{\times}} \Omega, \overset{n}{\underset{i=1}{\otimes}} \mathcal{A}, \{P_{n,\vartheta} := \overset{n}{\underset{i=1}{\otimes}} P_\vartheta : \vartheta \in \Theta\} \right)$$

with score $M_{n,\vartheta}$ in $\vartheta \in \Theta$ and information $I_{n,\vartheta} = n\, I_\vartheta$ as in Lemma 1.5(a). Let $\widehat{\vartheta}_n$ denote the maximum likelihood estimator for the unknown parameter in the product model $\mathcal{E}_n$, $n \geq 1$. We expect that the sequence $(\widehat{\vartheta}_n)_n$ is consistent as $n \to \infty$ (!!). Since $\widehat{\vartheta}_n$ is a point in Θ where the likelihood function attains its maximum, a Taylor expansion of the gradient of the log-likelihood function at $\widehat{\vartheta}_n$ reads as

$$(\nabla^\top \log f_n)(\xi) \approx \underbrace{(\nabla^\top \log f_n)(\widehat{\vartheta}_n)}_{=0} + (\xi - \widehat{\vartheta}_n)^\top (\nabla \nabla^\top \log f_n)(\widehat{\vartheta}_n)$$

for points ξ in Θ which are close to $\widehat{\vartheta}_n$. Inserting the true parameter value ϑ gives

(h3) $$(\nabla^\top \log f_n)(\vartheta) \overset{(!!)}{=} \left[\underbrace{(\nabla^\top \log f_n)(\widehat{\vartheta}_n)}_{=0} + (\vartheta - \widehat{\vartheta}_n)^\top (\nabla \nabla^\top \log f_n)(\widehat{\vartheta}_n) \right] (1 + o_{P_{n,\vartheta}}(1))_d \;.$$

Here $o_{P_{n,\vartheta}}(1)$ denotes remainder terms which vanish in $(P_{n,\vartheta})_n$-probability as n tends to ∞, and $(1 + o_{P_{n,\vartheta}}(1))_d$ means a diagonal matrix with diagonal entries $1 + o_{P_{n,\vartheta}}(1)$. We expect that second derivatives of the log-likelihoods should not vary much in the parameter (in the classical normal distribution model with unknown mean and known covariance, log-likelihoods are quadratic in the parameter, thus second derivatives do not depend on the parameter), i.e.

(h4) $$\frac{\partial}{\partial \vartheta_j} \frac{\partial}{\partial \vartheta_i} \log f_n(\widehat{\vartheta}_n) \approx \frac{\partial}{\partial \vartheta_j} \frac{\partial}{\partial \vartheta_i} \log f_n(\vartheta).$$

Inserting (h4) into the expansion (h3) we should have a final form

(h5) $$(\nabla^\top \log f_n)(\vartheta) \overset{(!!)}{=} (\vartheta - \widehat{\vartheta}_n)^\top (\nabla \nabla^\top \log f_n)(\vartheta) (1 + o_{P_{n,\vartheta}}(1))_d$$

of the expansion. In the product model $\mathcal{E}_n$ we have

$$\frac{1}{n}(\nabla\nabla^\top \log f_n)(\vartheta,(\omega_1,...,\omega_n)) = \frac{1}{n}\sum_{i=1}^n (\nabla\nabla^\top \log f)(\vartheta,\omega_i)$$

and the strong law of large numbers combined with (h2) gives almost sure convergence

$$(\diamond) \qquad \frac{1}{n}\left(\nabla\nabla^\top \log f_n\right)(\vartheta,\cdot) \longrightarrow E_\vartheta(\nabla\nabla^\top \log f(\vartheta,\cdot)) = -I_\vartheta$$

under ϑ as $n \to \infty$. According to Lemma 1.5, the score in product models

$$(\nabla \log f_n)(\vartheta,(\omega_1,...,\omega_n)) = M_{n,\vartheta}(\omega_1,...,\omega_n) = \sum_{i=1}^n M_\vartheta(\omega_i)$$

is asymptotically normal by definition of score and information:

$$(\diamond\diamond) \qquad \mathcal{L}\left(\frac{1}{\sqrt{n}} M_{n,\vartheta} \mid P_{n,\vartheta}\right) \longrightarrow \mathcal{N}(0, I_\vartheta)$$
(weak convergence in $\mathbb{R}^d$ as $n \to \infty$).

In combination with $(\diamond)$, expansion (h5) written as

$$\frac{1}{\sqrt{n}}(\nabla^\top \log f_n)(\vartheta) \stackrel{(!!)}{=} \sqrt{n}(\vartheta - \hat\vartheta_n)^\top \frac{1}{n}(\nabla\nabla^\top \log f_n)(\vartheta)\left(1 + o_{P_{n,\vartheta}}(1)\right)_d$$

takes the form

$$\frac{1}{\sqrt{n}} M_{n,\vartheta}^\top \stackrel{(!!)}{=} \left(\sqrt{n}(\hat\vartheta_n - \vartheta)\right)^\top I_\vartheta \left(1 + o_{P_{n,\vartheta}}(1)\right)_d .$$

By $(\diamond\diamond)$ the l.h.s. of this equation is tight in $\mathbb{R}^d$ under ϑ as $n \to \infty$; inverting the Fisher information matrix (!!) we get tightness in $\mathbb{R}^d$ of rescaled estimation errors under ϑ

(h6) $\qquad \left(\sqrt{n}(\hat\vartheta_n - \vartheta)\right) \stackrel{(!!)}{=} I_\vartheta^{-1}\left(1 + o_{P_{n,\vartheta}}(1)\right)_d \frac{1}{\sqrt{n}} M_{n,\vartheta}$

as $n \to \infty$. Combined with $(\diamond\diamond)$, (h6) thus implies asymptotic normality

(h7) $\qquad \mathcal{L}(\sqrt{n}(\hat\vartheta_n - \vartheta) \mid P_{n,\vartheta}) \longrightarrow \mathcal{N}(0, I_\vartheta^{-1})$ (weakly in $\mathbb{R}^d$ as $n \to \infty$)

of rescaled estimation errors. The limit variance at ϑ in (h7) is the inverse of the Fisher information in $\mathcal{E}$. According to Cramér–Rao asymptotics (Remark 1.10) or – assuming some uniformity in the parameter – van Trees asymptotics (Theorem 1.11) this is a lower bound for limit variances of 'good' estimators for the unknown parameter: hence ML estimators attain this bound asymptotically. □

Section 1.3 Heuristics on Maximum Likelihood Estimator Sequences

We turn to examples. There are two examples where all laws under consideration are normal distributions: in the first case, all approximations above are justified; in the second case we run into trouble. A third example illustrates that we can be extremely far from everything which seemed 'natural' in the above heuristics.

1.15 Example. Consider n-fold replication

$$\mathcal{E}_n := \left(\Omega_n := \underset{i=1}{\overset{n}{\times}} \mathbb{R}^k,\ \mathcal{A}_n := \underset{i=1}{\overset{n}{\otimes}} \mathcal{B}(\mathbb{R}^k),\ \left\{ P_{n,\vartheta} := \underset{i=1}{\overset{n}{\otimes}} \mathcal{N}(\vartheta, \Lambda) : \vartheta \in \Theta \right\} \right),$$

$$\Theta := \mathbb{R}^k$$

of the usual normal distribution model $\mathcal{E}$ with known $k \times k$-covariance matrix Λ, symmetric and strictly positive definite; the parameter of interest is the mean value $\vartheta \in \mathbb{R}^k$. Write $\omega = (\omega_1, \ldots, \omega_n)$ for the elements of Ω_n. The dimension of successive observations ω_i is $k \geq 1$, and we wish to estimate the unknown parameter.

(a) Based on observations $\omega_1, \ldots, \omega_n$ in $\mathbb{R}^k$, the likelihood function

$$\Theta \ni \vartheta \longrightarrow$$

$$f_n(\vartheta, (\omega_1, \ldots, \omega_n)) = \left(\frac{1}{(2\pi)^{\frac{k}{2}} (\det \Lambda)^{\frac{1}{2}}} \right)^n \exp\left(-\frac{1}{2} \sum_{i=1}^n (\omega_i - \vartheta)^\top \Lambda^{-1} (\omega_i - \vartheta) \right)$$

has a unique maximum at the empirical mean

$$\widehat{\vartheta}_n(\omega_1, \ldots, \omega_n) := \frac{1}{n} \sum_{i=1}^n \omega_i.$$

Using the notation $(X_1, \ldots, X_n) := id\,|_{\Omega_n}$, the ML estimator for the unknown parameter in $\mathcal{E}_n$ is thus

$$\widehat{\vartheta}_n = \frac{1}{n} \sum_{i=1}^n X_i.$$

(b) In all i.i.d. models where the single observation is a square integrable statistic, by the strong law of large numbers and the central limit theorem, the sequence of empirical means is a $\sqrt{n}$-consistent and asymptotically normal estimator sequence for $\vartheta = E_\vartheta(X_1)$. In the present example we have more: laws of rescaled estimation errors

(o) $\quad \mathcal{L}\left(\sqrt{n}(\widehat{\vartheta}_n - \vartheta) \mid P_{n,\vartheta} \right) = \mathcal{L}\left(\frac{1}{\sqrt{n}} \sum_{i=1}^n (X_i - \vartheta) \mid P_{n,\vartheta} \right) = \mathcal{N}(0_k, \Lambda)$

do not depend on $n \in \mathbb{N}$. It is easy to calculate score and information in $\mathcal{E}_n$:

$$M_{n,\vartheta} = \Lambda^{-1} \sum_{i=1}^n (X_i - \vartheta),\quad I_{n,\vartheta} = n\,\Lambda^{-1},\quad \vartheta \in \mathbb{R}^k.$$

Assertion (o) is a particular case of representation (h7), and

(oo) $\quad \sqrt{n}(\widehat{\vartheta}_n - \vartheta) = I_\vartheta^{-1} \frac{1}{\sqrt{n}} M_{n,\vartheta}$

a particular case of representation (h6), without $o_{P_{n,\vartheta}}(1)$-remainder terms. □

The log-likelihood function in the normal distribution model 1.15 is a negative quadratic polynomial

$$\vartheta \longrightarrow$$

$$-\frac{1}{2}\sum_{i=1}^{n}(X_i(\omega)-\vartheta)^\top \Lambda^{-1}(X_i(\omega)-\vartheta) + \text{expressions which do not depend on } \vartheta$$

in $\vartheta \in \mathbb{R}^k$, and is, for every $\omega \in \Omega$, centred at the maximum likelihood estimator $\widehat{\vartheta}_n = \frac{1}{n}\sum_{i=1}^{n} X_i$ which as in (o) and (oo) of Example 1.15 is close to the true parameter value. In this sense, the normal distribution model 1.15 serves as a prototype example for a broad class of statistical models where one can prove that log-likelihoods are locally – in small neighbourhoods of their unique argmax – approximatively quadratic in the parameter. Le Cam [83] amused himself in collecting some seemingly harmless examples where maximum likelihood goes wrong. In the next example – a $2k$-dimensional normal distribution model with unknown mean and unknown variance where the mean value parameter is restricted to a particular k-dimensional hyperplane in $\mathbb{R}^{2k}$ – maximum likelihood estimators are asymptotically normal, well concentrated, but not around the true parameter value.

1.16 Example ([83], going back to Neyman and Scott). Take $k \in \mathbb{N}$ arbitrarily large. On $(\Omega, \mathcal{A}) := (\mathbb{R}^{2k}, \mathcal{B}(\mathbb{R}^{2k}))$ where $(X_1, Y_1, X_2, Y_2, ..., X_k, Y_k) := id\,|_\Omega$ denotes the canonical variable, we consider the experiment $\{P_\vartheta : \vartheta \in \Theta\}$

$$\vartheta =: (\sigma^2, \xi_1, \xi_2, ..., \xi_k), \quad \Theta := (0, \infty) \times \mathbb{R}^k$$

defined by

$$\phi(\vartheta) := (\xi_1, \xi_1, \xi_2, \xi_2, ..., \xi_k, \xi_k) \in \mathbb{R}^{2k}, \quad P_\vartheta := \mathcal{N}(\phi(\vartheta), \sigma^2 I_{2k}), \quad \vartheta \in \Theta$$

together with two different estimators for the mapping $\gamma(\vartheta) = \sigma^2$.

(1) The random variables $\frac{X_i - Y_i}{\sqrt{2}}$, $i = 1, \ldots, k$, are i.i.d. and distributed according to $\mathcal{N}(0, \sigma^2)$, hence an empirical mean

$$T := \frac{1}{k}\sum_{i=1}^{k}\left(\frac{X_i - Y_i}{\sqrt{2}}\right)^2$$

is a reasonable estimator for $\gamma(\vartheta) = \sigma^2$. Since $\mathcal{L}(Z^2) = \Gamma(\frac{1}{2}, \frac{1}{2})$ for $Z \sim \mathcal{N}(0, 1)$, the usual properties of Gamma laws under convolution and scaling give

$$\mathcal{L}(T \mid P_\vartheta) = \Gamma\left(\frac{k}{2}, \frac{k}{2\sigma^2}\right), \quad E_\vartheta(T) = \sigma^2, \quad Var_\vartheta(T) = \frac{2\sigma^4}{k}.$$

In particular, this estimator T is unbiased for γ, square integrable, and well concentrated around the true σ^2 for large k.

(2) We prove that the likelihood function

$$\vartheta \longrightarrow \left(\frac{1}{\sqrt{2\pi\sigma^2}}\right)^{2k} \prod_{i=1}^{k} \exp\left(-\frac{1}{2\sigma^2}\{(X_i - \xi_i)^2 + (Y_i - \xi_i)^2\}\right)$$

admits a unique maximum in Θ. First, for $i = 1, ..., k$ fixed,

$$\xi_i \longrightarrow (X_i - \xi_i)^2 + (Y_i - \xi_i)^2 \quad \text{has a unique minimum at} \quad \widehat{\xi}_i := \frac{X_i + Y_i}{2},$$

second, inserting the $\widehat{\xi}_i$ into the likelihood and taking logarithms, it remains to consider

$$\sigma^2 \longrightarrow k \log\left(\frac{1}{\sigma^2}\right) - \sum_{i=1}^{k} \frac{1}{2\sigma^2} \left\{ 2 \left(\frac{X_i - Y_i}{2}\right)^2 \right\}$$

which has a unique maximum at

$$\widehat{\sigma}^2 := \frac{1}{k} \sum_{i=1}^{k} \left(\frac{X_i - Y_i}{2}\right)^2 = \frac{1}{2} T.$$

Hence there is a unique maximum likelihood estimator $\widehat{\vartheta}$ for ϑ, defined by its components $\widehat{\sigma}^2, \widehat{\xi}_1, ..., \widehat{\xi}_k$. Since $\widehat{\sigma}^2 = \frac{1}{2}T$, the above properties of T give

$$\mathcal{L}(\widehat{\sigma}^2 \mid P_\vartheta) = \Gamma\left(\frac{k}{2}, \frac{k}{\sigma^2}\right), \quad E_\vartheta(\widehat{\sigma}^2) = \frac{\sigma^2}{2}, \quad \text{Var}_\vartheta(\widehat{\sigma}^2) = \frac{\sigma^4}{2k}.$$

This means that the maximum likelihood estimator concentrates in its first component around one half of the true value, with obviously dramatic effects for large values of the parameter σ^2 once the model dimension k is large enough. □

The last example shows that the Heuristics 1.12–1.14 cannot pretend to serve as universal guidelines for good estimation. The next example, associated to classically well-behaving models, makes types of statistical experiments appear which are very far from all the considerations above.

1.16' Example. Consider a probability space $(\Omega, \mathcal{A}, P_0)$ carrying a one-dimensional Lévy process $(X_u)_{u \geq 0}$ (i.e.: stationary independent increments), starting from $X_0 = 0$, such that the Laplace transform of X_1 exists on some open interval D in $\mathbb{R}$. We can put Laplace transforms in the form

$$\vartheta \longrightarrow E_{P_0}(e^{\vartheta X_u}) = e^{u \psi(\vartheta)}, \quad \vartheta \in D, u \geq 0$$

which gives rise to two types of statistical models defined from the Laplace transforms.
(1) Fix $u = 1$ and define a family of probability laws $\{P_\vartheta : \vartheta \in D\}$ on $(\Omega, \mathcal{A})$ by

$$dP_\vartheta := e^{\vartheta X_1 - \psi(\vartheta)} dP_0, \quad \vartheta \in D.$$

As in Barra [4, Chap. X], this model is an exponential family, and the function ψ is $\mathcal{C}^\infty$ on D. In particular, for standard Brownian motion X we have $\psi(\vartheta) = \frac{1}{2}\vartheta^2$ and $D = \mathbb{R}$; for Poisson process X with parameter $\lambda > 0$ we have $\psi(\vartheta) = \lambda(e^\vartheta - 1)$ and $D = \mathbb{R}$.

(2) Fix a point ϑ_0 in D; w.l.o.g. we write $\vartheta_0 = 1$. Define probability laws $\{\widetilde{P}_u : u > 0\}$ on $(\Omega, \mathcal{A})$ by

$$d\widetilde{P}_u := e^{X_u - \psi(1)u} \, dP_0, \quad u > 0.$$

Now 'smoothness' of the log-likelihood function $u \to X_u - \psi(1)u$ is 'smoothness' of the path $u \to X_u$. In the case of standard Brownian motion X, the log-likelihood function is thus nowhere differentiable, P_0-a.s., and Hölder continuous of order $\frac{1}{2} - \varepsilon$ for every $\varepsilon > 0$ (e.g., see [112]). In the case of Poisson process X with parameter $\lambda > 0$, the log-likelihood function is piecewise linear, and has jumps of fixed size $+1$ at the jumps times of the Poisson path $u \to X_u$, i.e. after successive independent exponential waiting times with parameter λ. So we are very far from all assumptions in Definition 1.2.

(3) It is known that in models of type (2), under quadratic loss, maximum likelihood estimators are outperformed by squared-loss Bayesians. See Dachian [17] for an interesting class of examples of type (2); in the case of Brownian motion X, this has been established in successive steps by Golubev [32], Ibragimov and Khasminskii [60] and Rubin and Song [114]. □

1.4 Consistency of ML Estimators via Hellinger Distances

There are several ways to prove convergence of ML estimators for the unknown parameter. A rigorous route along the programme sketched in Heuristics 1.12–1.14 requires very strong conditions of type 'uniform integrability', see e.g. Witting and Müller-Funk [128, Chap. 6.1.3]); the proofs go back to Wald. In sharp contrast, the present section points towards a completely different way to pose the same problem; this can be applied successfully in a broad variety of cases, under weak and natural conditions. Ibragimov and Khasminskii [60, Thms. 19–20 in App. I.4, and Chaps. I.4–I.5] analysed the asymptotic behaviour of ML estimators in terms of Hellinger distances in the statistical model, and in terms of weak convergence of likelihoods – viewed as random fields in the (local) parameter – to a random field of limiting likelihoods. We explain only a first step on this route – consistency – in the present section, for i.i.d. models. The main result is in Theorem 1.24.

In general statistical models, likelihood functions $\vartheta \to f(\vartheta, \omega)$ do not necessarily achieve maxima on Θ, and if maxima exist, they need not be unique. When maxima exist, we shall be interested in the set of parameter values where the maximum is

Section 1.4 Consistency of ML Estimators via Hellinger Distances

attained. Thus we start with a careful definition of ML estimators and ML estimator sequences in arbitrary sequences of models.

1.17 Definition. (a) Consider a statistical model

$$\mathcal{E} = (\Omega, \mathcal{A}, \{P_\vartheta : \vartheta \in \Theta\}), \quad \Theta \subset \mathbb{R}^d$$

which is dominated by some σ-finite measure μ on $(\Omega, \mathcal{A})$, densities $f_\vartheta = \frac{dP_\vartheta}{d\mu}$ with respect to μ, and the likelihood function

$$\Theta \ni \vartheta \longrightarrow f(\vartheta, \omega) = f_\vartheta(\omega) \in [0, \infty)$$

based on the observation $\omega \in \Omega$.

(i) For $T : (\Omega, \mathcal{A}) \to (\mathbb{R}^d, \mathcal{B}(\mathbb{R}^d))$ measurable and $A \in \mathcal{A}$, we call T *maximum likelihood on the event A* if $\omega \in A$ implies

$$T(\omega) \in \Theta \quad \text{and} \quad f(T(\omega), \omega) = \sup\{f(\xi, \omega) : \xi \in \Theta\}.$$

(ii) We call $T : (\Omega, \mathcal{A}) \to (\mathbb{R}^d, \mathcal{B}(\mathbb{R}^d))$ a *maximum likelihood estimator* for the unknown parameter if there is some event $A \in \mathcal{A}$ such that

T is maximum likelihood on the event A, and $P_\vartheta(A) = 1$ for all $\vartheta \in \Theta$.

(b) Consider a sequence of experiments parameterised by $\Theta \in \mathbb{R}^d$

$$\mathcal{E}_n = (\Omega_n, \mathcal{A}_n, \{P_{n,\vartheta} : \vartheta \in \Theta\}), \quad n \geq 1$$

and assume that for every n, $\mathcal{E}_n$ is dominated by some μ_n with densities $f_n(\vartheta, \cdot) = \frac{dP_{n,\vartheta}}{d\mu_n}$, $\vartheta \in \Theta$. Consider a sequence of estimators $(T_n)_n$ in $(\mathcal{E}_n)_n$

$$T_n : (\Omega_n, \mathcal{A}_n) \longrightarrow (\mathbb{R}^d, \mathcal{B}(\mathbb{R}^d)), \quad n \geq 1$$

as in Definition 1.9. If there exist $A_n \in \mathcal{A}_n$ such that the following two conditions hold

$$\begin{cases} \text{for every } \vartheta \in \Theta : \quad \lim_{n \to \infty} P_{n,\vartheta}(A_n) = 1, \\ \text{for every } n \geq 1 : \quad T_n \text{ is maximum likelihood on the event } A_n, \end{cases}$$

then $(T_n)_n$ is called a *maximum likelihood estimator sequence* for the unknown parameter, and $(A_n)_n$ a *sequence of good sets* for $(T_n)_n$.

In most cases, explicit representations for ML estimators as in Examples 1.15 and 1.16 do not exist; usually in practical problems, even if there is a unique maximum in $\vartheta \to f_n(\vartheta, \omega)$ for every ω in the good set A_n it has to be evaluated numerically.

1.17' Exercise. Construct an ML estimator sequence in $(\mathcal{E}_n)_n$ of Definition 1.17(b) under the following set of conditions: (i) Θ is open in $\mathbb{R}^d$; (ii) for all $n \in \mathbb{N}$ and all $\omega \in \Omega_n$, densities $f_n(\vartheta, \omega)$ are continuous in $\vartheta \in \Theta$; (iii) for any compact exhaustion $(K_n)_n$ of Θ, events

$$A_n := \{\omega \in \Omega_n : \max\{f_n(\xi, \omega) : \xi \in K_n\} = \sup\{f_n(\xi, \omega) : \xi \in \Theta\}\}$$

are such that $\lim_{n\to\infty} P_{n,\vartheta}(A_n) = 1$ for every $\vartheta \in \Theta$. Hint: with respect to $(K_n)_n$ and $(A_n)_n$, for every n, associate to $\omega \in A_n$ a non-void compact set

$$M_n(\omega) := \{\xi^* \in K_n : f_n(\xi^*, \omega) = \max\{f_n(\xi, \omega) : \xi \in K_n\}\} \subset K_n$$

depending on ω, and specify a random variable $T_n : (\Omega_n, \mathcal{A}_n) \to (\mathbb{R}^d, \mathcal{B}(\mathbb{R}^d))$ such that

$$\omega \in A_n \quad \text{implies} \quad T_n(\omega) \in M_n(\omega) ;$$

here 'measurable selection' can be done e.g. as in step (2) of the proof of Proposition 2.10 in Chapter 2 below. □

In this section, we do not focus on the explicit construction of estimators (Chapter 2 below will cover this topic for a particular class of estimators): we are interested in the location of the random set of maxima of the likelihood function $\vartheta \to f_n(\vartheta, \omega)$, in order to show that this set – under suitable conditions – shrinks as $n \to \infty$ towards the true parameter value. The following example using Kullback divergence (from Pfanzagel [105, Chap. 6.5]) illustrates this point.

1.17" Example. Considering equivalent probability laws P_ϑ und $P_\xi \sim P_\vartheta$, Kullback divergence is defined (e.g. [124, Chap. 2.4]) as

$$K(P_\vartheta, P_\xi) := \int -\log\left(\frac{f_\xi}{f_\vartheta}\right) dP_\vartheta ;$$

by Jensen inequality for the convex function $-\log(\cdot)$ we have without further conditions

$$0 = -\log(1) = -\log E_\vartheta\left(\frac{f(\xi, \cdot)}{f(\vartheta, \cdot)}\right) \leq E_\vartheta\left(-\log \frac{f(\xi, \cdot)}{f(\vartheta, \cdot)}\right) \leq +\infty .$$

Consider an experiment

$$\mathcal{E} := (\Omega, \mathcal{A}, \{P_\vartheta : \vartheta \in \Theta\}), \quad \Theta \subset \mathbb{R}^d \text{ open}$$

with strictly positive densities $f(\vartheta, \omega)$ with respect to some dominating measure, and product models

$$\mathcal{E}_n = (\Omega_n, \mathcal{A}_n, \{P_{n,\vartheta} : \vartheta \in \Theta\}) = \left(\underset{i=1}{\overset{n}{\times}} \Omega, \underset{i=1}{\overset{n}{\otimes}} \mathcal{A}, \{P_{n,\vartheta} := \underset{i=1}{\overset{n}{\otimes}} P_\vartheta : \vartheta \in \Theta\}\right).$$

Assume that for every $\vartheta \in \Theta$ and every $\varepsilon > 0$ one can find a finite collection of open subsets $V_1, \ldots, V_l$ of Θ (with $l \in \mathbb{N}$ and $V_1, \ldots, V_l$ depending on ϑ and ε) such that $(*)$ and $(**)$ hold:

$$(*) \qquad \vartheta \notin \bigcup_{i=1}^{l} V_i \quad \text{and} \quad \{\xi \in \Theta : |\xi - \vartheta| > \varepsilon\} \subset \bigcup_{i=1}^{l} V_i$$

Section 1.4 Consistency of ML Estimators via Hellinger Distances

(**) $\quad \left(\sup_{\xi \in V_j} \log \frac{f(\xi, \cdot)}{f(\vartheta, \cdot)} \right) \in L^1(P_\vartheta) \quad$ and $\quad E_\vartheta \left(\sup_{\xi \in V_j} \log \frac{f(\xi, \cdot)}{f(\vartheta, \cdot)} \right) < 0,$

$$j = 1, \ldots, l.$$

Then every ML sequence for the unknown parameter is consistent.

Proof. (i) Given any ML sequence $(T_n)_n$ for the unknown parameter with 'good sets' $(A_n)_n$ in the sense of Definition 1.17 (b), fix $\theta \in \Theta$ and $\varepsilon > 0$ and select l and $V_1, \ldots, V_\ell$ according to (*). Then

$$\bigcap_{j=1}^{l} \left\{ \sup_{\xi \in V_j} \log f_n(\xi, \cdot) < \sup_{\xi \in \Theta : |\xi - \vartheta| \le \varepsilon} \log f_n(\xi, \cdot) \right\} \cap A_n \subset \{|T_n - \vartheta| \le \varepsilon\}$$

for every n fixed, by Definition 1.17(b) and since all densities are strictly positive. Inverting this,

$$\{|T_n - \vartheta| > \varepsilon\} \subset \bigcup_{j=1}^{l} \left\{ \sup_{\xi \in V_j} \log f_n(\xi, \cdot) \ge \sup_{\xi : |\xi - \vartheta| \le \varepsilon} \log f_n(\xi, \cdot) \right\} \cup A_n^c$$

$$\subset \bigcup_{j=1}^{l} \left\{ \sup_{\xi \in V_j} \log f_n(\xi, \cdot) \ge \log f_n(\vartheta, \cdot) \right\} \cup A_n^c .$$

(ii) In the sequence of product experiments $(\mathcal{E}_n)_n$, the $(A_n)_n$ being 'good sets' for $(T_n)_n$, we use the strong law of large numbers thanks to assumption (**) to show that

$$P_{n,\vartheta} \left(\left\{ (\omega_1, \ldots, \omega_n) : \frac{1}{n} \sum_{i=1}^{n} \left(\sup_{\xi \in V_j} \log \frac{f(\xi, \omega_i)}{f(\vartheta, \omega_i)} \right) \ge 0 \right\} \right) \longrightarrow 0$$

as $n \to \infty$ for every j, $1 \le j \le l$: this gives $\lim_{n \to \infty} P_{n,\vartheta}(|T_n - \vartheta| > \varepsilon) = 0.$ □

In order to introduce a geometry on the parametric statistical experiment, to be distinguished from Euclidian geometry on the parameter set $\Theta \subset \mathbb{R}^d$, we make use of *Hellinger distance* $H(\cdot, \cdot)$ which defines a metric on the space of all probability measures on $(\Omega, \mathcal{A})$; see e.g. [124, Chap. 2.4] or [121, Chap. I.2]. Recall that countable collections of probability measures on the same space are dominated.

1.18 Definition. The *Hellinger distance* $H(\cdot, \cdot)$ between probability measures Q_1 and Q_2 on $(\Omega, \mathcal{A})$ is defined as the square root of

$$H^2(Q_1, Q_2) := \frac{1}{2} \int \left| g_1^{1/2} - g_2^{1/2} \right|^2 d\mu \in [0, 1]$$

where $g_i = \frac{dQ_i}{d\mu}$ are densities with respect to a dominating measure μ, $i = 1, 2$. The *affinity* $A(\cdot, \cdot)$ between Q_1 and Q_2 is defined by

$$A(Q_1, Q_2) := 1 - H^2(Q_1, Q_2) = \int g_1^{1/2} g_2^{1/2} d\mu .$$

The integrals in Definition 1.18 do not depend on the choice of a dominating measure for Q_1 and Q_2 (similar to Remark 1.2'). We have $H(Q, Q') = 0$ if and only if probability measures Q, Q' coincide, and $H(Q, Q') = 1$ if and only if Q, Q' are mutually singular. Below, we focus on i.i.d. models $\mathcal{E}_n$ and follow Ibragimov and Khasminskii's route [60, Chap. 1.4] to consistency of maximum likelihood estimators under conditions on the Hellinger geometry in the single experiment $\mathcal{E}$.

For the remaining part of this section, the following assumptions will be in force. Note that we do not assume equivalence of probability laws, and do not assume continuity of densities in the parameter for fixed ω.

1.19 Notations and Assumptions. (a) (i) In the single experiment

$$\mathcal{E} = (\Omega, \mathcal{A}, \{P_\vartheta : \vartheta \in \Theta\}), \quad \Theta \subset \mathbb{R}^d \text{ open}$$

we have densities $f_\vartheta = \frac{dP_\vartheta}{d\mu}$ with respect to a dominating measure μ. For $\xi \neq \vartheta$, the likelihood ratio of P_ξ with respect to P_ϑ is

$$L^{\xi/\vartheta} = \frac{f_\xi}{f_\vartheta} 1_{\{f_\vartheta > 0\}} + \infty \cdot 1_{\{f_\vartheta = 0\}}.$$

(ii) We write $\mathcal{K}$ for the class of compact sets in $\mathbb{R}^d$ which are contained in Θ. A compact exhaustion of Θ is a sequence $(K_m)_m$ in $\mathcal{K}$ such that $K_m \subset \text{int}(K_{m+1})$ for all m, and $\bigcup_m K_m = \Theta$. We introduce the notations

$$\underline{h}(\xi, \gamma, K) := \inf_{\xi' \in K \setminus B_\gamma(\xi)} H^2(P_{\xi'}, P_\xi), \quad K \in \mathcal{K}, \xi \in K, \gamma > 0,$$

$$\overline{a}(\xi, K^c) := \int \sup_{\xi' \in K^c \cap \Theta} f_{\xi'}^{1/2} f_\xi^{1/2} \, d\mu, \quad K \in \mathcal{K}, \xi \in \text{int}(K),$$

$$\overline{\rho}(\xi, \delta) := \left(\int \sup_{\xi' \in B_\delta(\xi) \cap \Theta} \left| f_{\xi'}^{1/2} - f_\xi^{1/2} \right|^2 d\mu \right)^{1/2}, \quad \xi \in \Theta, \delta > 0.$$

(iii) For the single experiment $\mathcal{E}$, we assume a condition

(+) $\qquad\qquad\qquad \lim_{\delta \downarrow 0} \overline{\rho}(\xi, \delta) = 0 \quad \text{for every } \xi \in \Theta$

together with

(++) $\qquad\qquad\qquad \lim_{m \to \infty} \overline{a}(\xi, K_m^c) < 1$

for every $\xi \in \Theta$ and every compact exhaustion $(K_m)_m$ of Θ.

(b) For $n \geq 1$, we consider product models

$$\mathcal{E}_n = (\Omega_n, \mathcal{A}_n, \{P_{n,\vartheta} : \vartheta \in \Theta\}) = \left(\underset{i=1}{\overset{n}{\times}} \Omega, \underset{i=1}{\overset{n}{\otimes}} \mathcal{A}, \left\{ P_{n,\vartheta} := \underset{i=1}{\overset{n}{\otimes}} P_\vartheta : \vartheta \in \Theta \right\} \right).$$

Section 1.4 Consistency of ML Estimators via Hellinger Distances   35

In $\mathcal{E}_n$, we write $f_{n,\vartheta} = f_n(\vartheta, \cdot)$ for the density of $P_{n,\vartheta}$ with respect to $\otimes_{i=1}^n \mu$, and $L_n^{\xi/\vartheta}$ for the likelihood ratio of $P_{n,\xi}$ with respect to $P_{n,\vartheta}$.

In Assumptions 1.19(a), condition (+) implies continuity of the parameterisation in Hellinger distance, whereas (++) guarantees identifiability with respect to parameter values outside large compacts. The set of Assumptions 1.19 yields consistency of ML sequences, see Theorem 1.24 below. We proceed via some auxiliary results. The first is a control on likelihood ratios $L_n^{\xi/\vartheta}$ under $P_{n,\vartheta}$ when parameter values ξ are far away from ϑ.

1.20 Lemma. For $K \in \mathcal{K}$ and $\vartheta \in \text{int}(K)$, the condition $\bar{a}(\vartheta, K^c) < 1$ implies geometric decrease

$$E_{n,\vartheta}\left(\sup_{\xi \in \Theta \cap K^c} \sqrt{L_n^{\xi/\vartheta}}\right) \leq \left[\bar{a}(\vartheta, K^c)\right]^n, \quad n \geq 1.$$

Proof. Write for short $V := K^c \cap \Theta$. Then for $(\omega_1, \ldots, \omega_n) \in \{f_{n,\vartheta} > 0\}$

$$\left(\sup_{\xi \in V} \sqrt{L_n^{\xi/\vartheta}}\right)(\omega_1, \ldots, \omega_n) = \sup_{\xi \in V} \prod_{i=1}^n \frac{f^{1/2}(\xi, \omega_i)}{f^{1/2}(\vartheta, \omega_i)}$$

$$\leq \prod_{i=1}^n \left[\frac{1}{f^{1/2}(\vartheta, \omega_i)} \cdot \sup_{\xi \in V} f^{1/2}(\xi, \omega_i)\right]$$

and thus

$$E_{n,\vartheta}\left(\sup_{\xi \in V} \sqrt{L_n^{\xi/\vartheta}}\right) \leq \left[\int \sup_{\xi \in V} f_\vartheta^{1/2} f_\xi^{1/2} \, d\mu\right]^n \leq \left[\bar{a}(\vartheta, K^c)\right]^n$$

for every $n \geq 1$. □

1.21 Lemma. Condition 1.19 implies Hellinger continuity of the parameterisation

$$H(P_{\xi'}, P_\xi) \longrightarrow 0 \quad \text{whenever } \xi' \to \xi \text{ in } \Theta$$

which in turn guarantees identifiability within compacts contained in Θ:

$$\underline{h}(\xi, \gamma, K) > 0 \quad \text{for all } K \in \mathcal{K}, \xi \in K \text{ and } \gamma > 0.$$

Proof. By definition of $\bar{\rho}(\xi, \delta)$ and by (+) in Assumptions 1.19, the first assertion of the lemma holds by dominated convergence. From this, all mappings

(○) $\Theta \ni \xi' \longrightarrow H(P_{\xi'}, P_\xi) \in [0, 1], \quad \xi \in \Theta \text{ fixed}$

are continuous, by inverse triangular equality $|d(x, z) - d(y, z)| \leq d(x, y)$ for the metric $d(\cdot, \cdot) = H(\cdot, \cdot)$. Since we have always $P_{\xi'} \neq P_\xi$ when $\xi' \neq \xi$ (cf. Definition 1.1), the continuous mapping (○) has its unique zero at $\xi' = \xi$. As a consequence, fixing $K \in \mathcal{K}$, $\xi \in K$, $\gamma > 0$ and restricting ($\circ$) to the compact $K \setminus B_\gamma(\xi)$, the second assertion of the lemma follows. $\square$

This allows us to control likelihood ratios $L_n^{\xi/\vartheta}$ under $P_{n,\vartheta}$ when ξ ranges over small balls $B_\delta(\xi_0)$ which are distant from ϑ.

1.22 Lemma. Fix $K \in \mathcal{K}$, $\vartheta \in \text{int}(K)$, $\gamma > 0$. Under Assumptions 1.19, consider points
$$\xi_0 \in K \setminus B_\gamma(\vartheta)$$
and (using Lemma 1.21 and (+) in Assumptions 1.19) choose $\delta > 0$ small enough to have
$$\overline{p}(\xi_0, \delta) < \underline{h}(\vartheta, \gamma, K) .$$
Then we have geometric decrease
$$E_{n,\vartheta}\left(\sup_{\xi \in B_\delta(\xi_0) \cap K} \sqrt{L_n^{\xi/\vartheta}} \right) \leq [1 - \underline{h}(\vartheta, \gamma, K) + \overline{p}(\xi_0, \delta)]^n, \quad n \geq 1 .$$

Proof. Write for short $V := B_\delta(\xi_0) \cap K$. For $(\omega_1, \ldots, \omega_n) \in \{f_{n,\vartheta} > 0\}$, observe that
$$\left(\sup_{\xi \in V} \sqrt{L_n^{\xi/\vartheta}} \right) (\omega_1, \ldots, \omega_n) = \sup_{\xi \in V} \prod_{i=1}^{n} \frac{f^{1/2}(\xi, \omega_i)}{f^{1/2}(\vartheta, \omega_i)}$$
is smaller than
$$\prod_{i=1}^{n} \frac{1}{f^{1/2}(\vartheta, \omega_i)} \left[f^{1/2}(\xi_0, \omega_i) + \sup_{\xi \in V} \left| f^{1/2}(\xi, \omega_i) - f^{1/2}(\xi_0, \omega_i) \right| \right]$$
which yields
$$E_{n,\vartheta}\left(\sup_{\xi \in V} \sqrt{L_n^{\xi/\vartheta}} \right) \leq \left\{ \int_\Omega f_\vartheta^{1/2} \left[f_{\xi_0}^{1/2} + \sup_{\xi \in V} \left| f_\xi^{1/2} - f_{\xi_0}^{1/2} \right| \right] d\mu \right\}^n .$$
With Notations 1.19, we have by Lemma 1.21 and choice of ξ_0
$$\int f_\vartheta^{1/2} f_{\xi_0}^{1/2} d\mu = 1 - H^2(P_{\xi_0}, P_\vartheta) \leq 1 - \underline{h}(\vartheta, \gamma, K) < 1$$
whereas Cauchy–Schwarz inequality and choice of V imply
$$\int f_\vartheta^{1/2} \sup_{\xi \in V} \left| f_\xi^{1/2} - f_{\xi_0}^{1/2} \right| d\mu \leq 1 \cdot \overline{p}(\xi_0, \delta) = \overline{p}(\xi_0, \delta) .$$
Putting all this together, assumption (+) in Assumptions 1.19 and our choice of δ implies
$$E_{n,\vartheta}\left(\sup_{\xi \in B_\delta(\xi_0) \cap K} \sqrt{L_n^{\xi/\vartheta}} \right) \leq \{ 1 - \underline{h}(\vartheta, \gamma, K) + \overline{p}(\xi_0, \delta) \}^n$$
where the term $\{1 - \underline{h}(\vartheta, \gamma, K) + \overline{p}(\xi_0, \delta)\}$ is strictly between 0 and 1. $\square$

Section 1.4 Consistency of ML Estimators via Hellinger Distances

1.23 Lemma. For all $K \in \mathcal{K}$, $\vartheta \in \text{int}(K)$, $\gamma > 0$, we have under Assumptions 1.19 exponential bounds

$$E_{n,\vartheta}\left(\sup_{\xi \in K \setminus B_\gamma(\vartheta)} \sqrt{L_n^{\xi/\vartheta}}\right) \leq C\, e^{-n\frac{1}{2}\underline{h}(\vartheta,\gamma,K)}, \quad n \geq 1$$

where the constant $C < \infty$ depends on ϑ, γ, K.

Proof. Fix $K \in \mathcal{K}$, $\vartheta \in \text{int}(K)$, $\gamma > 0$. By Lemma 1.21 and (+) in Assumptions 1.19, we associate to every point ξ_0 in $K \setminus B_\gamma(\vartheta)$ some radius $\delta(\xi_0) > 0$ such that

(o) $$\overline{p}(\xi_0, \delta(\xi_0)) < \frac{1}{2}\underline{h}(\vartheta, \gamma, K)\,.$$

This defines an open covering $\{B_{\delta(\xi_0)}(\xi_0) : \xi_0 \in K \setminus B_\gamma(\vartheta)\}$ of the compact $K \setminus B_\gamma(\vartheta)$ in $\mathbb{R}^d$. We can select a finite subcovering $\{B_{\delta(\xi_{0,i})}(\xi_{0,i}) : i = 1, \ldots, \ell\}$. Applying Lemma 1.22 to every $V_i := B_{\delta(\xi_{0,i})}(\xi_{0,i}) \cap K$ we get

$$E_{n,\vartheta}\left(\sup_{\xi \in K \setminus B_\gamma(\vartheta)} \sqrt{L_n^{\xi/\vartheta}}\right) \leq \sum_{i=1}^{\ell} E_{n,\vartheta}\left(\sup_{\xi \in V_i} \sqrt{L_n^{\xi/\vartheta}}\right)$$

$$\leq \sum_{i=1}^{\ell} [1 - \underline{h}(\vartheta, \gamma, K) + \overline{p}(\xi_{0,i}, \delta(\xi_{0,i}))]^n$$

where by choice of the radii in (o), and by the elementary inequality $e^{-y} \geq 1 - y$ for $0 < y < 1$, the right-hand side is smaller than

$$\ell\left[1 - \frac{1}{2}\underline{h}(\vartheta,\gamma,K)\right]^n \leq \ell\, e^{-n\frac{1}{2}\underline{h}(\vartheta,\gamma,K)}\,.$$

This is the assertion, with $C := \ell$ the number of balls $B_{\delta(\xi_{0,i})}(\xi_{0,i})$ with radii (o) which were needed to cover the compact $K \setminus B_\gamma(\vartheta)$ subset of $\Theta \subset \mathbb{R}^d$. □

1.24 Theorem. As $n \to \infty$, consider independent replication of an experiment $\mathcal{E}$ which satisfies all of Assumptions 1.19. Then any maximum likelihood estimator sequence $(T_n)_n$ for the unknown parameter is consistent.

Proof. Let any ML estimator sequence $(T_n)_n$ with 'good sets' $(A_n)_n$ be given, as defined in Definition 1.17. Fix $\vartheta \in \Theta$ and $\gamma > 0$; we have to show

$$\lim_{n \to \infty} P_{n,\vartheta}(\{|T_n - \vartheta| \geq \gamma\}) = 0\,.$$

Put $U := \Theta \setminus B_\gamma(\vartheta)$. Since T_n is maximum likelihood on A_n, we have

$$\left\{\sup_{\xi \in U} f_{n,\xi} < f_{n,\vartheta}\right\} \cap A_n \subset \{T_n \notin U\}$$

for every $n \geq 1$. Taking complements we have
$$\{T_n \in U\} \subset \left\{\sup_{\xi \in U} f_{n,\xi} \geq f_{n,\vartheta}\right\} \cup A_n^c, \quad n \geq 1$$
and write for the first set on the right-hand side
$$P_{n,\vartheta}\left(\sup_{\xi \in U} \sqrt{\frac{f_{n,\xi}}{f_{n,\vartheta}}} \geq 1\right) \leq P_{n,\vartheta}\left(\sup_{\xi \in U} \sqrt{L_n^{\xi/\vartheta}} \geq 1\right) \leq E_{n,\vartheta}\left(\sup_{\xi \in U} \sqrt{L_n^{\xi/\vartheta}}\right).$$

Fix a compact exhaustion $(K_m)_m$ of Θ. Take m large enough for $B_\gamma(\vartheta) \subset K_m$ and large enough to have $\bar{a}(\vartheta, K_m^c) < 1$ in virtue of condition (++) in Assumptions 1.19. Then we decompose
$$U = \Theta \setminus B_\gamma(\vartheta) = U_1 \cup U_2, \quad U_1 := K_m \setminus B_\gamma(\vartheta), \quad U_2 := \Theta \cap K_m^c$$
where Lemma 1.20 applies to U_2 and Lemma 1.23 to U_1. Thus
$$P_{n,\vartheta}\left(\{|T_n - \vartheta| \geq \gamma\} \cap A_n\right) \leq E_{n,\vartheta}\left(\sup_{\xi \in \Theta \cap K_m^c} \sqrt{L_n^{\xi/\vartheta}}\right)$$
$$+ E_{n,\vartheta}\left(\sup_{\xi \in K_m \setminus B_\gamma(\vartheta)} \sqrt{L_n^{\xi/\vartheta}}\right)$$
$$\leq [\bar{a}(\vartheta, K_m^c)]^n + C\, e^{-n\frac{1}{2}\underline{h}(\vartheta,\gamma,K_m)}, \quad n \geq 1.$$

This right-hand side decreases exponentially fast as $n \to \infty$. By definition of the sequence of 'good sets' in Definition 1.17, we also have

(o) $$\lim_{n \to \infty} P_{n,\vartheta}(A_n^c) = 0$$

and are done. $\square$

Apart from the very particular case $A_n = \Omega_n$ for all n large enough, the preceding proof did *not* establish an exponential decrease of $P_{n,\vartheta}(|T_n - \vartheta| \geq \gamma)$ as $n \to \infty$. This is since Assumptions 1.19 do not provide a control on the speed of convergence in (o). Also, we are not interested in proving results 'uniformly in compact ϑ-sets'. Later, in more general sequences of statistical models, even the rate of convergence will vary from point to point on the parameter set, without any hope for results of such type. We shall work instead with contiguity and Le Cam's 'third lemma', see Chapters 3 and 7 below.

1.24" Exercise. Calculate the quantities defined in Assumption 1.19(a) for the single experiment $\mathcal{E}$ in the following case: $(\Omega, \mathcal{A}) = (\mathbb{R}, \mathcal{B}(\mathbb{R}))$; the dominating measure $\mu = \lambda$ is Lebesgue measure; Θ is some open interval in $\mathbb{R}$, bounded or unbounded; P_ϑ are uniform laws on intervals of unit length centred at ϑ
$$P_\vartheta := \mathcal{R}\left(\vartheta - \frac{1}{2}, \vartheta + \frac{1}{2}\right), \quad \vartheta \in \Theta.$$

Prove that the parameterisation is Hölder continuous of order $\frac{1}{2}$ in the sense of Hellinger distance:
$$H(P_{\xi'}, P_\xi) = \sqrt{|\xi' - \xi|} \quad \text{for } \xi' \text{ sufficiently close to } \xi.$$
Prove that condition (+) in Assumptions 1.19 holds with
$$\overline{\rho}(\xi, \delta) = 2\sqrt{\delta}$$
whenever $\delta > 0$ is sufficiently small (i.e. $0 < \delta \leq \delta_0(\xi)$ for $\xi \in \Theta$). Show also that it is impossible to satisfy condition (++) whenever $\text{diam}(\Theta) \leq 1$. In the case where $\text{diam}(\Theta) > 1$, prove
$$(1 - \text{dist}(\xi, \partial\Theta)) \leq \lim_{m\to\infty} \overline{a}(\xi, K_m^c) \leq 2(1 - \text{dist}(\xi, \partial\Theta)) \quad \text{when} \quad 0 < \text{dist}(\xi, \partial\Theta) \leq \frac{1}{2}$$
for any choice of a compact exhaustion of Θ, $\partial\Theta$ denoting the boundary of Θ: in this case condition (++) is satisfied. Hence, in the case where $\text{diam}(\Theta) > 1$, Theorem 1.24 establishes consistency of any sequence of ML estimators for the unknown parameter under independent replication of the experiment $\mathcal{E}$. □

We conclude this section by underlining the following. First, in i.i.d. models, rates of convergence need not be the well-known $\sqrt{n}$; second, in the case where ML estimation errors at ϑ are consistent at some rate $\varphi_n(\vartheta)$ as defined in Assumption 1.9(c), this is not more than just tightness of rescaled estimation errors $\mathcal{L}(\varphi_n(\vartheta)(\widehat{\vartheta}_n - \vartheta) \mid P_{n,\vartheta})$ as $n \to \infty$. Either there may be no weak convergence at all, or we may end up with a variety of limit laws whenever the model allows to define a variety of particular ML sequences.

1.25 Example. In the location model generated from the uniform law $\mathcal{R}(-\frac{1}{2}, \frac{1}{2})$
$$\mathcal{E} = \left(\mathbb{R}, \mathcal{B}(\mathbb{R}), \left\{ P_\vartheta := \mathcal{R}\left(\vartheta - \frac{1}{2}, \vartheta + \frac{1}{2}\right) : \vartheta \in \Theta = \mathbb{R} \right\} \right)$$
(extending exercise 1.24″), any choice of an ML estimator sequence for the unknown parameter is consistent at rate n. There is no unicity concerning limit distributions for rescaled estimation errors at ϑ to be attained simultaneously for any choice of an ML sequence.

Proof. We determine the density of P_0 with respect to λ as $f_0(x) := 1_{[-\frac{1}{2}, \frac{1}{2}]}(x)$, using the closed interval of length 1. This will simplify the representation since we may use two particular 'extremal' definitions – the sequences $(T_n^{(1)})_n$ and $(T_n^{(3)})_n$ introduced below – of an ML sequence.

(1) We start with a preliminary remark. If Y_i are i.i.d. random variables distributed according to $\mathcal{R}((0, 1))$, binomial trials show that the probability of an event
$$\left\{ \max_{1 \leq i \leq n} Y_i < 1 - \frac{u_1}{n}, \; \min_{1 \leq i \leq n} Y_i > \frac{u_2}{n} \right\}$$

tends to $e^{-(u_1+u_2)}$ as $n \to \infty$ for $u_1, u_2 > 0$ fixed. We thus have weak convergence in $\mathbb{R}^2$ as $n \to \infty$

$$\left(n(1 - \max_{1 \le i \le n} Y_i), \, n \cdot \min_{1 \le i \le n} Y_i \right) \longrightarrow (Z_1, Z_2)$$

where Z_1, Z_2 are independent and exponentially distributed with parameter 1.

(2) For $n \ge 1$, the likelihood function $\xi \to L_n^{\xi/\vartheta}$ coincides $P_{n,\vartheta}$-almost surely with the function

$$\mathbb{R} \ni \xi \longrightarrow 1_{\left[\max_{1 \le i \le n} X_i - \frac{1}{2}, \, \min_{1 \le i \le n} X_i + \frac{1}{2} \right]}(\xi) \in \{0, 1\}$$

in the product model $\mathcal{E}_n$. Hence any estimator sequence $(T_n)_n$ with 'good sets' $(A_n)_n$ such that

(×) $\qquad T_n \in \left[\max_{1 \le i \le n} X_i - \frac{1}{2}, \, \min_{1 \le i \le n} X_i + \frac{1}{2} \right] \quad \text{on} \quad A_n$

will be a maximum likelihood sequence for the unknown parameter. The random interval in (×) is of strictly positive length since $P_{n,\vartheta}$-almost surely

$$\max_{1 \le i \le n} X_i - \min_{1 \le i \le n} X_i < 1 \quad P_{n,\vartheta}\text{-almost surely.}$$

All random intervals above are closed by determination $1_{[-\frac{1}{2}+\xi, \frac{1}{2}+\xi]}$ of the density f_ξ for $\xi \in \Theta$.

(3) By (×) in step (2), the following $(T_n^{(i)})_n$ with good sets $(A_n)_n$ are ML sequences in $(\mathcal{E}_n)$:

$T_n^{(1)} := \min_{1 \le i \le n} X_i + \frac{1}{2}$ with $A_n := \Omega_n$

$T_n^{(2)} := \min_{1 \le i \le n} X_i + \frac{1}{2} - \frac{1}{n^2}$ with $A_n := \left\{ \max_{1 \le i \le n} X_i - \min_{1 \le i \le n} X_i < 1 - \frac{1}{n^2} \right\}$

$T_n^{(3)} := \max_{1 \le i \le n} X_i - \frac{1}{2}$ with $A_n := \Omega_n$

$T_n^{(4)} := \max_{1 \le i \le n} X_i - \frac{1}{2} + \frac{1}{n^2}$ with $A_n := \left\{ \max_{1 \le i \le n} X_i - \min_{1 \le i \le n} X_i < 1 - \frac{1}{n^2} \right\}$

$T_n^{(5)} := \frac{1}{2} \left(\min_{1 \le i \le n} X_i + \max_{1 \le i \le n} X_i \right)$ with $A_n := \Omega_n$.

Without the above convention on closed intervals in the density f_0 we would remove the sequences $(T_n^{(1)})_n$ and $(T_n^{(3)})_n$ from this list.

(4) Fix $\vartheta \in \Theta$ and consider at stage n of the asymptotics the model $\mathcal{E}_n$ locally at ϑ, via a reparameterisation $\xi = \vartheta + u/n$ according to rate n suggested by step (1). Reparameterised, the likelihood function $u \to L_n^{(\vartheta + u/n)/\vartheta}$ coincides $P_{n,\vartheta}$-almost surely with

$$\mathbb{R} \ni u \longrightarrow 1_{\left[n\left(\max_{1 \le i \le n} (X_i - \vartheta) - \frac{1}{2} \right), \, n\left(\min_{1 \le i \le n} (X_i - \vartheta) + \frac{1}{2} \right) \right]}(u) \in \{0, 1\}.$$

Section 1.4 Consistency of ML Estimators via Hellinger Distances

As $n \to \infty$, under $P_{n,\vartheta}$, we obtain by virtue of step (1) the 'limiting likelihood function'

(××) $$\mathbb{R} \ni u \longrightarrow 1_{(-Z_1, Z_2)}(u) \in \{0, 1\}$$

where Z_1, Z_2 are independent exponentially distributed.

(5) Step (4) yields the following list of convergences as $n \to \infty$:

$$\mathcal{L}\left(n\left(T_n^{(i)} - \vartheta\right) \mid P_{n,\vartheta}\right) \longrightarrow \begin{cases} -Z_1 & \text{for } i = 3, 4 \\ Z_2 & \text{for } i = 1, 2 \\ \frac{1}{2}(Z_2 - Z_1) & \text{for } i = 5. \end{cases}$$

Clearly, we can also realise convergences

$$\mathcal{L}(n(T_n - \vartheta) \mid P_{n,\vartheta}) \longrightarrow \alpha Z_2 - (1 - \alpha) Z_1$$

for any $0 < \alpha < 1$ if we consider $T_n := \alpha T_n^{(2)} + (1 - \alpha) T_n^{(4)}$, or we can realise

$$\mathcal{L}(n(T_n - \vartheta) \mid P_{n,\vartheta}) \quad \text{does not converge weakly in } \mathbb{R} \text{ as } n \to \infty$$

if we define T_n as $T_n^{(2)}$ when n is odd, and by $T_n^{(4)}$ if n is even. Thus, for a maximum likelihood estimation in the model $\mathcal{E}$ as $n \to \infty$, it makes no sense to put forward the notion of limit distribution, the relevant property of ML estimator sequences being n-consistency (and nothing more). □

Chapter 2

Minimum Distance Estimators

Topics for Chapter 2:

2.1 Measurable Stochastic Processes with Paths in $L^p(T,\mathcal{T},\mu)$
Measurable stochastic processes and their paths 2.1
Measurable stochastic processes with paths in $L^p(T,\mathcal{T},\mu)$ 2.2
Characterising weak convergence in $L^p(T,\mathcal{T},\mu)$ 2.3
Main theorem: sufficient conditions for weak convergence in $L^p(T,\mathcal{T},\mu)$ 2.4
Auxiliary results on integrals along the path of a process 2.5–2.5'
Proving the main theorem 2.6–2.6"

2.2 Minimum Distance Estimator Sequences
Example: fitting the empirical distribution function to a parametric family 2.7
Assumptions and notations 2.8
Defining minimum distance (MD) estimator sequences 2.9
Measurable selection 2.10
Strong consistency of MD estimator sequences 2.11
Some auxiliary results 2.12–2.13
Main theorem: representation of rescaled estimation errors in MD estimator sequences 2.14
Variant: weakly consistent MD estimator sequence 2.15

2.3 Some Comments on Gaussian Processes
Gaussian and μ-Gaussian processes 2.16
Some examples (time changed Brownian bridge) 2.17
Existence of μ-Gaussian processes 2.18
μ-integrals along the path of a μ-Gaussian process 2.19

2.4 Example: Asymptotic Normality of Minimum Distance Estimator Sequences
Assumptions and notations 2.20–2.21
Main theorem: asymptotic normality for MD estimator sequences 2.22
Empirical distribution functions and Brownian bridges 2.23
Example: MD estimators defined from the empirical distribution function 2.24
Example: MD estimator sequences for symmetric stable i.i.d. observations 2.25

Exercises: 2.1', 2.3', 2.24', 2.25'.

This chapter is devoted to the study of one class of estimators in parametric families which – without aiming at any notion of optimality – do have reasonable asymptotic properties under assumptions which are weak and easy to verify. Most examples will consider i.i.d. models, but the setting is more general: sequences of experiments where empirical objects $\widehat{\Psi}_n$, calculated from the data at level n of the asymptotics, are compared to theoretical counterparts Ψ_ϑ under ϑ, for all values of the parameter $\vartheta \in \Theta$, which are deterministic and independent of n. Our treatment of asymptotics of minimum distance (MD) estimators follows Millar [100], for an outline see Kutoyants [78–80]. Below, Sections 2.1 and 2.3 contain the mathematical tools and are of auxiliary character; the statistical part is concentrated in Sections 2.2 and 2.4. The main statistical results are Theorem 2.11 (almost sure convergence of MD estimators), Theorem 2.14 (representation of rescaled MD estimator errors) and Theorem 2.22 (asymptotic normality of rescaled MD estimation errors). We conclude with an example where the parameters of a symmetric stable law are estimated by means of MD estimators based on the empirical characteristic function of the first n observations.

2.1 Stochastic Processes with Paths in $L^p(T, \mathcal{T}, \mu)$

In preparation for asymptotics of MD estimator sequences, we characterise weak convergence in relevant path spaces. We follow Cremers and Kadelka [16], see also Grinblat [37]. Throughout this section, the following set 2.1 of assumptions will be in force.

2.1 Assumptions and Notations for Section 2.1. (a) Let μ denote a finite measure on a measurable space $(T, \mathcal{T})$ with countably generated σ-field $\mathcal{T}$. For $1 \leq p < \infty$ fixed, the space $L^p(\mu) = L^p(T, \mathcal{T}, \mu)$ of (μ-equivalence classes of) p-integrable functions $f : (T, \mathcal{T}) \to (\mathbb{R}, \mathcal{B}(\mathbb{R}))$ is equipped with its norm

$$\|f\| = \|f\|_{L^p(T,\mathcal{T},\mu)} = \left(\int_T |f(t)|^p \mu(dt) \right)^{\frac{1}{p}} < \infty$$

and its Borel-σ-field $\mathcal{B}(L^p(\mu))$. $\mathcal{T}$ being countably generated, the space $L^p(T, \mathcal{T}, \mu)$ is separable ([127, p. 138], [117, pp. 269–270]): there is a countable subset $S \subset L^p(\mu)$ which is dense in $L^p(T, \mathcal{T}, \mu)$, and $\mathcal{B}(L^p(\mu))$ is generated by the countable collection of open balls

$$B_r(g) := \{ f \in L^p(\mu) : \|f - g\|_{L^p(\mu)} < r \}, \quad r \in \mathbb{Q}^+, \, g \in S.$$

(b) With parameter set T as in (a), a real valued stochastic process $X = (X_t)_{t \in T}$ on a probability space $(\Omega, \mathcal{A}, P)$ is a collection of random variables $X_t, t \in T$, defined on $(\Omega, \mathcal{A})$ and taking values in $(\mathbb{R}, \mathcal{B}(\mathbb{R}))$. This process is termed *measurable* if $(t, \omega) \to X(t, \omega)$ is a measurable mapping from $(T \times \Omega, \mathcal{T} \otimes \mathcal{A})$ to $(\mathbb{R}, \mathcal{B}(\mathbb{R}))$. In a measurable

process $X = (X_t)_{t \in T}$, every *path*
$$X_{\bullet}(\omega) : \quad T \ni t \longrightarrow X(t, \omega) \in \mathbb{R}, \quad \omega \in \Omega$$
is a measurable mapping from $(T, \mathcal{T})$ to $(\mathbb{R}, \mathcal{B}(\mathbb{R}))$.

(c) Points in $\mathbb{R}^m$ are written as $t = (t_1, ..., t_m)$. We write $(s_1, ..., s_m) \leq (t_1, ..., t_m)$ if $s_j \leq t_j$ for all $1 \leq j \leq m$; in this case, we put $(s, t] := \mathsf{X}_{j=1}^m (s_j, t_j]$ or $[s, t) := \mathsf{X}_{j=1}^m [s_j, t_j)$ etc. and speak of intervals in $\mathbb{R}^m$ instead of rectangles.

2.1' Exercise. Consider i.i.d. random variables $Y_1, Y_2, \ldots$ defined on some $(\Omega, \mathcal{A}, P)$, and taking values in $(T, \mathcal{T}) := (\mathbb{R}^m, \mathcal{B}(\mathbb{R}^m))$.

(a) For any $n \geq 1$, the empirical distribution function associated to the first n observations
$$\widehat{F}_n(t, \omega) := \frac{1}{n} \sum_{i=1}^n \mathbf{1}_{(-\infty, t]}(Y_i(\omega)) = \frac{1}{n} \sum_{i=1}^n \mathbf{1}_{\{Y_i \leq t\}}(\omega), \quad t \in \mathbb{R}^m, \omega \in \Omega$$
is a stochastic process $\widehat{F}_n = (\widehat{F}_n(t))_{t \in \mathbb{R}^m}$ on $(\Omega, \mathcal{A})$. Prove that $\widehat{F}_n$ is a measurable process in the sense of 2.1.

Hint: with respect to grids $2^{-k} \mathbb{Z}^m$, $k \geq 1$, $l = (l_1, \ldots, l_m) \in \mathbb{Z}^m$, write
$$l^+(k) := \left(\frac{l_1+1}{2^k}, \ldots, \frac{l_m+1}{2^k} \right), \quad A_l(k) := \mathsf{X}_{j=1}^m \left[\frac{l_j}{2^k}, \frac{l_j+1}{2^k} \right) ;$$
then for every k, the mappings
$$(t, \omega) \longrightarrow \sum_{l \in \mathbb{Z}^m} \mathbf{1}_{A_l(k)}(t) \, \widehat{F}_n(l^+(k), \omega)$$
are measurable from $(\mathbb{R}^m \times \Omega, \mathcal{B}(\mathbb{R}^m) \otimes \mathcal{A})$ to $(\mathbb{R}, \mathcal{B}(\mathbb{R}))$, so the same holds for their pointwise limit as $k \to \infty$ which is $(t, \omega) \to \widehat{F}_n(t, \omega)$.

(b) Note that the reasoning in (a) did not need more than continuity from the right – in every component of the argument $t \in \mathbb{R}^m$ – of the paths $\widehat{F}_n(\cdot, \omega)$ which for fixed $\omega \in \Omega$ are distribution functions on $\mathbb{R}^m$. Deduce the following: every real valued stochastic process $(Y_t)_{t \in \mathbb{R}^m}$ on $(\Omega, \mathcal{A})$ whose paths are continuous from the right is a measurable process.

(c) Use the argument of (b) to show that rescaled differences $\sqrt{n}(\widehat{F}_n - F)$
$$(t, \omega) \longrightarrow \sqrt{n}(\widehat{F}_n(t, \omega) - F(t)), \quad t \in \mathbb{R}^m, \omega \in \Omega$$
are measurable stochastic processes, for every $n \geq 1$.

(d) Prove that every real valued stochastic process $(Y_t)_{t \in \mathbb{R}^m}$ on $(\Omega, \mathcal{A})$ whose paths are continuous from the left is a measurable process: proceed in analogy to (a) and (b), replacing the intervals $A_\ell(k)$ used in (a) by $\widetilde{A}_l(k) := \mathsf{X}_{j=1}^m (\frac{l_j}{2^k}, \frac{l_j+1}{2^k}]$, and the points $\ell^+(k)$ by $2^{-k}\ell$. □

2.2 Lemma. For a measurable stochastic process $X = (X_t)_{t \in T}$ on $(\Omega, \mathcal{A}, P)$, introduce the condition
$$(*) \qquad \int_T |X(t, \omega)|^p \, \mu(dt) < \infty \quad \text{for all } \omega \in \Omega.$$

Then the mapping

$$X_\bullet : \Omega \ni \omega \longrightarrow X_\bullet(\omega) \in L^p(\mu)$$

is a well-defined random variable on $(\Omega, \mathcal{A})$ taking values in the space $(L^p(T, \mathcal{T}, \mu), \mathcal{B}(L^p(T, \mathcal{T}, \mu)))$, and we call X a *measurable stochastic process with paths in* $L^p(\mu)$.

Proof. For arbitrary $g \in L^p(T, \mathcal{T}, \mu)$ fixed, for measurable processes $X = (X_t)_{t \in T}$ on $(\Omega, \mathcal{A})$ which satisfy the condition $(*)$, consider

$$Z_g : \omega \longrightarrow \|X_\bullet(\omega) - g\| = \left(\int_T |X(t, \omega) - g(t)|^p \mu(dt) \right)^{\frac{1}{p}} < \infty .$$

Considering the right-hand side of this expression, our assumptions guarantee that Z_g is a measurable mapping from $(\Omega, \mathcal{A})$ to $(\mathbb{R}, \mathcal{B}(\mathbb{R}))$. For balls $B_r(g)$ as defined in Assumption 2.1(a), this gives

$$X_\bullet^{-1}(B_r(g)) = \{\omega : X_\bullet(\omega) \in B_r(g)\} = \{Z_g < r\} \in \mathcal{A}$$

where the collection of balls $B_r(g)$ generates $\mathcal{B}(L^p(\mu))$. Hence $X_\bullet$ is a measurable mapping from $(\Omega, \mathcal{A})$ to $(L^p(T, \mathcal{T}, \mu), \mathcal{B}(L^p(T, \mathcal{T}, \mu)))$. □

We give a characterisation of weak convergence in $L^p(T, \mathcal{T}, \mu)$:

2.3 Proposition. *Let* $(X_t^n)_{t \in T}$, $n \geq 1$, *and* $(X_t)_{t \in T}$ *be measurable stochastic processes with paths in* $L^p(T, \mathcal{T}, \mu)$. *A necessary and sufficient condition for weak convergence in* $L^p(T, \mathcal{T}, \mu)$

$$X_\bullet^n \xrightarrow{\mathcal{L}} X_\bullet, \quad n \to \infty$$

is the following:

$$\begin{cases} \text{for arbitrary } l \geq 1 \text{ and arbitrary choice of functions } g_1, ..., g_l \text{ in } L^p(T, \mathcal{T}, \mu), \\ (\|X_\bullet^n - g_1\|, ..., \|X_\bullet^n - g_l\|) \xrightarrow{\mathcal{L}} (\|X_\bullet - g_1\|, ..., \|X_\bullet - g_l\|) \\ \text{weakly in } \mathbb{R}^l \text{ as } n \to \infty . \end{cases}$$

Proof. Write for short $L^p = L^p(T, \mathcal{T}, \mu)$. Mappings

$$h : L^p \ni f \longrightarrow (\|f - g_1\|, ..., \|f - g_l\|) \in \mathbb{R}^l$$

being continuous, the condition stated above is a necessary condition, by the continuous mapping theorem. We prove that the condition stated above is sufficient.

Let X^n be defined on spaces $(\Omega_n, \mathcal{A}_n, P_n)$, $n \geq 1$, and X on $(\Omega, \mathcal{A}, P)$. We write Q^n for the law $\mathcal{L}(X_\bullet^n | P_n)$ on $(L^p, \mathcal{B}(L^p))$, $n \geq 1$, and Q for $\mathcal{L}(X_\bullet | P)$; according to the Portmanteau theorem (e.g. Billingsley [11]), we shall prove

(+) $\qquad \liminf_{n \to \infty} Q^n(G) \geq Q(G) \quad$ for every open set G in L^p .

By separability of L^p according to Assumption 2.1(a) we can write G as a countable union of open balls

$$G = \bigcup_{i=1}^{\infty} B_{r_i}(g_i) \quad \text{for suitable } g_i \in S, \; r_i \in \mathbb{Q}^+.$$

From this representation of G, associate to every $\varepsilon > 0$ some $\ell = \ell(\varepsilon)$ such that

$$Q(G) \leq Q\left(\bigcup_{i=1}^{l} B_{r_i}(g_i)\right) + \varepsilon.$$

Thus we have as $n \to \infty$

$$\liminf_n Q^n(G) \geq \liminf_n Q^n\left(\bigcup_{i=1}^{l} B_{r_i}(g_i)\right) = \liminf_n P_n\left(X_\bullet^n \in \bigcup_{i=1}^{l} B_{r_i}(g_i)\right)$$

$$= \liminf_n P_n \left(\text{there is some } 1 \leq i \leq l \text{ such that } \|X_\bullet^n - g_i\| < r_i\right)$$

$$= \liminf_n \left(1 - P_n\left((\|X_\bullet^n - g_1\|, \ldots, \|X_\bullet^n - g_l\|) \in \bigtimes_{i=1}^{l} [r_i, \infty)\right)\right).$$

Using the condition which ensures weak convergence of $(\|X_\bullet^n - g_1\|, \ldots, \|X_\bullet^n - g_l\|)$ in $\mathbb{R}^l$ as $n \to \infty$, and the Portmanteau theorem with closed sets in $\mathbb{R}^l$, we can continue

$$\geq 1 - P\left((\|X_\bullet - g_1\|, \ldots, \|X_\bullet - g_l\|) \in \bigtimes_{i=1}^{l} [r_i, \infty)\right)$$

$$= P \left(\text{there is some } 1 \leq i \leq l \text{ such that } \|X_\bullet - g_i\| < r_i\right)$$

$$= P\left(X_\bullet \in \bigcup_{i=1}^{l} B_{r_i}(g_i)\right) = Q\left(\bigcup_{i=1}^{l} B_{r_i}(g_i)\right)$$

$$\geq Q(G) - \varepsilon$$

by choice of $l = \ell(\varepsilon)$. Since $\varepsilon > 0$ was arbitrary, we have (+). This finishes the proof. □

2.3' Exercise. Under the assumptions of Proposition 2.3, use the continuous mapping theorem to show that weak convergence $X_\bullet^n \longrightarrow X_\bullet$ in $L^p(T, \mathcal{T}, \mu)$ as $n \to \infty$ implies weak convergence of all integrals

$$\int_T g(s) X_s^n \mu(ds) \longrightarrow \int_T g(s) X_s \mu(ds) \quad (\text{weakly in } \mathbb{R}, \text{ as } n \to \infty)$$

for functions $g : T \to \mathbb{R}$ which belong to $L^q(\mu)$, $\frac{1}{p} + \frac{1}{q} = 1$, and thus also weak convergence

$$\int_T g_n(s) X_s^n \mu(ds) \longrightarrow \int_T g(s) X_s \mu(ds) \quad (\text{weakly in } \mathbb{R}, \text{ as } n \to \infty)$$

for arbitrary sequences $(g_n)_n$ in $L^q(\mu)$ with the property $g_n \to g$ in $L^q(\mu)$. □

The following Theorem 2.4, the main result of this section, gives sufficient conditions for weak convergence in $L^p(T, \mathcal{T}, \mu)$ – of type 'convergence of finite dimensional distributions plus uniform integrability' – from which weak convergence in $L^p(T, \mathcal{T}, \mu)$ can be checked quite easily.

2.4 Theorem (Cremers and Kadelka [16]). Consider measurable stochastic processes

$(X_t^n)_{t \in T}$ defined on $(\Omega_n, \mathcal{A}_n, P_n)$, $n \geq 1$, and $(X_t)_{t \in T}$ defined on $(\Omega, \mathcal{A}, P)$

with paths in $L^p(T, \mathcal{T}, \mu)$, under all assumptions of 2.1.
(a) In order to establish

$$X_\bullet^n \xrightarrow{\mathcal{L}} X_\bullet \quad \text{(weak convergence in } L^p(T, \mathcal{T}, \mu)\text{, as } n \to \infty\text{)},$$

a sufficient condition is that the following properties (i) and (ii) hold simultaneously:
(i) convergence of finite dimensional distributions up to some exceptional set $N \in \mathcal{T}$ such that $\mu(N) = 0$: for arbitrary $l \geq 1$ and any choice of $t_1, \ldots, t_l$ in $T \setminus N$, one has

$$\mathcal{L}\left((X_{t_1}^n, \ldots, X_{t_l}^n) \mid P_n\right) \xrightarrow{\mathcal{L}} \mathcal{L}\left((X_{t_1}, \ldots, X_{t_l}) \mid P\right)$$

(weak convergence in $\mathbb{R}^l$, as $n \to \infty$);

(ii) uniform integrability of $\{|X^n(\cdot, \cdot)|^p : n \geq 0\}$ for the random variables (including $X^0 := X$ and $P_0 := P$)

X^n defined on $(T \times \Omega_n, \mathcal{T} \otimes \mathcal{A}_n, \mu \otimes P_n)$ with values in $(\mathbb{R}, \mathcal{B}(\mathbb{R}))$, $n \geq 0$

in the following sense: for every $\varepsilon > 0$ there is some $K = K(\varepsilon) < \infty$ such that

$$\sup_{n \geq 0} \int_{T \times \Omega_n} 1_{\{|X^n| > K\}} |X^n|^p \, d(\mu \otimes P_n) < \varepsilon.$$

(b) Whenever condition (a.i) is satisfied, any one of the following two conditions is sufficient for (a.ii):

(2.4′) $\begin{cases} X^n(\cdot, \cdot), n \geq 1, X(\cdot, \cdot) \text{ are elements of } L^p(T \times \Omega_n, \mathcal{T} \otimes \mathcal{A}_n, \mu \otimes P_n), \text{ and} \\ \limsup_{n \to \infty} \int_{T \times \Omega_n} |X^n|^p \, d(\mu \otimes P_n) \leq \int_{T \times \Omega} |X|^p \, d(\mu \otimes P), \end{cases}$

(2.4″) $\begin{cases} \text{there is some function } f \in L^1(T, \mathcal{T}, \mu) \text{ such that for } \mu\text{-almost all } t \in T \\ E\left(|X_t^n|^p\right) \leq f(t) \text{ for all } n \geq 1, \text{ and } \lim_{n \to \infty} E\left(|X_t^n|^p\right) = E\left(|X_t|^p\right). \end{cases}$

The remaining parts of this section contain the proof of Theorem 2.4, to be completed in Proofs 2.6 and 2.6', and some auxiliary results; we follow Cremers and Kadelka [16]. Recall that the set of Assumptions 2.1 is in force. W.l.o.g., we take the finite measure μ on $(T, \mathcal{T})$ as a probability measure $\mu(T) = 1$.

2.5 Lemma. Write $\mathcal{H}$ for the class of bounded measurable functions $\varphi : T \times \mathbb{R} \to \mathbb{R}$ such that

for every $t \in T$ fixed, the mapping $\varphi(t, \cdot) : x \to \varphi(t, x)$ is continuous.

Then condition (a.i) of Theorem 2.4 gives for every $\varphi \in \mathcal{H}$ convergence in law of the integrals

$$\int_T \varphi(s, X_s^n) \mu(ds) \longrightarrow \int_T \varphi(s, X_s) \mu(ds) \quad \text{(weak convergence in } \mathbb{R}\text{, as } n \to \infty\text{)}.$$

Proof. For real valued measurable processes $(X'_t)_{t \in T}$ defined on some $(\Omega', \mathcal{A}', P')$, the mapping $(t, \omega) \to (t, X'(t, \omega))$ is measurable from $T \otimes \mathcal{A}$ to $T \otimes \mathcal{B}(\mathbb{R})$, hence composition with φ gives a mapping $(t, \omega) \to \varphi(t, X'(t, \omega))$ which is $T \otimes \mathcal{A}$-$\mathcal{B}(\mathbb{R})$-measurable. As a consequence,

$$\omega \longrightarrow \int_T \varphi(s, X'(s, \omega)) \mu(ds)$$

is a well-defined random variable on $(\Omega', \mathcal{A}')$ taking values in $(\mathbb{R}, \mathcal{B}(\mathbb{R}))$. In order to prove convergence in law of integrals $\int_T \varphi(s, X_s^n) \mu(ds)$ as $n \to \infty$, we shall show

$$(\diamond) \quad E_{P_n}\left(g\left(\int_T \varphi(s, X_s^n) \mu(ds)\right)\right) \longrightarrow E_P\left(g\left(\int_T \varphi(s, X_s) \mu(ds)\right)\right)$$

for arbitrary $g \in \mathcal{C}_b(\mathbb{R})$, the class of bounded continuous functions $\mathbb{R} \to \mathbb{R}$.

Fix any constant $M < \infty$ such that $\sup_{T \times \mathbb{R}} |\varphi| \le M$. Thanks to the convention $\mu(T) = 1$ above, for functions $g \in \mathcal{C}_b(\mathbb{R})$ to be considered in $(\diamond)$, only the restriction $g|_{[-M,M]}$ to the interval $[-M, M]$ is relevant. According to Weierstrass, approximating g uniformly on $[-M, M]$ by polynomials, it will be sufficient to prove $(\diamond)$ for polynomials. By additivity, it remains to consider the special case $g(x) := x^l, l \in \mathbb{N}$, and to prove $(\diamond)$ in this case. We fix $l \in \mathbb{N}$ and show

$$E_{P_n}\left(\left(\int_T \varphi(s, X_s^n) \mu(ds)\right)^l\right) \longrightarrow E_P\left(\left(\int_T \varphi(s, X_s) \mu(ds)\right)^l\right), \quad n \to \infty.$$

Put $X^0 := X$ and $P_0 := P$, and write left- and right-hand sides in the form

$$E_{P_n}\left(\left(\int_T \varphi(s, X_s^n) \mu(ds)\right)^l\right)$$

$$= \int_T \cdots \int_T E_{P_n}\left(\prod_{i=1}^l \varphi(s_i, X_{s_i}^n)\right) \mu(ds_1) \ldots \mu(ds_l)$$

$$= \int_T \cdots \int_T E_{P_n}\left(\psi_{(s_1,\ldots,s_l)}(X_{s_1}^n, \ldots, X_{s_l}^n)\right) \mu(ds_1) \ldots \mu(ds_l).$$

Here, at every point $(s_1, \ldots, s_l)$ of the product space $\times_{i=1}^{l} T$, a function

$$\psi(x_1, \ldots, x_l) = \psi_{(s_1, \ldots, s_l)}(x_1, \ldots, x_l) := \prod_{i=1}^{l} \varphi(s_i, x_i) \quad \in \mathcal{C}_b(\mathbb{R}^l)$$

arises which indeed is bounded and continuous: for $\varphi \in \mathcal{H}$, we exploit at this point of the proof the defining property of class $\mathcal{H}$. Condition (a.i) of Theorem 2.4 guarantees convergence of finite dimensional distributions of X^n to those of X up to some exceptional μ-null set $\mathcal{T} \in \mathcal{T}$, thus we have

$$E_{P_n}\left(\psi(X_{s_1}^n, \ldots, X_{s_l}^n)\right) \longrightarrow E_P\left(\psi(X_{s_1}, \ldots, X_{s_l})\right), \quad n \to \infty$$

for any choice of $(s_1, \ldots, s_l)$ such that $s_1, \ldots, s_l \in T \setminus N$. Going back to the above integrals on the product space $\times_{i=1}^{l} T$, note that all expressions in the last convergence are bounded by M^l, uniformly in $(s_1, \ldots, s_l)$ and n: hence

$$\int_T \cdots \int_T E_{P_n}\left(\psi_{(s_1, \ldots, s_l)}(X_{s_1}^n, \ldots, X_{s_l}^n)\right) \mu(ds_1) \ldots \mu(ds_l)$$

will tend as $n \to \infty$ by dominated convergence to

$$\int_T \cdots \int_T E_P\left(\psi_{(s_1, \ldots, s_l)}(X_{s_1}, \ldots, X_{s_l})\right) \mu(ds_1) \ldots \mu(ds_l) .$$

This proves $(\diamond)$ in the case where $g(x) = x^l$, for arbitrary $l \in \mathbb{N}$. This finishes the proof. $\square$

2.5' Lemma. For $g \in L^p(T, \mathcal{T}, \mu)$ fixed, consider the family of random variables

$$Z^n(g) := \|X_\bullet^n - g\|^p, \quad n \geq 0$$

as in the proof of Lemma 2.2 (with $X^0 := X$, $P_0 := P$), and the class $\mathcal{H}$ as defined in Lemma 2.5. Under condition (a.ii) of Theorem 2.4, we can approximate $Z^n(g)$ uniformly in $n \in \mathbb{N}_0$ by integrals $\int_T \varphi(s, X_s^n) \mu(ds)$ for suitable choice of $\varphi \in \mathcal{H}$: for every $g \in L^p(\mu)$ and every $\delta > 0$ there is some $\varphi = \varphi_{g,\delta} \in \mathcal{H}$ such that

$$\sup_{n \geq 0} P_n\left(\left|\|X_\bullet^n - g\|^p - \int_T \varphi_{g,\delta}(s, X_s^n) \mu(ds)\right| > \delta\right) < \delta .$$

Proof. (1) Fix $g \in L^p(T, \mathcal{T}, \mu)$ and $\delta > 0$. Exploiting the uniform integrability condition (a.ii) of Theorem 2.4, we select $C = C(\delta) < \infty$ large enough for

$(\diamond 1)$ $$\sup_{n \geq 0} \int_{T \times \Omega_n} 1_{\{|X^n| > C\}} |X^n|^p \, d(\mu \otimes P_n) < \frac{1}{4} 2^{-(p+1)} \delta^2$$

and $\eta = \eta(\delta) > 0$ small enough such that

($\diamond 2$)
$$G_n \in \mathcal{T} \otimes \mathcal{A}_n \,, \, (\mu \otimes P_n)(G_n) < \eta \implies$$
$$\int_{T \times \Omega_n} 1_{G_n} |X^n|^p \, d(\mu \otimes P_n) < \frac{1}{4} 2^{-p} \delta^2$$

independently of n. Assertion ($\diamond 2$) amounts to the usual ε-δ-characterisation of uniform integrability, in the case where random variables $(t, \omega) \to |X_n(t, \omega)|^p$ are defined on different probability spaces $(T \times \Omega_n, \mathcal{T} \otimes \mathcal{A}_n, \mu \otimes P_n)$. In the same way, we view $(t, \omega) \to |g(t)|^p$ as a family of random variables on $(T \times \Omega_n, \mathcal{T} \otimes \mathcal{A}_n, \mu \otimes P_n)$ for $n \geq 1$ which is uniformly integrable. Increasing $C < \infty$ of ($\diamond 1$) and decreasing $\eta > 0$ of ($\diamond 2$) if necessary, we obtain in addition

($\diamond 3$)
$$\int_T 1_{\{|g|>C\}} |g|^p \, d\mu < \frac{1}{4} 2^{-(p+1)} \delta^2$$

together with

($\diamond 4$)
$$G_n \in \mathcal{T} \otimes \mathcal{A}_n \,, \, (\mu \otimes P_n)(G_n) < \eta \implies$$
$$\int_{T \times \Omega_n} 1_{G_n} |g|^p \, d(\mu \otimes P_n) < \frac{1}{4} 2^{-p} \delta^2 \,.$$

A final increase of $C < \infty$ makes sure that in all cases $n \geq 0$ the sets

($\diamond 5$)
$$G_n := \{|X^n| > C\} \quad \text{or} \quad G_n := \{|g| > C\}$$

satisfy the condition $(\mu \otimes P_n)(G_n) < \eta$, and thus can be inserted in ($\diamond 2$) and ($\diamond 4$).

(2) With g, δ, C of step (1), introduce a truncated identity $h(x) = (-C) \vee x \wedge (+C)$ and define a function $\varphi = \varphi_{g,\delta} : T \times \mathbb{R} \to \mathbb{R}$ by

$$\varphi(s, x) := |h(x) - h(g(s))|^p \,, \quad s \in T \,, \, x \in \mathbb{R} \,.$$

Then φ belongs to class $\mathcal{H}$ as defined in Lemma 2.5, and is such that

(+) $\quad |X_s^n(\omega) - g(s)|^p = \varphi(s, X_s^n(\omega))$ on $\{(s, \omega) : |X^n(s, \omega)| \leq C \,, \, |g(s)| \leq C\}$.

As a consequence of (+), for $\omega \in \Omega$ fixed, the difference

$$\left| \|X_\bullet^n(\omega) - g\|^p - \int_T \varphi(s, X_s^n(\omega)) \, \mu(ds) \right| \leq$$
($*, n$)
$$\int_T ||X_s^n(\omega) - g(s)|^p - \varphi(s, X_s^n(\omega))| \, \mu(ds)$$

admits, for $\omega \in \Omega$ fixed, the upper bound

$$\int_T 1_{\{|X^n(s,\omega)|>C\} \cup \{|g(s)|>C\}} ||X_s^n(\omega) - g(s)|^p - \varphi(s, X_s^n(\omega))| \, \mu(ds)$$

which is (using the elementary $|a+b|^p \leq (|a|+|b|)^p \leq 2^p(|a|^p+|b|^p)$, and definition of φ) smaller than

$$\int_T 1_{\{|X^n(s,\omega)|>C\} \cup \{|g(s)|>C\}} \left(2^p |X_s^n(\omega)|^p + 2^p |g(s)|^p + 2^p C^p\right) \mu(ds)$$

$$\leq 2^p \bigg\{ \int_T 1_{\{|X^n(s,\omega)|>C\}} (|X^n(s,\omega)|^p + C^p) \mu(ds)$$

$$+ \int_T 1_{\{|X^n(s,\omega)|>C\}} |g(s)|^p \mu(ds) + \int_T 1_{\{|g(s)|>C\}} |X^n(s,\omega)|^p \mu(ds)$$

$$+ \int_T 1_{\{|g(s)|>C\}} (|g(s)|^p + C^p) \mu(ds) \bigg\}$$

and finally smaller than

$$2^p \bigg\{ 2 \int_T 1_{\{|X^n(s,\omega)|>C\}} |X^n(s,\omega)|^p \mu(ds) + \int_T 1_{\{|X^n(s,\omega)|>C\}} |g(s)|^p \mu(ds)$$

$$+ \int_T 1_{\{|g(s)|>C\}} |X^n(s,\omega)|^p \mu(ds) + 2 \int_T 1_{\{|g(s)|>C\}} |g(s)|^p \mu(ds) \bigg\} .$$

The last right-hand side is the desired bound for $(*,n)$, for $\omega \in \Omega$ fixed.

(3) Integrating this bound obtained in step (2) with respect to P_n, we obtain

$$E_{P_n} \left(\left| \|X_\bullet^n(\omega) - g\|^p - \int_T \varphi(s, X_s^n(\omega)) \mu(ds) \right| \right)$$

$$\leq 2^{p+1} \int_{T \times \Omega_n} 1_{\{|X^n|>C\}} |X^n|^p \, d(\mu \otimes P_n) + 2^p \int_{T \times \Omega_n} 1_{\{|X^n|>C\}} |g|^p \, d(\mu \otimes P_n)$$

$$+ 2^p \int_{T \times \Omega_n} 1_{\{|g|>C\}} |X^n|^p \, d(\mu \otimes P_n) + 2^{p+1} \int_T 1_{\{|g|>C\}} |g|^p \, d\mu$$

where $(\diamond 1)$–$(\diamond 5)$ make every term on the right-hand side smaller than $\frac{1}{4} \delta^2$, independently of n. Thus

$$\sup_{n \geq 0} E_{P_n} \left(\left| \|X_\bullet^n(\omega) - g\|^p - \int_T \varphi(s, X_s^n(\omega)) \mu(ds) \right| \right) < \delta^2$$

for $\varphi = \varphi_{\delta, g} \in \mathcal{H}$, and application of the Markov inequality gives

$$\sup_{n \geq 0} P_n \left(\left| \|X_\bullet^n - g\|^p - \int_T \varphi(s, X_s^n) \mu(ds) \right| > \delta \right) < \delta$$

as desired. □

2.6 Proof of Theorem 2.4(a). (1) We shall prove part (a) of Theorem 2.4 using 'accompanying sequences'. We explain this for some sequence $(\widetilde{Y}_n)_n$ of real valued random variables whose convergence in law to $\widetilde{Y}_0$ we wish to establish. Write $\mathcal{C}_u(\mathbb{R})$ for the class of uniformly continuous and bounded functions $\mathbb{R} \to \mathbb{R}$; for $f \in \mathcal{C}_u(\mathbb{R})$ put $M_f := 2 \sup |f|$. A sequence $(\widetilde{Z}^n)_n$ – where for every $n \geq 0$, $\widetilde{Z}_n$ and $\widetilde{Y}_n$ live on the same probability space – is called a δ-*accompanying sequence for* $(\widetilde{Y}^n)_n$ if

$$\sup_{n \geq 0} P_n \left(|\widetilde{Y}^n - \widetilde{Z}^n| > \delta \right) < \delta .$$

Selecting for every $f \in \mathcal{C}_u(\mathbb{R})$ and every $\eta > 0$ some $\delta = \delta(f, \eta) > 0$ such that

$$\sup_{|x-x'|<\delta} |f(x) - f(x')| < \eta ,$$

we have for $\delta = \delta(f, \eta)$-accompanying sequences $(\widetilde{Z}^n)_n$ the inequality

(o) $\quad \left| E(f(\widetilde{Y}^n)) - E(f(\widetilde{Y}^0)) \right| \leq \left| E(f(\widetilde{Z}^n)) - E(f(\widetilde{Z}^0)) \right| + 2\eta + 2M_f \delta .$

For a given sequence $(\widetilde{Y}^n)_{n \geq 0}$ we thus have the following: whenever we are able to associate *for every* $\delta > 0$ *some sequence* $(\widetilde{Z}^n(\delta))_{n \geq 0}$ which is δ-accompanying and such that $\widetilde{Z}^n(\delta)$ converges in law to $\widetilde{Z}^0(\delta)$ as $n \to \infty$, we do have convergence in law of $(\widetilde{Y}^n)_n$ to $\widetilde{Y}^0$, thanks to (o).

(2) We start the proof of Theorem 2.4(a). In order to show

$$X_\bullet^n \longrightarrow X_\bullet \quad \text{(weak convergence in } L^p(T, \mathcal{T}, \mu), n \to \infty)$$

it is sufficient by Proposition 2.3 to consider arbitrary $g_1, ..., g_l$ in $L^p(T, \mathcal{T}, \mu), l \geq 1$, and to prove

$$(\|X_\bullet^n - g_1\|, ..., \|X_\bullet^n - g_l\|) \longrightarrow (\|X_\bullet - g_1\|, ..., \|X_\bullet - g_l\|)$$
(weakly in $\mathbb{R}^l, n \to \infty$)

or equivalently

$$(\|X_\bullet^n - g_1\|^p, ..., \|X_\bullet^n - g_l\|^p) \longrightarrow (\|X_\bullet - g_1\|^p, ..., \|X_\bullet - g_l\|^p)$$
(weakly in $\mathbb{R}^l, n \to \infty$) .

According to Cramér–Wold (or to P. Lévy's continuity theorem for characteristic functions), to establish the last assertion we have to prove for all $\alpha = (\alpha_1, ..., \alpha_\ell) \in \mathbb{R}^l$

(++) $\quad Y^n := \sum_{i=1}^{l} \alpha_i \|X_\bullet^n - g_i\|^p \longrightarrow \sum_{i=1}^{l} \alpha_i \|X_\bullet - g_i\|^p =: Y^0$

(weakly in $\mathbb{R}, n \to \infty$)

where w.l.o.g. we can assume first $\alpha_i \neq 0$ for $1 \leq i \leq l$, and second $\frac{1}{l} \sum_{i=1}^{l} \frac{1}{|\alpha_i|} = 1$, by multiplication of $(\alpha_1, ..., \alpha_\ell)$ with some constant.

Section 2.1 Stochastic Processes with Paths in $L^p(T, \mathcal{T}, \mu)$ 53

(3) With $(\alpha_1, \ldots, \alpha_\ell)$ and $g_1, \ldots, g_\ell$ of (++), for $\delta > 0$ arbitrary, select functions $\varphi_{g_i, \frac{\delta}{l|\alpha_i|}}$ in $\mathcal{H}$ such that

$$\sup_{n \geq 0} P_n \left(\left| \|X^n_\bullet - g_i\|^p - \int_T \varphi_{g_i, \frac{\delta}{l|\alpha_i|}}(s, X^n_s) \mu(ds) \right| > \frac{\delta}{l|\alpha_i|} \right) < \frac{\delta}{l|\alpha_i|}$$

with notations of Lemma 2.5'. Then also

$$\varphi_{\alpha, g_1, \ldots, g_l, \delta}(t, x) := \sum_{i=1}^{l} \alpha_i \, \varphi_{g_i, \frac{\delta}{l|\alpha_i|}}(t, x), \quad t \geq 0, \, x \in \mathbb{R}$$

is a function in class $\mathcal{H}$. Introducing the random variables

$$Z^n := \int_T \varphi_{\alpha, g_1, \ldots, g_l, \delta}(s, X^n_s) \, \mu(ds), \quad n \geq 1,$$

$$Z^0 := \int_T \varphi_{\alpha, g_1, \ldots, g_l, \delta}(s, X_s) \, \mu(ds),$$

by Lemma 2.5, condition (a.i) of Theorem 2.4 gives convergence in law of the integrals

(∗) $\qquad Z^n \longrightarrow Z^0 \quad$ (weak convergence in $\mathbb{R}$), $n \to \infty$.

Now, triangle inequality and norming $\frac{1}{l} \sum_{i=1}^{l} \frac{1}{|\alpha_i|} = 1$ show that independently of $n \geq 0$

$$P_n \left(|Y^n - Z^n| > \delta \right)$$

$$\leq \sum_{i=1}^{l} P_n \left(\left| \alpha_i \|X^n_\bullet - g_i\|^p - \alpha_i \int_T \varphi_{g_i, \frac{\delta}{l|\alpha_i|}}(s, X^n_s) \mu(ds) \right| > \frac{\delta}{l} \right)$$

$$= \sum_{i=1}^{l} P_n \left(\left| \|X^n_\bullet - g_i\|^p - \int_T \varphi_{g_i, \frac{\delta}{l|\alpha_i|}}(s, X^n_s) \mu(ds) \right| > \frac{\delta}{l|\alpha_i|} \right) < \delta$$

which identifies $(Z^n)_{n \in \mathbb{N}_0}$ as a δ-accompanying sequence for $(Y^n)_{n \in \mathbb{N}_0}$

(∗∗) $\qquad \sup_{n \geq 0} P_n \left(\left| Y^n - \int_T \varphi_{\alpha, g_1, \ldots, g_l, \delta}(s, X^n_s) \mu(ds) \right| > \delta \right) < \delta.$

(4) Combining (∗) and (∗∗), we have constructed a δ-accompanying sequence for $(Y^n)_{n \in \mathbb{N}_0}$ which is weakly convergent, for $\delta > 0$ arbitrary. By step (1), we thus have proved convergence in law of Y_n to Y_0 as $n \to \infty$. This is (++). According to step (2), part (a) of Theorem 2.4 is proved. □

2.6' Proof of Theorem 2.4(b). By dominated convergence with respect to μ, condition (2.4") obviously implies condition (2.4'). We have to prove that condition (2.4') combined with convergence of finite dimensional distributions up to some exceptional set of μ-measure 0 (condition (a.i) of Theorem 2.4) guarantees that uniform integrability as required in (a.ii) of Theorem 2.4 holds.

Condition (2.4') implies $X(.,.) \in L^p(T \times \Omega, \mathcal{T} \otimes \mathcal{A}, \mu \otimes P)$, hence for $\varepsilon > 0$ there is $C < \infty$ such that

$$\int_{T \times \Omega} 1_{\{|X|>C\}} |X|^p \, d(\mu \otimes P) < \varepsilon.$$

Write $f_0(x) := |x|^p$ on $\mathbb{R}^d$, introduce a function $f_1 \in \mathcal{C}_b(\mathbb{R})$ which coincides with f_0 on $\{|x| \leq C\}$, vanishes on $\{|x| > 2C\}$, and satisfies $0 \leq f_1 \leq f_0$ everywhere on $\mathbb{R}^d$. Define $f_2 := f_0 - f_1$: then f_2 coincides with f_0 on $\{|x| > 2C\}$, vanishes on $\{|x| \leq C\}$, and satisfies $0 \leq f_2 \leq f_0$ on $\mathbb{R}^d$. By choice of $C = C(\varepsilon)$ we have in particular

(o) $$E_P \left(\int_T f_2(X_s) \mu(ds) \right) < \varepsilon.$$

Next, the function $\phi(s,x) := f_1(x)$ belongs to class $\mathcal{H}$ as defined in Lemma 2.5. Thanks to Lemma 2.5, condition (a.i) of Theorem 2.4 which is assumed here yields convergence in law

$$\int_T f_1(X_s^n) \mu(ds) \longrightarrow \int_T f_1(X_s) \mu(ds) \quad (\text{weakly in } \mathbb{R}, n \to \infty)$$

which by definition of weak convergence is

(+) $$E_{P_n} \left(g \left(\int_T f_1(X_s^n) \mu(ds) \right) \right) \longrightarrow E_P \left(g \left(\int_T f_1(X_s) \mu(ds) \right) \right), \quad n \to \infty$$

for arbitrary $g \in \mathcal{C}_b(\mathbb{R})$. Put $M := \sup |f_1|$ and consider in particular functions $g \in \mathcal{C}_b(\mathbb{R})$ which on $[-M, +M]$ coincide with the identity. Since $\mu(T) = 1$, (+) for such g yields

(++) $$E_{P_n} \left(\int_T f_1(X_s^n) \mu(ds) \right) \longrightarrow E_P \left(\int_T f_1(X_s) \mu(ds) \right), \quad n \to \infty.$$

Since $f_1 + f_2 = f_0$, $f_0(x) = |x|^p$, our assumption (2.4') combined with (++) and (o) gives

$$\limsup_{n \to \infty} E_{P_n} \left(\int_T f_2(X_s^n) \mu(ds) \right) \leq E_P \left(\int_T f_2(X_s) \mu(ds) \right) < \varepsilon.$$

The function f_2 coinciding with f_0 on $\{|x| > 2C\}$, we have a fortiori

$$\int_{T \times \Omega_n} 1_{\{|X^n|>2C\}} |X^n|^p \, d(\mu \otimes P_n) < \varepsilon \quad \text{for } n \text{ sufficiently large}.$$

By assumption (2.4'), all $X^n(.,.)$ being in $L^p(T\times\Omega_n, \mathcal{T}\otimes\mathcal{A}_n, \mu\otimes P_n)$, we can increase this constant $2C$ to some $K = K(\varepsilon) < \infty$ which satisfies

$$\sup_{n\geq 0} \int_{T\times\Omega_n} 1_{\{|X^n|>K\}} |X^n|^p \, d(\mu\otimes P_n) < \varepsilon \, .$$

This is uniform integrability as stated in condition (a.ii) of Theorem 2.4. The proof of Theorem 2.4(b) is finished. □

2.2 Minimum Distance Estimator Sequences

Millar [100] has shown that minimum distance (MD) estimator sequences are consistent and asymptotically normal under rather weak conditions, and have interesting robustness properties under small deviations from the statistical model. Hence such estimators – without aiming at optimality within a specified statistical model – are interesting under the practical aspect that in all applications a statistical model is at best an approximately accurate one. Le Cam [83, Ex. 1 on p. 154] presents a striking example of how ML estimation of the parameters in a normal distribution model can switch to a notion void of sense once certain – arbitrarily small – 'contaminations' are added: the likelihood surface then develops singularities which are in one-to-one correspondence to the observations themselves. The present section introduces MD estimator sequences. The main result is strong consistency (Proposition 2.11 below) and a representation of rescaled MD estimation errors (Theorem 2.14); on this basis, asymptotic normality for MD estimator sequences will be dealt with in Section 2.4. Our approach follows [100].

2.7 Example. Observe i.i.d. random variables $Y_1, Y_2, \ldots, Y_n$ taking values in $\mathbb{R}^m$ whose law depends on an unknown parameter $\vartheta \in \Theta$. Assume that Θ is an open subset of $\mathbb{R}^d$, and let the single observation under P_ϑ have continuous distribution function $F_\vartheta : \mathbb{R}^m \to [0, 1]$. Consider the family $\{F_\vartheta : \vartheta \in \Theta\}$ as a subset of $\mathcal{C}_b(\mathbb{R}^m)$ equipped with uniform convergence, and let the mapping

(+) $\qquad\qquad\qquad \Theta \ni \vartheta \longrightarrow F_\vartheta(\cdot) \in \mathcal{C}_b(\mathbb{R}^m)$

be continuous. Let $\widehat{F}_n(\cdot, \omega)$ denote the empirical distribution function

$$\widehat{F}_n(t, \omega) := \frac{1}{n}\sum_{i=1}^n 1_{(-\infty,t]}(Y_i(\omega)) = \frac{1}{n}\sum_{i=1}^n 1_{\{Y_i \leq t\}}(\omega), \quad t \in \mathbb{R}^m \, .$$

Considering as in Exercise 2.1'(a) the mapping $(t, \omega) \to \widehat{F}_n(t, \omega)$, $\widehat{F}_n(\cdot, \cdot)$ is a measurable stochastic process in the sense of Assumption 2.1(b). For any choice of a finite measure μ on $(\mathbb{R}^m, \mathcal{B}(\mathbb{R}^m))$, we write for short $L^2(\mu) = L^2(\mathbb{R}^m, \mathcal{B}(\mathbb{R}^m), \mu)$,

$\widehat{F}_n(\cdot,\cdot)$ is a process with paths in $(L^2(\mu), \mathcal{B}(L^2(\mu)))$ as defined in Lemma 2.2. Assumption (+) guarantees that for fixed ω the mapping

$$(++) \qquad \xi \longrightarrow \|\widehat{F}_n(\cdot,\omega) - F_\xi(\cdot)\|_{L^2(\mu)}$$

is continuous on Θ. Thinking of (++) as a surface over Θ which is a function of the observed $Y_1(\omega), \ldots, Y_n(\omega)$, 'best approximations' (later we will consider all the details which are necessary here)

$$\vartheta_n^*(\omega) = \underset{\xi \in \Theta}{\arg\inf}\ \|\widehat{F}_n(\cdot,\omega) - F_\xi(\cdot)\|_{L^2(\mu)}$$

provides an estimator for the unknown parameter $\vartheta \in \Theta$. This estimator is called a minimum distance (MD) estimator. It compares the empirical quantity $\widehat{F}_n(\cdot,\omega)$ to its theoretical counterparts $F_\xi(\cdot)$ under possible values ξ of the parameter, in the Hilbert space $L^2(\mu)$, where Glivenko–Cantelli

$$(*) \qquad \lim_{n\to\infty}\ \sup_{t\in\mathbb{R}^m} |\widehat{F}_n(t) - F_\vartheta(t)| = 0 \quad P_\vartheta\text{-almost surely, for every } \vartheta \in \Theta$$

makes random surfaces (++) stabilise under P_ϑ as $n \to \infty$ at a deterministic limit

$$(+++) \qquad \Theta \ni \xi \longrightarrow \|F_\vartheta(\cdot) - F_\xi(\cdot)\|_{L^2(\mu)} \in \mathbb{R},$$

for every $\vartheta \in \Theta$. This allows to obtain strong convergence of MD estimators under very weak – in contrast to ML estimation, cf. Assumptions 1.19 and Theorem 1.24 – assumptions. We sketch the core of this argument, leaving questions of existence or measurability to Propositions 2.10–2.11 below. Starting from

$$\left\{\min_{\xi:|\xi-\vartheta|\le\varepsilon} \|\widehat{F}_n - F_\xi\|_{L^2(\mu)} < \inf_{\xi:|\xi-\vartheta|>\varepsilon} \|\widehat{F}_n - F_\xi\|_{L^2(\mu)}\right\} \subset \{|\vartheta_n^* - \vartheta| \le \varepsilon\}$$

we take complements and obtain the inclusions

$$\{|\vartheta_n^* - \vartheta| > \varepsilon\} \subset \left\{\min_{\xi:|\xi-\vartheta|\le\varepsilon} \|\widehat{F}_n - F_\xi\|_{L^2(\mu)} \ge \inf_{\xi:|\xi-\vartheta|>\varepsilon} \|\widehat{F}_n - F_\xi\|_{L^2(\mu)}\right\}$$

$$\subset \left\{\|\widehat{F}_n - F_\vartheta\|_{L^2(\mu)} \ge \inf_{\xi:|\xi-\vartheta|>\varepsilon} \|\widehat{F}_n - F_\xi\|_{L^2(\mu)}\right\}$$

$$\subset \left\{\|\widehat{F}_n - F_\vartheta\|_{L^2(\mu)} \ge \inf_{\xi:|\xi-\vartheta|>\varepsilon} \left(\|F_\vartheta - F_\xi\| - \|\widehat{F}_n - F_\vartheta\|\right)\right\}$$

$$\subset \left\{2\|\widehat{F}_n - F_\vartheta\|_{L^2(\mu)} \ge \inf_{\xi:|\xi-\vartheta|>\varepsilon} \|F_\vartheta - F_\xi\|_{L^2(\mu)}\right\}$$

for n large enough, using inverse triangular inequality. Glivenko–Cantelli ($*$) in the last expression shows that an identifiability condition

$$(**) \qquad \inf_{\xi\in\Theta:|\xi-\vartheta|>\varepsilon} \|F_\xi - F_\vartheta\|_{L^2(\mu)} > 0, \quad \text{for every } \vartheta \in \Theta \text{ and every } \varepsilon > 0$$

will guarantee consistency of the estimator sequence $(\vartheta_n^*)_n$ for the unknown parameter as $n \to \infty$. Due to the structure of the last right-hand side in the chain of inclusions above, the convergence $\vartheta_n^* \to \vartheta$ under P_ϑ will necessarily be almost sure convergence, for every $\vartheta \in \Theta$. Thanks to the continuity (+), the identifiability condition (**) is usually easy to satisfy in restriction to compact subsets of Θ; difficulties may arise for ξ at large distances from ϑ, in the case where Θ is unbounded and of large dimension. □

We give a list of assumptions and notations to be used in this subsection. The assumptions in 2.8(I) will always be in force; out of the list 2.8(III), we will indicate separately for each result what we assume.

2.8 Assumptions and Notations for Section 2.2. $\{P_\vartheta : \vartheta \in \Theta\}$ is a family of probability measures on some space $(\Omega, \mathcal{A})$. The parameter space $\Theta \subset \mathbb{R}^d$ is open. $(\mathcal{F}_n)_n$ is an increasing family of sub-σ-fields in $\mathcal{A}$ where $\mathcal{F}_n$ represents stage n of the asymptotics (e.g. n-fold independent replication of an experiment). We write $P_{n,\vartheta} := P_\vartheta | \mathcal{F}_n$ for the restriction of P_ϑ to $\mathcal{F}_n$, and consider the sequence of experiments

$$\mathcal{E}_n := (\Omega, \mathcal{F}_n, \{P_{n,\vartheta} : \vartheta \in \Theta\}) , \quad n \geq 1 .$$

(I) H is a Hilbert space, with scalar product $\langle \cdot, \cdot \rangle_H$ and norm $\|\cdot\|_H$, equipped with its Borel-σ-field $\mathcal{B}(H)$. We consider a sequence $\widehat{\Psi}_n$ of $\mathcal{F}_n$-measurable H-valued random variables

$$\widehat{\Psi}_n : \quad (\Omega, \mathcal{F}_n) \longrightarrow (H, \mathcal{B}(H)) , \quad n \geq 1$$

and a deterministic family $\{\Psi_\xi : \xi \in \Theta\}$ of objects in H such that

(+) \qquad the mapping $\Theta \ni \xi \to \Psi_\xi \in H$ is continuous.

At stage n of the asymptotics, $\widehat{\Psi}_n$ is a statistic in $\mathcal{E}_n$, and Ψ_ξ a theoretical counterpart – deterministic and independent of n – for $\widehat{\Psi}_n$ under a true value ξ of the unknown parameter.

(II) By continuity (+), for $\omega \in \Omega$ fixed, the mappings

$$\Theta \ni \xi \to \left\| \widehat{\Psi}_n(\omega) - \Psi_\xi \right\|_H \in [0, \infty)$$

are continuous: hence for open (or closed) sets B or compact sets F which are contained in Θ,

$$\inf_{\xi \in B} \left\| \widehat{\Psi}_n - \Psi_\xi \right\|_H , \quad \sup_{\xi \in B} \left\| \widehat{\Psi}_n - \Psi_\xi \right\|_H , \quad \min_{\xi \in F} \left\| \widehat{\Psi}_n - \Psi_\xi \right\|_H , \quad \ldots$$

are well-defined random variables on $(\Omega, \mathcal{F}_n)$.

(III) We introduce a list of assumptions concerning the objects in (I) under parameter values from Θ:

(a) *Strong law of large numbers* **SLLN(ϑ)**: we require

$$P_\vartheta\text{-almost surely:} \quad \|\widehat{\Psi}_n - \Psi_\vartheta\|_H \longrightarrow 0 \text{ as } n \to \infty;$$

(b) *Identifiability condition* **I(ϑ)**: we require

$$\inf_{\xi\in\Theta,\,|\xi-\vartheta|>\delta} \|\Psi_\xi - \Psi_\vartheta\|_H > 0 \quad \text{for } \delta > 0 \text{ arbitrarily small;}$$

(c) *Tightness condition* **T(ϑ)**: there is a sequence $\varphi_n = \varphi_n(\vartheta) \uparrow \infty$ of norming constants such that

$$\left\{ \mathcal{L}\left(\varphi_n \|\widehat{\Psi}_n - \Psi_\vartheta\|_H \mid P_\vartheta \right) : n \geq 1 \right\} \quad \text{is tight in } \mathbb{R};$$

(d) *Differentiability condition* **D(ϑ)** (including *non-singularity* of the derivative): the mapping

(+) $\qquad\qquad\qquad\qquad \Theta \ni \xi \longrightarrow \Psi_\xi \in H$

is Fréchet differentiable at $\xi = \vartheta$, i.e. there is a derivative

$$D\Psi_\vartheta := \begin{pmatrix} D_1\Psi_\vartheta \\ \cdots \\ D_d\Psi_\vartheta \end{pmatrix}, \quad D_j\Psi_\vartheta \in H, \quad 1 \leq j \leq d$$

at ϑ such that

$$\frac{1}{|\xi-\vartheta|} \left\| \Psi_\xi - \Psi_\vartheta - (\xi-\vartheta)^\top D\Psi_\vartheta \right\|_H \longrightarrow 0 \quad \text{as } |\xi-\vartheta| \to 0.$$

We require in addition that the components $D_j\Psi_\vartheta$, $1 \leq j \leq d$, be linearly independent in H.

2.9 Definition. A sequence of estimators $\vartheta_n^* : (\Omega, \mathcal{F}_n) \to (\mathbb{R}^d, \mathcal{B}(\mathbb{R}^d))$ for the unknown parameter $\vartheta \in \Theta$ is called a *minimum distance* (MD) sequence if there is a sequence of events $A_n \in \mathcal{F}_n$ such that

$$P_\vartheta\left(\liminf_{n\to\infty} A_n \right) = 1$$

for all $\vartheta \in \Theta$ fixed, and such that for $n \geq 1$

$$\vartheta_n^*(\omega) \in \Theta \quad \text{and} \quad \left\|\widehat{\Psi}_n(\omega) - \Psi_{\vartheta_n^*(\omega)}\right\|_H = \inf_{\xi\in\Theta} \left\|\widehat{\Psi}_n(\omega) - \Psi_\xi\right\|_H \quad \text{for all } \omega \in A_n.$$

Whenever the symbolic notation

$$\vartheta_n^* = \arginf_{\xi\in\Theta} \left\|\widehat{\Psi}_n - \Psi_\xi\right\|_H, \quad n \to \infty$$

will be used below, we assume that the above properties do hold.

Section 2.2 Minimum Distance Estimator Sequences

2.10 Proposition. Consider a compact exhaustion $(K_n)_n$ of Θ. Define with respect to K_n the event

$$A_n := \left\{ \min_{\xi \in K_n} \left\| \widehat{\Psi}_n - \Psi_\xi \right\|_H = \inf_{\xi \in \Theta} \left\| \widehat{\Psi}_n - \Psi_\xi \right\|_H \right\} \in \mathcal{F}_n, \quad n \geq 1.$$

Then one can construct a sequence $(T_n)_n$ such that the following holds for all $n \geq 1$:

$$T_n : (\Omega, \mathcal{F}_n) \longrightarrow (\mathbb{R}^d, \mathcal{B}(\mathbb{R}^d)) \quad \text{is measurable,}$$

$$\forall \, \omega \in A_n : T_n(\omega) \in K_n, \quad \left\| \widehat{\Psi}_n(\omega) - \Psi_{T_n(\omega)} \right\|_H = \inf_{\xi \in \Theta} \left\| \widehat{\Psi}_n(\omega) - \Psi_\xi \right\|_H.$$

Proof. (1) We fix a compact exhaustion $(K_n)_n$ of Θ:

$$K_n \text{ compact in } \mathbb{R}^d, \quad K_n \subset \text{int}(K_{n+1}), \quad K_n \uparrow \Theta.$$

For every n and every $\omega \in \Omega$, define

$$M_n(\omega) := \left\{ \zeta \in K_n : \left\| \widehat{\Psi}_n(\omega) - \Psi_\zeta \right\|_H = \inf_{\xi \in \Theta} \left\| \widehat{\Psi}_n(\omega) - \Psi_\xi \right\|_H \right\}$$

the set of all points in K_n where the mapping $\xi \to \|\widehat{\Psi}_n(\omega) - \Psi_\xi\|_H$ attains its global minimum. By continuity of this mapping and definition of A_n, $M_n(\omega)$ is a non-void closed subset of K_n when $\omega \in A_n$, hence a non-void compact set. Thus, for $\omega \in A_n$, out of arbitrary sequences in $M_n(\omega)$, we can select convergent subsequences having limits in $M_n(\omega)$.

(2) For fixed $n \geq 1$ and fixed $\omega \in A_n$ we specify one particular point $\alpha(\omega) \in M_n(\omega)$ as follows. Put

$$\alpha_1(\omega) := \inf\{\zeta_1 : \text{there are points } \zeta = (\zeta_1, \zeta_2, ..., \zeta_d) \in M_n(\omega)\}.$$

Selecting convergent subsequences in the non-void compact $M_n(\omega)$, we see that $M_n(\omega)$ contains points of the form $(\alpha_1(\omega), \zeta_2, ..., \zeta_d)$. Next we consider

$$\alpha_2(\omega) := \inf\{\zeta_2 : \text{there are points } \zeta = (\alpha_1(\omega), \zeta_2, ..., \zeta_d) \in M_n(\omega)\}.$$

Again, selecting convergent subsequences in $M_n(\omega)$, we see that $M_n(\omega)$ contains points of the form $(\alpha_1(\omega), \alpha_2(\omega), \zeta_3, ..., \zeta_d)$. Continuing in this way, we end up with a point

$$\alpha(\omega) := (\alpha_1(\omega), \alpha_2(\omega), ..., \alpha_d(\omega)) \in M_n(\omega)$$

which has the property

$$\alpha_j(\omega) = \min\{\zeta_j : \text{there are points } \zeta = (\alpha_1(\omega), \ldots, \alpha_{j-1}(\omega), \zeta_j, \ldots, \zeta_d) \in M_n(\omega)\}$$

for all components $1 \leq j \leq d$.

(3) For $\omega \in A_n$ and for the particular point $\alpha(\omega) \in M_n(\omega)$ selected in step (2), write $\alpha^{(n)}(\omega) := \alpha(\omega)$ for clarity, and define (fixing some default value $\vartheta_0 \in \Theta$)

$$T_n(\omega) := \vartheta_0 \, 1_{A_n^c}(\omega) + \alpha^{(n)}(\omega) \, 1_{A_n}(\omega) \,, \quad \omega \in \Omega \,.$$

Then $T_n(\omega)$ represents *one* point in K_n such that the mapping $\xi \to \|\widehat{\Psi}_n(\omega) - \Psi_\xi\|_H$ attains its global minimum on Θ at $T_n(\omega)$, provided $\omega \in A_n$. It remains to show that T_n is $\mathcal{F}_n$-measurable.

(4) Write for short $\rho := \inf_{\xi \in \Theta} \|\widehat{\Psi}_n - \Psi_\xi\|_H$. By construction of the sequence $(\alpha^{(n)})_n$, fixing arbitrary $(b_1, \ldots, b_d) \in \mathbb{R}^d$ and selecting convergent subsequences in compacts $K_n \cap (\times_{j=1}^m (-\infty, b_j] \times \mathbb{R}^{d-m})$, the following is seen to hold successively in $1 \leq m \leq d$:

$$A_n \cap \left\{\alpha_1^{(n)} \leq b_1, \ldots, \alpha_m^{(n)} \leq b_m\right\} =$$

$$\bigcap_{r > 0 \text{ rational}} \bigcup_{\substack{\xi \in \mathbb{Q}^d \cap K_n \\ \xi_j \leq b_j, 1 \leq j \leq m}} \left\{\|\widehat{\Psi}_n - \Psi_\xi\|_H < \rho + r\right\} \in \mathcal{F}_n$$

(where again we have used continuity (+) of the mapping $\xi \to \|\widehat{\Psi}_n(\omega) - \Psi_\xi\|_H$ on Θ for ω fixed). As a consequence of this equality, taking $m = d$, the mapping $\omega \to \alpha^{(n)}(\omega) 1_{A_n}(\omega)$ is $\mathcal{F}_n$-measurable. Thus, with constant ϑ_0 and $A_n \in \mathcal{F}_n$, T_n defined in step (3) is a $\mathcal{F}_n$-measurable random variable, and all assertions of the proposition are proved. □

The core of the last proof was 'measurable selection' out of the set of points where $\xi \to \|\widehat{\Psi}_n(\omega) - \Psi_\xi\|_H$ attains its global minimum on Θ. In asymptotic statistics, problems of this type arise in almost all cases where one wishes to construct estimators through minima, maxima or zeros of suitable mappings $\xi \to H(\xi, \omega)$. See [43, Thm. A.2 in App. A] for a general and easily applicable result to solve 'measurable selection problems' in parametric models, see also [105, Thm. 6.7.22 and Lem. 6.7.23].

2.11 Proposition. Assume **SLLN**(ϑ) and $\mathbf{I}(\vartheta)$ for all $\vartheta \in \Theta$. Then the sequence $(T_n)_n$ constructed in Proposition 2.10 is a minimum distance estimator sequence for the unknown parameter. Moreover, *arbitrary* minimum distance estimator sequences $(\vartheta_n^*)_n$ for the unknown parameter as defined in Definition 2.9 are strongly consistent:

$$\forall \, \vartheta \in \Theta: \quad \vartheta_n^* \longrightarrow \vartheta \quad P_\vartheta\text{-almost surely as } n \to \infty.$$

Proof. (1) For the particular sequence $(T_n)_n$ which has been constructed in Proposition 2.10, using a compact exhaustion $(K_n)_n$ of Θ and measurable selection on events

$$A_n := \left\{\min_{\xi \in K_n} \|\widehat{\Psi}_n - \Psi_\xi\|_H = \inf_{\xi \in \Theta} \|\widehat{\Psi}_n - \Psi_\xi\|_H\right\} \in \mathcal{F}_n$$

Section 2.2 Minimum Distance Estimator Sequences

defined with respect to K_n, we have to show that the conditions of our proposition imply

(o) $$P_\vartheta \left(\liminf_{n \to \infty} A_n \right) = 1$$

for all $\vartheta \in \Theta$. Then, by Proposition 2.10, the sequence $(T_n)_n$ with 'good sets' $(A_n)_n$ will have all properties required in Definition 2.9 of an MD estimator sequence.

Fix $\vartheta \in \Theta$. Since $(K_n)_n$ is a compact exhaustion of the open parameter set Θ, there is some n_0 and some $\varepsilon_0 > 0$ such that $B_{2\varepsilon_0}(\vartheta) \subset K_n$ for all $n \geq n_0$. Consider $0 < \varepsilon < \varepsilon_0$ arbitrarily small. By the definition of A_n and T_n in Proposition 2.10, we have for $n \geq n_0$

$$\left\{ \min_{\xi:|\xi-\vartheta|\leq\varepsilon} \left\| \widehat{\Psi}_n - \Psi_\xi \right\|_H < \inf_{\xi:|\xi-\vartheta|>\varepsilon} \left\| \widehat{\Psi}_n - \Psi_\xi \right\|_H \right\} \subset A_n \cap \{|T_n - \vartheta| \leq \varepsilon\}.$$

Passing to complements this reads for $n \geq n_0$

$$\{|T_n - \vartheta| > \varepsilon\} \cup A_n^c$$

$$\subset \left\{ \min_{\xi:|\xi-\vartheta|\leq\varepsilon} \left\| \widehat{\Psi}_n - \Psi_\xi \right\|_H \geq \inf_{\xi:|\xi-\vartheta|>\varepsilon} \left\| \widehat{\Psi}_n - \Psi_\xi \right\|_H \right\}$$

$$\subset \left\{ \left\| \widehat{\Psi}_n - \Psi_\vartheta \right\|_H \geq \inf_{\xi:|\xi-\vartheta|>\varepsilon} \left\| \widehat{\Psi}_n - \Psi_\xi \right\|_H \right\}$$

$$\subset \left\{ \| \widehat{\Psi}_n - \Psi_\vartheta \|_H \geq \inf_{\xi:|\xi-\vartheta|>\varepsilon} \left(\| \Psi_\vartheta - \Psi_\xi \|_H - \| \widehat{\Psi}_n - \Psi_\vartheta \|_H \right) \right\}$$

$$\subset C_n := \left\{ 2 \cdot \left\| \widehat{\Psi}_n - \Psi_\vartheta \right\|_H \geq \inf_{\xi:|\xi-\vartheta|>\varepsilon} \| \Psi_\xi - \Psi_\vartheta \|_H \right\}$$

(in the third line, we use inverse triangular inequality). For the event C_n defined by the right-hand side of this chain of inclusions, **SLLN**(ϑ) combined with **I**(ϑ) yields

$$P_\vartheta \left(\limsup_{n \to \infty} C_n \right) = P_\vartheta \left(\{ \omega : \omega \in C_n \text{ for infinitely many } n \} \right) = 0.$$

But A_n^c is a subset of C_n for $n \geq n_0$, hence we have $P_\vartheta (\limsup_{n \to \infty} A_n^c) = 0$ and thus (o).

(2) Next we consider an arbitrary MD estimator sequence $(\vartheta_n^*)_n$ for the unknown parameter according to Definition 2.9, and write $(\widetilde{A}_n)_n$ for its sequence of 'good sets': thus $\widetilde{A}_n \in \mathcal{F}_n$, for $\omega \in \widetilde{A}_n$ the mapping $\xi \to \| \widehat{\Psi}_n(\omega) - \Psi_\xi \|_H$ attains its global minimum on Θ at $\vartheta_n^*(\omega)$, and we have

(oo) $$P_\vartheta \left(\liminf_{n \to \infty} \widetilde{A}_n \right) = 1.$$

By Definition 2.9, we have necessarily

$$\left\{ \min_{\xi:|\xi-\vartheta|\leq\varepsilon} \left\| \widehat{\Psi}_n - \Psi_\xi \right\|_H < \inf_{\xi:|\xi-\vartheta|>\varepsilon} \left\| \widehat{\Psi}_n - \Psi_\xi \right\|_H \right\} \subset \left(\widetilde{A}_n^c \cup \{|\vartheta_n^* - \vartheta| \leq \varepsilon\} \right)$$

(note the different role played by general 'good sets' $\widetilde{A}_n$ in comparison to the particular construction of Proposition 2.10 and step (1) above). If we transform the left-hand side by the chain of inclusions of step (1) with C_n defined as there, we now obtain

$$\{|\vartheta_n^* - \vartheta| > \varepsilon\} \cap \widetilde{A}_n \subset C_n$$

for all $n \geq 1$, and thus

$$\{|\vartheta_n^* - \vartheta| > \varepsilon\} \subset C_n \cup \widetilde{A}_n^c \ , \quad n \geq 1 \ .$$

Again **SLLN**(ϑ) and **I**(ϑ) guarantee $P_\vartheta(\limsup_{n\to\infty} C_n) = 0$ as in step (1) above. Since property (∞) yields $P_\vartheta(\limsup_{n\to\infty} \widetilde{A}_n^c) = 0$, we arrive at

$$P_\vartheta\left(\limsup_{n\to\infty} \{|\vartheta_n^* - \vartheta| > \varepsilon\}\right) = 0 \ .$$

This holds for all $\varepsilon > 0$, and we have proved P_ϑ-almost sure convergence of $(\vartheta_n^*)_n$ to ϑ. □

Two auxiliary results prepare for the proof of our main result – the representation of rescaled MD estimator errors, Lemma 2.13 below – in this section. The first is purely analytical.

2.12 Lemma. Under **I**(ϑ) and **D**(ϑ) we have

$$\lim_{c\uparrow\infty} \liminf_{n\to\infty} \inf_{|h|>c} \left\| \varphi_n \left(\Psi_{\vartheta+h/\varphi_n} - \Psi_\vartheta\right) \right\|_H = \infty$$

for any sequence $\varphi_n \uparrow \infty$ of real numbers.

Proof. By assumption **I**(ϑ), we have for $\varepsilon > 0$ fixed as $n \to \infty$

$$(+) \quad \inf_{h:|h/\varphi_n|\geq\varepsilon} \left\| \varphi_n \left(\Psi_{\vartheta+h/\varphi_n} - \Psi_\vartheta\right) \right\|_H = \varphi_n \cdot \inf_{\xi:|\xi-\vartheta|\geq\varepsilon} \left\| \Psi_\xi - \Psi_\vartheta \right\|_H \longrightarrow \infty \ .$$

Assumption **D**(ϑ) includes linear independence in H of the components $D_1\Psi_\vartheta, \ldots, D_d\Psi_\vartheta$ of the derivative $D\Psi_\vartheta$. Since $u \to \|u^\top D\Psi_\vartheta\|_H$ is continuous, we have on the unit sphere in $\mathbb{R}^d$

$$(++) \quad \rho := \min_{|u|=1} \|u^\top D\Psi_\vartheta\|_H > 0 \ .$$

Assumption **D**(ϑ) shows also

$$(+++) \quad \sup_{\xi:|\xi-\vartheta|<\varepsilon} \frac{1}{|\xi-\vartheta|} \left\| \Psi_\xi - \Psi_\vartheta - (\xi-\vartheta)^\top D\Psi_\vartheta \right\|_H < \frac{\rho}{2}$$

provided $\varepsilon > 0$ is small enough. For any $0 < c < \infty$, because of (+), all what remains to consider in

$$\inf_{|h|>c} \left\| \varphi_n \left(\Psi_{\vartheta+h/\varphi_n} - \Psi_\vartheta\right) \right\|_H$$

Section 2.2 Minimum Distance Estimator Sequences

are contributions

$$\inf_{|h|>c,\,|h/\varphi_n|<\varepsilon} \|\varphi_n(\Psi_{\vartheta+h/\varphi_n} - \Psi_\vartheta)\|_H, \quad \varepsilon > 0 \text{ arbitrarily small}$$

which by (++) and (+++) allow for lower bounds

$$\inf_{|h|>c,\,|h/\varphi_n|<\varepsilon} |h| \, \frac{\|\Psi_{\vartheta+h/\varphi_n} - \Psi_\vartheta\|_H}{|h/\varphi_n|}$$

$$\geq c \cdot \inf_{\xi:0<|\xi-\vartheta|<\varepsilon} \frac{\|\Psi_\xi - \Psi_\vartheta\|_H}{|\xi-\vartheta|}$$

$$\geq c \cdot \left(\inf_{\xi:0<|\xi-\vartheta|<\varepsilon} \frac{\|(\xi-\vartheta)^\top D\Psi_\vartheta\|_H}{|\xi-\vartheta|} \right.$$

$$\left. - \sup_{\xi:|\xi-\vartheta|<\varepsilon} \frac{\|\Psi_\xi - \Psi_\vartheta - (\xi-\vartheta)^\top D\Psi_\vartheta\|_H}{|\xi-\vartheta|} \right)$$

$$\geq c \cdot (\rho - \frac{\rho}{2}) = c \cdot \frac{\rho}{2}$$

which increase to ∞ together with c. □

2.13 Lemma. In addition to $\mathbf{I}(\vartheta)$ and $\mathbf{D}(\vartheta)$, assume $\mathbf{T}(\vartheta)$ with norming constants $\varphi_n = \varphi_n(\vartheta) \uparrow \infty$. Then arbitrary MD sequences $(\vartheta_n^*)_n$ according to Definition 2.9 are $(\varphi_n)_n$-consistent at ϑ:

$$\{ \mathcal{L}(\varphi_n(\vartheta_n^* - \vartheta) \mid P_\vartheta) : n \geq 1 \} \quad \text{is tight in } \mathbb{R}^d.$$

Proof. Let $(\vartheta_n^*)_n$ with 'good sets' $(A_n)_n$ denote any choice of a MD estimator sequence for the unknown parameter according to Definition 2.9.

(1) For $K < \infty$ arbitrarily large but fixed, we repeat the reasoning of step (2) in the proof of Proposition 2.11, except that we insert K/φ_n in place of the ε there, for n large enough. Writing

$$C_n(K) := \left\{ 2 \cdot \|\widehat{\Psi}_n - \Psi_\vartheta\|_H \geq \inf_{\xi:|\xi-\vartheta|>K/\varphi_n} \|\Psi_\xi - \Psi_\vartheta\|_H \right\}$$

we thus obtain

(o) $$\{|\varphi_n(\vartheta_n^* - \vartheta)| > K\} \subset C_n(K) \cup A_n^c$$

for n large enough, with 'good sets' A_n satisfying $P_\vartheta(\liminf_{n\to\infty} A_n) = 1$.

(2) Under assumptions $\mathbf{I}(\vartheta)$ and $\mathbf{D}(\vartheta)$, Lemma 2.12 shows that deterministic quantities

$$\liminf_{n\to\infty} \inf_{|h|>K} \|\varphi_n(\Psi_{\vartheta+h/\varphi_n} - \Psi_\vartheta)\|_H$$

can be made arbitrarily large by choosing K large, whereas the tightness condition $\mathbf{T}(\vartheta)$ yields

$$\lim_{M\uparrow\infty}\sup_{n\geq 1} P_\vartheta\left(\|\varphi_n(\widehat{\Psi}_n - \Psi_\vartheta)\|_H > M\right) = 0.$$

Combining both statements, we obtain for the events $C_n(K)$ in step (1)

for every $\varepsilon > 0$ there is $K = K(\varepsilon) < \infty$ such that $\limsup\limits_{n\to\infty} P_\vartheta(C_n(K(\varepsilon))) < \varepsilon$.

The 'good sets' A_n of step (1) satisfy a fortiori

$$\lim_{n\to\infty} P_\vartheta(A_n) = 1$$

(view $\liminf\limits_{n\to\infty} A_n = \bigcup_m \bigcap_{n\geq m} A_n$ as increasing limit of events $\bigcap_{n\geq m} A_n$ as m tends to ∞). Combining both last assertions with ($\circ$), we have for every $\varepsilon > 0$ some $K(\varepsilon) < \infty$ such that

$$\limsup_{n\to\infty} P_\vartheta\left(|\varphi_n(\vartheta_n^* - \vartheta)| > K(\varepsilon)\right) < \varepsilon$$

holds. This finishes the proof. □

We arrive at the main result of this section.

2.14 Theorem (Millar [100]). Assume **SLLN**(ϑ), **I**(ϑ), **D**(ϑ), and **T**(ϑ) with a sequence of norming constants $\varphi_n = \varphi_n(\vartheta) \uparrow \infty$. With $d \times d$ matrix

$$\Lambda_\vartheta := \left(\langle D_i\Psi_\vartheta, D_j\Psi_\vartheta\rangle\right)_{1\leq i,j\leq d}$$

(invertible by assumption **D**(ϑ)) define a linear mapping $\Pi_\vartheta : H \longrightarrow \mathbb{R}^d$

$$\Pi_\vartheta(f) := \Lambda_\vartheta^{-1}\begin{pmatrix} \langle D_1\Psi_\vartheta, f\rangle \\ \ldots \\ \langle D_d\Psi_\vartheta, f\rangle \end{pmatrix}, \quad f \in H.$$

Then rescaled estimation errors at ϑ of arbitrary minimum distance estimator sequences $(\vartheta_n^*)_n$ as in Definition 2.9 admit the representation

$$\varphi_n(\vartheta_n^* - \vartheta) = \Pi_\vartheta(\varphi_n(\widehat{\Psi}_n - \Psi_\vartheta)) + o_{P_\vartheta}(1), \quad n \to \infty.$$

Proof. (1) We begin with a preliminary remark. By assumption **D**(ϑ), see Assumptions 2.8(III), the components $D_1\Psi_\vartheta, \ldots, D_d\Psi_\vartheta$ of the derivative $D\Psi_\vartheta$ are linearly independent in H. Hence

$$V_\vartheta := \text{span}(D_i\Psi_\vartheta : 1 \leq i \leq d)$$

is a d-dimensional closed linear subspace of H. For points $h \in \mathbb{R}^d$ with components $h_1, \ldots, h_d$ and for elements $f \in H$, the following two statements are equivalent:

Section 2.2 Minimum Distance Estimator Sequences

(i) the orthogonal projection of f on V_ϑ takes the form $\sum_{i=1}^n h_i D_i \Psi_\vartheta = h^\top D\Psi_\vartheta$;

(ii) one has $\Pi_\vartheta(f) = h$.

Note that the orthogonal projection of f on V_ϑ corresponds to the unique h in $\mathbb{R}^d$ such that $(f - h^\top D\Psi_\vartheta) \perp V_\vartheta$. This can be rewritten as

$$0 = \left\langle f - \sum_{i=1}^d h_i D_i \Psi_\vartheta, D_j \Psi_\vartheta \right\rangle = \langle f, D_j \Psi_\vartheta \rangle - \sum_{i=1}^d h_i (\Lambda_\vartheta)_{i,j}, \quad \text{for all } 1 \leq j \leq d$$

by definition of Λ_ϑ, and thus in the form $(\langle D_1 \Psi_\vartheta, f \rangle, \ldots, \langle D_d \Psi_\vartheta, f \rangle) = h^\top \Lambda_\vartheta$; transposing and using symmetry of Λ_ϑ the last line gives

$$\Lambda_\vartheta h = \begin{pmatrix} \langle D_1 \Psi_\vartheta, f \rangle \\ \ldots \\ \langle D_d \Psi_\vartheta, f \rangle \end{pmatrix}.$$

Inverting Λ_ϑ we have the assertion.

(2) Inserting the H-valued random variable $\varphi_n(\widehat{\Psi}_n - \Psi_\vartheta)$ under P_ϑ in place of f in step (1), we define

$$\widehat{h}_n := \Pi_\vartheta \left(\varphi_n (\widehat{\Psi}_n - \Psi_\vartheta) \right)$$

which is a random variable taking values in $\mathbb{R}^d$. Then by step (1),

(∗) $\widehat{h}_n^\top D\Psi_\vartheta$ is the orthogonal projection of $\varphi_n(\widehat{\Psi}_n - \Psi_\vartheta)$ on the subspace V_ϑ.

Moreover, assumption $\mathbf{T}(\vartheta)$ imposing tightness of $\|\varphi_n(\widehat{\Psi}_n - \Psi_\vartheta)\|_H$ under P_ϑ as $n \to \infty$ guarantees

the family of laws $\mathcal{L}(\widehat{h}_n \mid P_\vartheta), n \geq 1$, is tight in $\mathbb{R}^d$

since Π_ϑ is a linear mapping.

(3) Let $(\vartheta_n^*)_n$ denote any MD estimator sequence for the unknown parameter with 'good sets' $(A_n)_n$ as in Definition 2.9: for every $n \geq 1$, on the event A_n, the mapping

$$\Theta \ni \xi \longrightarrow \left\| \widehat{\Psi}_n - \Psi_\xi \right\|_H \in [0, \infty)$$

attains its global minimum at ϑ_n^*, and we have (in particular, cf. proof of Lemma 2.13) $\lim_{n \to \infty} P_\vartheta(A_n) = 1$. With the norming sequence of assumption $\mathbf{T}(\vartheta)$, put

$$h_n^* := \varphi_n(\vartheta_n^* - \vartheta), \quad \Theta_{\vartheta,n} := \{h \in \mathbb{R}^d : \vartheta + h/\varphi_n \in \Theta\}.$$

Then the sets $\Theta_{\vartheta,n}$ increase to $\mathbb{R}^d$ as $n \to \infty$ since Θ is open. Rewriting points $\xi \in \Theta$ relative to ϑ in the form $\vartheta + h/\varphi_n$, $h \in \Theta_{\vartheta,n}$, we have the following: for every $n \geq 1$, on the event A_n, the mapping

(∗∗) $$\Theta_{\vartheta,n} \ni h \longrightarrow \left\| \varphi_n(\widehat{\Psi}_n - \Psi_{\vartheta + h/\varphi_n}) \right\| =: \Delta_{n,\vartheta}(h)$$

attains a global minimum at h_n^*. By Lemma 2.13,

the family of laws $\mathcal{L}(h_n^* \mid P_\vartheta), n \geq 1$, is tight in $\mathbb{R}^d$.

(4) In order to prove our theorem, we have to show
$$h_n^* = \widehat{h}_n + o_{P_\vartheta}(1) \quad \text{as } n \to \infty$$
with the notation of steps (2) and (3). The idea is as follows. On the one hand, the function $\Delta_{n,\vartheta}(\cdot)$ defined in (**) attains a global minimum at h_n^* (on the event A_n), on the other hand, by (**),
$$\Delta_{n,\vartheta}(h) = \| \varphi_n(\widehat{\Psi}_n - \Psi_\vartheta) - \varphi_n(\Psi_{\vartheta+h/\varphi_n} - \Psi_\vartheta) \|$$
$$= \| \varphi_n(\widehat{\Psi}_n - \Psi_\vartheta) - h^\top D\Psi_\vartheta - \varphi_n(\Psi_{\vartheta+h/\varphi_n} - \Psi_\vartheta - (h/\varphi_n)^\top D\Psi_\vartheta) \|$$
will be close (for n large) to the function

(* * *) $\qquad D_{n,\vartheta}(h) := \| \varphi_n(\widehat{\Psi}_n - \Psi_\vartheta) - h^\top D\Psi_\vartheta \|$

which admits a unique minimum at $\widehat{h}_n$ according to (*) in step (2), up to small perturbations of order
$$\| \varphi_n (\Psi_{\vartheta+h/\varphi_n} - \Psi_\vartheta - (h/\varphi_n)^\top D\Psi_\vartheta) \|$$
which have to be negligible as $n \to \infty$, uniformly in compact h-intervals, in virtue of the differentiability assumption $D(\vartheta)$. We will make this precise in steps (5)–(7) below.

(5) By definition of the random functions $\Delta_{n,\vartheta}(\cdot)$ and $D_{n,\vartheta}(\cdot)$, an inverse triangular inequality in H establishes
$$|\Delta_{n,\vartheta}(h) - D_{n,\vartheta}(h)| \leq \| \varphi_n (\Psi_{\vartheta+h/\varphi_n} - \Psi_\vartheta - (h/\varphi_n)^\top D\Psi_\vartheta) \|$$
for all $\omega \in \Omega$, all $h \in \Theta_{n,\vartheta}$ and all $n \geq 1$. The bound on the right-hand side is deterministic. The differentiability assumption $D(\vartheta)$ shows that for arbitrarily large constants $C < \infty$,
$$\sup_{|h| \leq C} \| \varphi_n (\Psi_{\vartheta+h/\varphi_n} - \Psi_\vartheta - (h/\varphi_n)^\top D\Psi_\vartheta) \| \leq$$
$$C \sup_{|h| \leq C} \frac{\| \Psi_{\vartheta+h/\varphi_n} - \Psi_\vartheta - (h/\varphi_n)^\top D\Psi_\vartheta \|}{|h/\varphi_n|}$$
vanishes as $n \to \infty$. For n sufficiently large, $\Theta_{n,\vartheta}$ includes $\{|h| \leq C\}$. For all $\omega \in \Omega$, we thus have

($\diamond$) $\qquad \sup_{h \in K} |\Delta_{n,\vartheta}(h) - D_{n,\vartheta}(h)| \longrightarrow 0 \quad \text{as } n \to \infty$

on arbitrary compacts K in $\mathbb{R}^d$.

(6) Squaring the random function $D_{n,\vartheta}(\cdot)$ defined in (* * *), we find a quadratic lower bound

(+) $\qquad D_{n,\vartheta}^2(h) \geq D_{n,\vartheta}^2(\widehat{h}_n) + |h - \widehat{h}_n|^2 \rho^2, \quad \text{for all } \omega \in \Omega, h \in \mathbb{R}^d, n \geq 1$

Section 2.2 Minimum Distance Estimator Sequences

around its unique minimum at $\widehat{h}_n$, with

$$\rho = \min_{|u|=1} \| u^T D\Psi_\vartheta \| > 0$$

introduced in (++) in the proof of Lemma 2.12. Assertion (+) is proved as follows. From (∗) in step (2)

$\widehat{h}_n^T D\Psi_\vartheta$ is the orthogonal projection of $\varphi_n(\widehat{\Psi}_n - \Psi_\vartheta)$ on the subspace V_ϑ

or equivalently

$$\left(\varphi_n(\widehat{\Psi}_n - \Psi_\vartheta) - \widehat{h}_n^T D\Psi_\vartheta \right) \perp V_\vartheta$$

we have for the function $D_{n,\vartheta}(\cdot)$ according to Pythagoras

$$D_{n,\vartheta}^2(h) = \left\| (h - \widehat{h}_n)^T D\Psi_\vartheta \right\|^2 + D_{n,\vartheta}^2(\widehat{h}_n) \geq \left| h - \widehat{h}_n \right|^2 \rho^2 + D_{n,\vartheta}^2(\widehat{h}_n) .$$

This assertion holds for all $\omega \in \Omega$, all $h \in \mathbb{R}^d$ and all $n \geq 1$.

(7) To conclude the proof, we fix $\varepsilon > 0$ arbitrarily small. By tightness of $(\widehat{h}_n)_n$ and $(h_n^*)_n$ under P_ϑ, see steps (2) and (3) above, there is a compact $K = K(\varepsilon)$ in $\mathbb{R}^d$ such that

$$\sup_{n \geq 1} P_\vartheta \left(\widehat{h}_n \notin K \right) < \frac{\varepsilon}{2}, \quad \sup_{n \geq 1} P_\vartheta \left(h_n^* \notin K \right) < \frac{\varepsilon}{2} .$$

Then for the MD estimator sequence $(\vartheta_n^*)_n$ of step (3), with 'good sets' $(A_n)_n$,

$$P_\vartheta \left(|h_n^* - \widehat{h}_n| > \varepsilon \right)$$
$$\leq P_\vartheta \left(\widehat{h}_n \notin K \right) + P_\vartheta \left(h_n^* \notin K \right) + P_\vartheta \left(A_n^c \right)$$
$$\quad + P_\vartheta \left(\{\widehat{h}_n \in K\} \cap \{h_n^* \in K\} \cap A_n \cap \{|h_n^* - \widehat{h}_n| > \varepsilon\} \right)$$
$$\leq \frac{\varepsilon}{2} + \frac{\varepsilon}{2} + P_\vartheta \left(A_n^c \right)$$
$$\quad + P_\vartheta \left(\{\widehat{h}_n \in K\} \cap \{h_n^* \in K\} \cap A_n \cap \{D_{n,\vartheta}^2(h_n^*) \geq D_{n,\vartheta}^2(\widehat{h}_n) + \varepsilon^2 \rho^2\} \right)$$

where we have used the quadratic lower bound (+) for $D_{n,\vartheta}^2(\cdot)$ around $\widehat{h}_n$ from step (6). Recall that $\lim_{n \to \infty} P_\vartheta(A_n) = 1$, as in step (2) in the proof of Lemma 2.13. Thanks to the approximation (◇) from step (5), we can replace – uniformly on K as $n \to \infty$ – the random function $D_{n,\vartheta}^2(\cdot)$ by $\Delta_{n,\vartheta}^2(\cdot)$. This gives

$$\limsup_{n \to \infty} P_\vartheta \left(|h_n^* - \widehat{h}_n| > \varepsilon \right)$$
$$< \varepsilon + \limsup_{n \to \infty} P_\vartheta \left(A_n \cap \{\Delta_{n,\vartheta}^2(h_n^*) \geq \Delta_{n,\vartheta}^2(\widehat{h}_n) + \frac{1}{2}\varepsilon^2 \rho^2\} \right) .$$

We shall show that the last assertion implies

$$(\diamond\diamond) \qquad \limsup_{n\to\infty} P_\vartheta\left(|h_n^* - \widehat{h}_n| > \varepsilon\right) < \varepsilon.$$

For n large enough, the compact K is contained in $\Theta_{\vartheta,n}$. For $\omega \in A_n$, by $(**)$ in step (3), the global minimum $\underline{M}$ of the mapping $\Theta_{\vartheta,n} \ni h \to \Delta_{n,\vartheta}^2(h)$ is attained at h_n^*. Hence for n large enough, the intersection of the sets

$$A_n \quad \text{and} \quad \left\{\Delta_{n,\vartheta}^2(\widehat{h}_n) \le \underline{M} - \frac{1}{2}\varepsilon^2\rho^2\right\}$$

must be void: this proves $(\diamond\diamond)$. By $(\diamond\diamond)$, $\varepsilon > 0$ being arbitrary, the sequences $(h_n^*)_n$ and $(\widehat{h}_n)_n$ under P_ϑ are asymptotically equivalent. Thus we have proved

$$h_n^* = \widehat{h}_n + o_{P_\vartheta}(1) \quad \text{as } n \to \infty$$

which – as stated at the start of step (4) – concludes the proof. $\square$

2.15 Remark. We indicate a variant of our approach. Instead of almost sure convergence of the MD sequence $(\vartheta_n^*)_n$ to the true parameter one might be interested only in convergence in probability (consistency in the usual sense of Definition 1.9). For this, it is sufficient to work with Definition 2.9 of MD estimator sequences with good sets $(A_n)_n$ which satisfy the weaker condition $\lim_{n\to\infty} P_\vartheta(A_n) = 1$, and to weaken the condition **SLLN**(ϑ) to $\|\widehat{\Psi}_n - \Psi_\vartheta\|_H \longrightarrow 0$ in P_ϑ-probability, i.e. to a *weak law of large numbers* **WLLN**(ϑ). With these changes, Proposition 2.12, Lemma 2.13 and Theorem 2.14 remain valid, and Proposition 2.11 changes to convergence in P_ϑ-probability instead of P_ϑ-almost sure convergence.

2.3 Some Comments on Gaussian Processes

In the case where the Hilbert space of Section 2.2 is $H = L^2(T, \mathcal{T}, \mu)$ as considered in Assumption 2.1(a), Gaussian processes arise as limits of (weakly convergent subsequences of) $\varphi_n(\widehat{\Psi}_n - \Psi_\vartheta)$ under P_ϑ.

2.16 Definition. Consider a measurable space $(T, \mathcal{T})$ with countably generated σ-field $\mathcal{T}$, and a symmetric mapping $K(\cdot,\cdot): T \times T \to \mathbb{R}$.

(a) For finite measures μ on $(T, \mathcal{T})$, a real valued measurable process $(X_t)_{t \in T}$ with the property

$$\begin{cases} \text{there is an exceptional set } N \in \mathcal{T} \text{ of } \mu\text{-measure zero such that} \\ \mathcal{L}(X_{t_1}, \ldots, X_{t_l}) = \mathcal{N}\left(0_l, (K(t_i, t_j))_{i,j=1,\ldots,l}\right) \\ \text{for arbitrary } t_1, \ldots, t_l \in T \setminus N, \ \ell \ge 1 \end{cases}$$

is called (centred) μ-*Gaussian with covariance kernel* $K(\cdot,\cdot)$.

Section 2.3 Some Comments on Gaussian Processes

(b) A real valued measurable process $(X_t)_{t \in T}$ with the property

$$\begin{cases} \mathcal{L}(X_{t_1}, \ldots, X_{t_l}) = \mathcal{N}\left(0_l, (K(t_i, t_j))_{i,j=1,\ldots,l}\right) \\ \text{for arbitrary } t_1, \ldots, t_l \in T, \ \ell \geq 1 \end{cases}$$

is called (centred) *Gaussian with covariance kernel* $K(\cdot, \cdot)$.

2.17 Examples. (a.i) For $T = [0, \infty)$ or $T = [0, 1]$, consider standard Brownian motion $(B_t)_{t \in T}$ with $B_0 \equiv 0$. By continuity of all paths, B defined on any $(\Omega', \mathcal{A}', P')$ is a measurable stochastic process (cf. Exercise 2.1'(b)). By independence of increments, $B_{t_2} - B_{t_1}$ having law $\mathcal{N}(0, t_2 - t_1)$, we have $E(B_{t_1} B_{t_2}) = E(B_{t_1}^2) + E(B_{t_1}(B_{t_2} - B_{t_1})) = t_1$ for $t_1 < t_2$. Hence, writing

$$K(t_1, t_2) := E(B_{t_1} B_{t_2}) = t_1 \wedge t_2 \quad \text{for all } t_1, t_2 \text{ in } T,$$

Brownian motion $(B_t)_{t \in T}$ is a Gaussian process in the sense of Definition 2.16(b) with covariance kernel $K(\cdot, \cdot)$.
(ii) If a finite measure μ on $(T, \mathcal{T})$ satisfies the condition

$$E\left(\int_T B_t^2 \mu(dt)\right) = \int_T K(t, t) \mu(dt) = \int_T t \mu(dt) < \infty,$$

we can modify the paths of B on the P'-null set $A := \{\int_T B_t^2 \mu(dt) = \infty\}$ in $\mathcal{A}'$ (put $B(t, \omega) \equiv 0$ for all $t \in T$ if $\omega \in A$): then Brownian motion $(B_t)_{t \in T}$ is a measurable process with paths in $L^2(T, \mathcal{B}(T), \mu)$ according to Lemma 2.2.
(b.i) Put $T = [0, 1]$ and consider Brownian bridge

$$B^0 = (B_t^0)_{0 \leq t \leq 1}, \quad B_t^0 := B_t - t \cdot B_1, \quad 0 \leq t \leq 1.$$

Transformation properties of multidimensional normal laws under linear transformations and

$$\begin{pmatrix} B_{t_1}^0 \\ \vdots \\ B_{t_l}^0 \end{pmatrix} = \widetilde{A} \cdot \begin{pmatrix} B_{t_1} \\ \vdots \\ B_{t_l} \\ B_1 \end{pmatrix}, \quad \widetilde{A} := \begin{pmatrix} 1 & 0 & \cdots & 0 & -t_1 \\ 0 & 1 & \cdots & 0 & -t_2 \\ & & \vdots & & \\ 0 & 0 & \cdots & 1 & -t_l \end{pmatrix}$$

yield the finite dimensional distributions of B^0: hence Brownian bridge is a Gaussian process in the sense of Definition 2.16(b) with covariance kernel

$$K(t_1, t_2) = t_1 \wedge t_2 - t_1 t_2, \quad t_1, t_2 \in [0, 1].$$

(ii) The paths of B^0 being bounded functions on $[0, 1]$, Brownian bridge is a measurable stochastic process with paths in $L^2([0, 1], \mathcal{B}([0, 1]), \mu)$ (apply Lemma 2.2) for every finite measure μ on $([0, 1], \mathcal{B}([0, 1]))$.

(c.i) Fix a distribution function F, associated to an arbitrary probability distribution on $(\mathbb{R}, \mathcal{B}(\mathbb{R}))$. Brownian bridge time-changed by F is the process

$$B^{0,F} = \left(B_t^{0,F}\right)_{t \in \mathbb{R}}, \quad B_t^{0,F} := B_{F(t)}^0, \, t \in \mathbb{R}.$$

All paths of $B^{0,F}$ being càdlàg (right continuous with left-hand limits: this holds by construction since F is càdlàg), $B^{0,F}$ is a measurable stochastic process (cf. Exercise 2.1'(b)). Using (b), $B^{0,F}$ is a Gaussian process in the sense of Definition 2.16(b) with covariance kernel

$$K(t_1, t_2) = F(t_1) \wedge F(t_2) - F(t_1)F(t_2), \quad t_1, t_2 \in \mathbb{R}.$$

(ii) The paths of $B^{0,F}$ being bounded functions on $\mathbb{R}$, Brownian bridge time-changed by F is a process with paths in $L^2(\mathbb{R}, \mathcal{B}(\mathbb{R}), \mu)$ by Lemma 2.2, for arbitrary choice of a finite measure μ. □

Gaussian processes have been considered since about 1940, together with explicit orthogonal representations of the process in terms of eigenfunctions and eigenvalues of the covariance kernel $K(\cdot,\cdot)$ (Karhunen–Loève expansions). See Loeve [89, vol. 2, Sects. 36–37, in particular p. 144] or [89, 3rd. ed., p. 478], see also Gihman and Skorohod [28, vol. II, pp. 229–230].

2.18 Theorem. Consider T compact in $\mathbb{R}^k$, equipped with its Borel σ-field $\mathcal{T} = \mathcal{B}(T)$. Consider a mapping $K(\cdot,\cdot) : T \times T \to \mathbb{R}$ which is symmetric, continuous, and non-negative definite in the sense

$$\sum_{i,j=1,\ldots,\ell} \alpha_i K(t_i, t_j) \alpha_j \geq 0 \quad \text{for all } \ell \geq 1, t_1, \ldots, t_\ell \in T, \alpha_1, \ldots, \alpha_\ell \in \mathbb{R}.$$

Then for every finite measure μ on $(T, \mathcal{T})$, there is a real valued measurable process $(X_t)_{t \in T}$ which is μ-Gaussian with covariance kernel $K(\cdot,\cdot)$ as in Definition 2.16(a).

Proof. (1) Since the kernel $K(.,.)$ is symmetric and non-negative definite, for arbitrary choice of $t_1, \ldots, t_\ell$ in T, $\ell \geq 1$, there is a centred normal law $P_{t_1,\ldots,t_\ell}$ with covariance matrix $(K(t_i, t_j))_{i,j=1,\ldots,\ell}$ on $(\mathbb{R}^\ell, \mathcal{B}(\mathbb{R}^\ell))$, with characteristic function

$$\mathbb{R}^\ell \ni \lambda \longrightarrow e^{-\frac{1}{2}\lambda^T \Sigma \lambda}, \quad \Sigma := (K(t_i, t_j))_{i,j=1,\ldots,\ell}.$$

By the consistency theorem of Kolmogorov, the family of laws

$P_{t_1,\ldots,t_\ell}$ probability measure on $\left(\bigtimes_{i=1}^{\ell} \mathbb{R}, \bigotimes_{i=1}^{\ell} \mathcal{B}(\mathbb{R})\right)$, $t_1, \ldots, t_\ell \in T$, $\ell \geq 1$

being consistent, there exists a unique probability measure

P on $(\Omega, \mathcal{A}) := \left(\bigtimes_{t \in T} \mathbb{R}, \bigotimes_{t \in T} \mathcal{B}(\mathbb{R})\right)$

Section 2.3 Some Comments on Gaussian Processes

such that the canonical process $X = (X_t)_{t \in T}$ on $(\Omega, \mathcal{A}, P)$ – the process of coordinate projections – has finite dimensional distributions

(+) $$\mathcal{L}\left((X_{t_1}, \ldots, X_{t_\ell}) \mid P\right) = \mathcal{N}\left((0)_{i=1,\ldots,\ell}, (K(t_i, t_j))_{i,j=1,\ldots,\ell}\right),$$
$$t_1, \ldots, t_\ell \in T, \ \ell \geq 1.$$

Since $K(.,.)$ is continuous, any convergent sequence $t_n \to t$ in T makes

$$E_P\left((X_{t_n} - X_t)^2\right) = K(t_n, t_n) - 2K(t_n, t) + K(t, t)$$

vanish as $n \to \infty$. In this sense, the process $(X_t)_{t \in T}$ under P is 'mean square continuous'. As a consequence, X under P is continuous in probability:

(++) for convergent sequences $t_n \to t$: $X_{t_n} = X_t + o_P(1), \quad n \to \infty$.

So far, being a canonical process on a product space, X has no path properties.

(2) We define $\mathcal{T} \otimes \mathcal{A}$–measurable approximations $(X^m)_m$ to the process X. As in Exercise 2.1', we cover $\mathbb{R}^k$ with half-open cubes $A_l(m)$

$$A_l(m) := \bigtimes_{j=1}^{k} \left[\frac{l_j}{2^m}, \frac{l_j+1}{2^m}\right), \quad l = (l_1, \ldots, l_k) \in \mathbb{Z}^k$$

according to a k-dimensional dyadic grid with step size 2^{-m} in each dimension, and define $\Lambda(m, T)$ as the set of all indices $l \in \mathbb{Z}^k$ such that $A_l(m) \cap T \neq \emptyset$. For $l \in \Lambda(m, T)$ we select some point $t_l^*(m)$ in $A_l(m) \cap T$. Note that for $t \in T$ fixed and $m \in \mathbb{N}$, there is exactly one $l \in \mathbb{Z}^k$ such that $t \in A_l(m)$ holds: we write $l(t, m)$ for this index l. Define $\mathcal{T} \otimes \mathcal{A}$-measurable processes

$$X^m(t, \omega) := \sum_{l \in \Lambda(m,T)} 1_{A_l(m) \cap T}(t) \, X(t_l^*(m), \omega), \quad t \in T, \ \omega \in \Omega, \ m \geq 1.$$

Then by (+) in step (1), arbitrary finite dimensional distributions of X^m are Gaussian with covariances

$$E\left(X^m(t_1) X^m(t_2)\right) = K\left(t_{l(t_1,m)}^*(m), t_{l(t_2,m)}^*(m)\right), \quad t_1, t_2 \in T.$$

By continuity of $K(\cdot, \cdot)$ and convergence $t_{l(t,m)}^*(m) \to t$, the finite dimensional distributions of X^m converge as $m \to \infty$ to those of the process X constructed in step (1): the last equation combined with (+) gives for arbitrary $t_1, \ldots, t_\ell \in T$ and $\ell \geq 1$

(*) $$\mathcal{L}\left((X_{t_1}^m, \ldots, X_{t_\ell}^m) \mid P\right) \longrightarrow \mathcal{L}\left((X_{t_1}, \ldots, X_{t_\ell}) \mid P\right)$$
(weak convergence in $\mathbb{R}^\ell$, as $m \to \infty$).

(3) We fix any finite measure μ on $(T, \mathcal{T})$ and show that the sequence X^m converges in $L^2(T \times \Omega, \mathcal{T} \otimes \mathcal{A}, \mu \otimes P)$ as $m \to \infty$ to some limit process $\widetilde{X}$. Renormalising μ, it is sufficient to consider probability measures $\widetilde{\mu}$ on $(T, \mathcal{T})$.

(i) We prove that $(X^m)_m$ is a Cauchy sequence in $L^2(T\times\Omega, \mathcal{T}\otimes\mathcal{A}, \widetilde{\mu}\otimes P)$. First, every X^m belongs to $L^2(T\times\Omega, \mathcal{T}\otimes\mathcal{A}, \widetilde{\mu}\otimes P)$ since $K(\cdot,\cdot)$ is bounded on the compact $T\times T$:

$$\int_{T\times\Omega} (X^m(t,\omega))^2 \, (\widetilde{\mu}\otimes P)(dt,d\omega) = \int_T E\left((X_t^m)^2\right) \widetilde{\mu}(dt)$$
$$\leq [\sup_{t'\in T} K(t',t')] \, \widetilde{\mu}(T) < \infty.$$

Consider now pairs (m,m') where $m' > m$. By construction of the covering of $\mathbb{R}^k$ in step (2) at stages m and m', we have for any two indices l' and l either $A_{l'}(m') \subset A_l(m)$ or $A_{l'}(m') \cap A_l(m) = \emptyset$. Thus

$$\int_{T\times\Omega} |X^{m'} - X^m|^2 \, d(\widetilde{\mu}\otimes P) = \int_T E\left((X_t^{m'} - X_t^m)^2\right) \widetilde{\mu}(dt)$$

takes the form

(○) $$\sum_{\substack{l'\in\Lambda(m',T)\\ l\in\Lambda(m,T)}} \widetilde{\mu}\left(A_{l'}(m') \cap A_l(m) \cap T\right) \left(K(t_{l'}^*(m'), t_{l'}^*(m'))\right.$$
$$\left. - 2K(t_{l'}^*(m'), t_l^*(m)) + K(t_l^*(m), t_l^*(m))\right).$$

Since $K(\cdot,\cdot)$ is uniformly continuous on the compact $T\times T$ and since $|t_{l'}^*(m') - t_l^*(m)| \leq \sqrt{k}\, 2^{-m}$ for indices l' and l such that $A_{l'}(m') \subset A_l(m)$, integrands in (○)

$$\left(K(t_{l'}^*(m'), t_{l'}^*(m')) - 2K(t_{l'}^*(m'), t_l^*(m)) + K(t_l^*(m), t_l^*(m))\right)$$
for $t \in A_{l'}(m') \cap A_l(m)$

vanish as $m \to \infty$ uniformly in $t \in T$ and uniformly in $m' > m$. Hence, $\widetilde{\mu}$ being a finite measure, integrals (○) vanish as $m \to \infty$ uniformly in $m' > m$, and we have proved

$$\lim_{m\to\infty} \sup_{m'>m} \int_{T\times\Omega} |X^{m'} - X^m|^2 \, d(\widetilde{\mu}\otimes P) = 0.$$

(ii) For the Cauchy sequence $(X^m)_m$ under $\widetilde{\mu}\otimes P$ there is some $\widetilde{X} \in L^2(T\times\Omega, \mathcal{T}\otimes\mathcal{A}, \widetilde{\mu}\otimes P)$ such that

(+++) $\qquad X^m \longrightarrow \widetilde{X}$ in $L^2(T\times\Omega, \mathcal{T}\otimes\mathcal{A}, \widetilde{\mu}\otimes P)$ as $m \to \infty$.

In particular, $\widetilde{X}$ is measurable, and we deduce from (+++) convergence $X_t^m \to \widetilde{X}_t$ in $L^2(P)$ as $m \to \infty$ for $\widetilde{\mu}$-almost all $t \in T$; the exceptional $\widetilde{\mu}$-null set in $\mathcal{T}$ arising here can in general not be avoided.

(4) In (+++), we select a subsequence $(m_k)_k$ along which $\widetilde{\mu}\otimes P$-almost sure convergence holds:

$$X^{m_k} \longrightarrow \widetilde{X} \quad \widetilde{\mu}\otimes P\text{-almost surely on } (T\times\Omega, \mathcal{T}\otimes\mathcal{A}) \text{ as } k\to\infty.$$

Thus there is some set $M \in \mathcal{T} \otimes \mathcal{A}$ of full measure under $\widetilde{\mu} \otimes P$ such that

(**) $\qquad 1_M X^{m_k} \longrightarrow 1_M \widetilde{X}$ pointwise on $T \times \Omega$ as $k \to \infty$.

With notation M_t for t-sections through M

$$M_t = \{\omega \in \Omega : (t, \omega) \in M\} \in \mathcal{A}, \quad t \in T$$

we have $1 = (\widetilde{\mu} \otimes P)(M) = \int_T P(M_t) \widetilde{\mu}(dt)$. Hence there is some $\widetilde{\mu}$-null set $N \in \mathcal{T}$ such that

(***) $\qquad\qquad P(M_t) = 1$ for all t in $T \setminus N$.

Now the proof is finished: for arbitrary $\ell \geq 1$ and $t_1, \ldots, t_\ell$ in $T \setminus N$, we have pointwise convergence

($\diamond$) $\quad \left(1_{M_{t_1}} X_{t_1}^{m_k}, \ldots, 1_{M_{t_\ell}} X_{t_\ell}^{m_k}\right) \longrightarrow \left(1_{M_{t_1}} \widetilde{X}_{t_1}, \ldots, 1_{M_{t_\ell}} \widetilde{X}_{t_\ell}\right), \quad k \to \infty$

for all $\omega \in \Omega$ by (**), and at the same time –combining (***) and (*)– weak convergence in $\mathbb{R}^\ell$

($\diamond\diamond$) $\quad \mathcal{L}\left(1_{M_{t_1}} X_{t_1}^{m_k}, \ldots, 1_{M_{t_\ell}} X_{t_\ell}^{m_k}\right) \longrightarrow \mathcal{N}\left(0_\ell, (K(t_i, t_j))_{i,j=1,\ldots,l}\right), \quad k \to \infty$.

This shows that the (real valued and measurable) process $1_M \widetilde{X}$ is μ-Gaussian with covariance kernel $K(\cdot, \cdot)$ in the sense of Definition 2.16(a). $\square$

Frequently in what follows, we will consider μ-integrals along the path of a μ-Gaussian process.

2.19 Proposition. Consider a μ-Gaussian process $(X_t)_{t \in T}$ with covariance kernel $K(\cdot, \cdot)$, defined on some $(\Omega', \mathcal{A}', P')$. Assume that T is compact in $\mathbb{R}^k$ and that $K(\cdot, \cdot)$ is continuous on $T \times T$. Then, modifying paths of X on some P'-null set $N' \in \mathcal{A}'$ if necessary, X is a process with paths in $L^2(T, \mathcal{T}, \mu)$, and we have

$$\int_T g(t) X_t \, \mu(dt) \sim \mathcal{N}(0, \tau^2)$$

where $\quad \tau^2 := \int_T \int_T g(t_1) K(t_1, t_2) g(t_2) \mu(dt_1) \mu(dt_2)$

for functions $g \in L^2(T, \mathcal{T}, \mu)$.

Proof. For X μ-Gaussian, fix an exceptional μ-null set $N \in \mathcal{T}$ such that whenever $t_1, \ldots, t_r$ do not belong to N, finite dimensional laws $\mathcal{L}(X_{t_1}, \ldots, X_{t_r} \mid P')$ are normal laws with covariance matrix $(K(t_i, t_j))_{i,j=1,\ldots,r}$.
(1) X on $(\Omega', \mathcal{A}', P')$ being real valued and measurable, the set

$$N' := \left\{\omega \in \Omega' : \int_T X^2(t, \omega) \mu(dt) = +\infty\right\} \in \mathcal{A}'$$

has P-measure zero since under our assumptions
$$\int_{T'\times\Omega} X^2\, d(\mu\otimes P') = \int_T E_{P'}(X_t^2)\,\mu(dt) \leq [\sup_{t\in T} K(t,t)]\,\mu(T) < \infty.$$
Redefining $X(t,\omega) = 0$ for all $t \in T$ if $\omega \in N'$, all paths of X are in $L^2(T,\mathcal{T},\mu)$.

(2) Cover $\mathbb{R}^k$ with half-open intervals $A_l(m) := \times_{j=1}^k [\frac{l_j}{2^m}, \frac{l_j+1}{2^m})$, $l = (\ell_1, \ldots, l_k)$ $\in \mathbb{Z}^k$, as in step (2) of the proof of Theorem 2.18. Differently from the proof of Theorem 2.18, define $\Lambda(m,T)$ as the set of all l such that $\mu(A_l(m) \cap T) > 0$. For $l \in \Lambda(m,T)$, select $t_l^*(m)$ in $A_l(m) \cap T$ such that $t_l^*(m)$ does not belong to the exceptional set $N \subset T$, and define
$$X^m(t,\omega) := \sum_{l\in\Lambda(m,T)} 1_{A_l(m)\cap T}(t)\, X(t_l^*(m),\omega), \quad t\in T,\ \omega\in\Omega,\ m\geq 1.$$
Then all finite dimensional distributions of X^m, $m \geq 1$, are normal distributions; $\Lambda(m,T)$ being finite since T is compact, all X^m have paths in $L^2(T,\mathcal{T},\mu)$. For functions $g \in L^2(T,\mathcal{T},\mu)$, μ-integrals
$$\int_T g(t)\, X_t^m\, \mu(dt) = \sum_{l\in\Lambda(m,T)} X(t_l^*(m))\, \mu\left(g\, 1_{A_l(m)\cap T}\right)$$
are random variables on $(\Omega', \mathcal{A}', P')$ following normal laws
$$\mathcal{N}\left(0, \tau_m^2\right) \quad \text{where}$$
$$\tau_m^2 := \sum_{l,l'\in\Lambda(m,T)} \mu\left(g\, 1_{A_l(m)\cap T}\right) K\left(t_l^*(m), t_{l'}^*(m)\right) \mu\left(g\, 1_{A_{l'}(m)\cap T}\right).$$
Under our assumptions, $|g|(t)\mu(dt)$ is a finite measure on $(T,\mathcal{T})$, and the kernel $K(\cdot,\cdot)$ is continuous and bounded on $T\times T$. Thus by dominated convergence
$$\tau_m^2 \longrightarrow \int_{T\times T} g(t_1)\, K(t_1,t_2)\, g(t_2)\, (\mu\otimes\mu)(dt_1, dt_2) = \tau^2, \quad m\to\infty$$
which yields

(+) $\qquad \int_T g(t)\, X_t^m\, \mu(dt) \xrightarrow{\mathcal{L}} \mathcal{N}\left(0, \tau^2\right)$ as $m \to \infty$.

(3) As in the proof of Theorem 2.18 we have for points $t_1, \ldots, t_r$ in $T \setminus N$
$$\mathcal{L}\left((X_{t_1}^m, \ldots, X_{t_r}^m) \mid P'\right) \xrightarrow{\mathcal{L}} \mathcal{L}\left((X_{t_1}, \ldots, X_{t_r}) \mid P'\right)$$
(weak convergence in $\mathbb{R}^r$, as $m \to \infty$)

where all laws on the right-hand side are normal laws, and for points t in $T \setminus N$
$$\sup_m E(|X_t^m|^2) \leq \sup_{t'\in T} K(t',t') < \infty \quad \text{and} \quad \lim_{m\to\infty} E(|X_t^m|^2) = E(|X_t|^2).$$

Thus the assumptions of Theorem 2.4(a) are satisfied (here we use the sufficient condition (2.4") with $p = 2$ and constant f to establish uniform integrability). Applying Theorem 2.4(a) we obtain

$$X_\bullet^m \xrightarrow{\mathcal{L}} X_\bullet \quad \text{(weak convergence in } L^2(T, \mathcal{T}, \mu) \text{ as } m \to \infty),$$

from which the continuous mapping theorem (see Exercise 2.3') gives weak convergence of integrals

(++)
$$\int_T g(t) X_t^m \, \mu(dt) \xrightarrow{\mathcal{L}} \int_T g(t) X_t \, \mu(dt)$$

(weak convergence in $\mathbb{R}$ as $m \to \infty$).

Comparing (++) to (+), the assertion is proved. □

2.4 Asymptotic Normality for Minimum Distance Estimator Sequences

In this section, we continue the approach of Section 2.2: based on the representation of Theorem 2.14 of rescaled MD estimation errors, the results of Sections 2.3 and 2.1 – with $H = L^2(T, \mathcal{T}, \mu)$ as Hilbert space – will allow to prove asymptotic normality. Our set of Assumptions 2.20 provides additional structure for the empirical quantities $\widehat{\Psi}_n$ from which MD estimators were defined, merging the Assumptions 2.1 of Section 2.1 with the Assumptions 2.8(I) of Section 2.2. We complete the list of Assumptions 2.8(III) by introducing an asymptotic normality condition $\mathbf{AN}(\vartheta)$. The main result of this subsection is Theorem 2.22.

2.20 Assumptions and Notations for Section 2.4. (a) T is compact in $\mathbb{R}^k$, $\mathcal{T} = \mathcal{B}(T)$, and

$$H := L^2(T, \mathcal{T}, \mu)$$

for some finite measure μ on $(T, \mathcal{T})$. For $\Theta \subset \mathbb{R}^d$ open, $\{P_\vartheta : \vartheta \in \Theta\}$ is a family of probability measures on some $(\Omega, \mathcal{A})$; we have an increasing family of sub-σ-fields $(\mathcal{F}_n)_n$ in $\mathcal{A}$, and $P_{n,\vartheta} := P_\vartheta | \mathcal{F}_n$ is the restriction of P_ϑ to $\mathcal{F}_n$. We consider the sequence of experiments

$$\mathcal{E}_n := (\Omega, \mathcal{F}_n, \{P_{n,\vartheta} : \vartheta \in \Theta\}), \quad n \geq 1,$$

and have a sequence $\widehat{\Psi}_n$ of $\mathcal{F}_n$-measurable H-valued random variables

$$\widehat{\Psi}_n : (\Omega, \mathcal{F}_n) \longrightarrow (H, \mathcal{B}(H)), \quad n \geq 1$$

together with a deterministic family $\{\Psi_\xi : \xi \in \Theta\}$ in H such that

(+)   the mapping $\Theta \ni \xi \to \Psi_\xi \in H$ is continuous.

Specialising to compacts in $\mathbb{R}^k$, this unites the Assumptions 2.1(a) and 2.8(I).

(b) We now impose additional spatial structure on $\widehat{\Psi}_n$, $n \geq 1$: there is a measurable process

$$X^n : (T \times \Omega, \mathcal{T} \otimes \mathcal{F}_n) \longrightarrow (\mathbb{R}, \mathcal{B}(\mathbb{R})) \quad \text{with paths in } H = L^2(T, \mathcal{T}, \mu)$$

(cf. Lemma 2.2) such that $\widehat{\Psi}_n$ in (a) arises as path

$$\widehat{\Psi}_n(\omega) := X_\bullet^n(\omega) = X^n(\cdot, \omega), \quad \omega \in \Omega$$

of X^n, at every stage $n \geq 1$ of the asymptotics.

The spatial structure assumed in (b) was already present in Example 2.7. We complete the list of Assumptions 2.8(III) by strengthening the tightness condition $\mathbf{T}(\vartheta)$ in 2.8(III):

2.21 Asymptotic Normality Condition AN(ϑ). Following Notations and Assumptions 2.20, there is a sequence of norming constants $\varphi_n = \varphi_n(\vartheta) \uparrow \infty$ such that processes

$$W^n := \varphi_n \left(X^n - \Psi_\vartheta \right), \quad n \geq 1, \quad \text{under } P_\vartheta$$

have the following properties: there is a kernel

$$K(\cdot, \cdot) : T \times T \longrightarrow \mathbb{R} \quad \text{symmetric, continuous and non-negative definite,}$$

an exceptional set $N \in \mathcal{T}$ of μ-measure zero such that for arbitrary points $t_1, \ldots, t_l$ in $T \setminus N$, $l \geq 1$,

$$\mathcal{L}\left((W_{t_1}^n, \ldots, W_{t_l}^n) \mid P_\vartheta \right) \xrightarrow{\mathcal{L}} \mathcal{N}\left(0, (K(t_i, t_j))_{1 \leq i, j \leq l} \right)$$

(weak convergence in $\mathbb{R}^l$, $n \to \infty$),

and some function $f \in L^1(T, \mathcal{T}, \mu)$ such that for μ-almost all t in T

$$E(|W_t^n|^2 \mid P_\vartheta) \leq f(t) < \infty, \quad n \geq 1, \quad \lim_{n \to \infty} E(|W_t^n|^2 \mid P_\vartheta) = K(t, t).$$

Now we state the main theorem on minimum distance estimators. It relies directly on Proposition 2.11 and on the representation of rescaled estimation errors in Theorem 2.14.

2.22 Theorem. Under Assumptions 2.20, let for every $\vartheta \in \Theta$ the set of conditions

$$\mathbf{SLLN}(\vartheta), \mathbf{I}(\vartheta), \mathbf{D}(\vartheta), \mathbf{AN}(\vartheta), \text{ with } \varphi_n = \varphi_n(\vartheta) \uparrow \infty$$

Section 2.4 Asymptotic Normality for Minimum Distance Estimator Sequences

hold. Then any minimum distance estimator sequence $(\vartheta_n^*)_n$ for the unknown parameter $\vartheta \in \Theta$ defined according to Definition 2.9 is (strongly consistent and) asymptotically normal. We have for every $\vartheta \in \Theta$

$$\mathcal{L}\left(\varphi_n(\vartheta_n^* - \vartheta) \mid P_\vartheta\right) \longrightarrow \mathcal{N}\left(0, \Lambda_\vartheta^{-1} \Xi_\vartheta \Lambda_\vartheta^{-1}\right)$$

(weak convergence in $\mathbb{R}^d$, as $n \to \infty$), with a $d \times d$ matrix Ξ_ϑ having entries

(+) $$(\Xi_\vartheta)_{i,j} = \int_T \int_T D_i \Psi_\vartheta(t_1) \, K(t_1, t_2) \, D_j \Psi_\vartheta(t_2) \, \mu(dt_1) \mu(dt_2)$$
$$1 \leq i, j \leq d,$$

and with Λ_ϑ as defined in Theorem 2.14:

$$(\Lambda_\vartheta)_{i,j} = \langle D_i \Psi_\vartheta, D_j \Psi_\vartheta \rangle, \quad 1 \leq i, j \leq d.$$

Proof. Fix $\vartheta \in \Theta$. T being compact in $\mathbb{R}^k$, for any finite measure μ on $(T, \mathcal{T})$ and for any covariance kernel $K(\cdot, \cdot)$ on $T \times T$ which is symmetric, continuous and nonnegative definite, there exists a real valued measurable process W which is μ-Gaussian with covariance kernel $K(\cdot, \cdot)$, by Theorem 2.18. Depending on ϑ, $W = W(\vartheta)$ is defined on some space $(\Omega', \mathcal{A}', P')$. As in the proof of Proposition 2.19, W has paths in $L^2(T, \mathcal{T}, \mu)$ (after modification of paths on some P'-null set in $\mathcal{A}'$).

(1) For $\widehat{\Psi}_n = X_\bullet^n$ as in Assumption 2.20(b), and $W^n = \varphi_n(X^n - \Psi_\vartheta)$ under P_ϑ as in Condition 2.21, we have an exceptional set $N \in \mathcal{T}$ of μ-measure zero such that for arbitrary points $t_1, \ldots, t_l$ in $T \setminus N$, $l \geq 1$,

$$\mathcal{L}\left((W_{t_1}^n, \ldots, W_{t_l}^n) \mid P_\vartheta\right) \xrightarrow{\mathcal{L}} \mathcal{N}\left(0, (K(t_i, t_j))_{1 \leq i, j \leq l}\right)$$

(weak convergence in $\mathbb{R}^l$, $n \to \infty$)

by assumption **AN**(ϑ). Defining $N' \in \mathcal{T}$ as the union of N with the exceptional μ-null set which is contained in the definition of W as a μ-Gaussian process (cf. Definition 2.16), we have for $t_1, \ldots, t_l$ in $T \setminus N'$

$$\mathcal{L}\left((W_{t_1}^n, \ldots, W_{t_l}^n) \mid P_\vartheta\right) \xrightarrow{\mathcal{L}} \mathcal{L}\left((W_{t_1}, \ldots, W_{t_l}) \mid P_\vartheta\right)$$

(weak convergence in $\mathbb{R}^l$, $n \to \infty$).

Thus we have checked assumption (a.i) of Theorem 2.4. Next, again by assumption **AN**(ϑ), there is some function $f \in L^1(T, \mathcal{T}, \mu)$ such that for μ-almost all $t \in T$

$$E(|W_t^n|^2 \mid P_\vartheta) \leq f(t) < \infty, \quad n \geq 1, \quad \lim_{n \to \infty} E(|W_t^n|^2 \mid P_\vartheta) = K(t, t).$$

This is assumption (a.ii) of Theorem 2.4, via condition (2.4"); hence Theorem 2.4 establishes

$$W_\bullet^n \xrightarrow{\mathcal{L}} W_\bullet \quad \text{(weak convergence in } L^2(T, \mathcal{T}, \mu), \text{ as } n \to \infty).$$

Write $\langle .,. \rangle$ for the scalar product in $L^2(T, \mathcal{T}, \mu)$. Then the continuous mapping theorem gives

$$(\diamond) \qquad \langle g, W_\bullet^n \rangle \xrightarrow{\mathcal{L}} \langle g, W_\bullet \rangle \quad \text{(weakly in } \mathbb{R}, \text{ as } n \to \infty\text{)}$$

for any $g \in L^2(T, \mathcal{T}, \mu)$ where according to Proposition 2.19

$$(\diamond\diamond) \qquad \mathcal{L}\left(\langle g, W_\bullet \rangle \mid P'\right) = \mathcal{N}\left(0, \int_T \int_T g(t_1) \, K(t_1, t_2) \, g(t_2) \, \mu(dt_1)\mu(dt_2)\right).$$

(2) By assumption $\mathbf{D}(\vartheta)$, the components $D_1 \Psi_\vartheta, \ldots, D_d \Psi_\vartheta$ of the derivative $D\Psi_\vartheta$ are elements of $H = L^2(T, \mathcal{T}, \mu)$. If we apply $(\diamond)+(\diamond\diamond)$ to $g := h^\top D\Psi_\vartheta$, $h \in \mathbb{R}^d$ arbitrary, Cramér–Wold yields

$$\mathcal{L}\left(\begin{pmatrix} \langle D_1\Psi_\vartheta, W_\bullet^n \rangle \\ \vdots \\ \langle D_d\Psi_\vartheta, W_\bullet^n \rangle \end{pmatrix} \middle| P_\vartheta\right) \to \mathcal{L}\left(\begin{pmatrix} \langle D_1\Psi_\vartheta, W_\bullet \rangle \\ \vdots \\ \langle D_d\Psi_\vartheta, W_\bullet \rangle \end{pmatrix} \middle| P'\right) = \mathcal{N}(0, \Xi_\vartheta)$$

(weak convergence in $\mathbb{R}^d$, as $n \to \infty$) since the covariance matrix Ξ_ϑ in (+) has entries

$$\int_T \int_T D_i \Psi_\vartheta(s) \, K(s, t) \, D_j \Psi_\vartheta(t) \, \mu(ds)\mu(dt) = E\left(\langle D_i\Psi_\vartheta, W_\bullet \rangle, \langle D_j\Psi_\vartheta, W_\bullet \rangle\right),$$

$$i, j = 1, \ldots, d.$$

To conclude the proof, it is sufficient to combine the last convergence with the representation of Theorem 2.14

$$\varphi_n(\vartheta_n^* - \vartheta) = \Pi_\vartheta\left(W_\bullet^n\right) + o_{P_\vartheta}(1), \quad n \to \infty$$

of rescaled MD estimation errors. To apply Theorem 2.14, note that assumption $\mathbf{AN}(\vartheta)$ – through step (1) above and the continuous mapping theorem – establishes weak convergence $\|W_\bullet^n\| \xrightarrow{\mathcal{L}} \|W_\bullet\|$ in $\mathbb{R}$ as $n \to \infty$, and thus in particular tightness as required in condition $\mathbf{T}(\vartheta)$. $\square$

In Example 2.7, we started to consider MD estimator sequences based on the empirical distribution function for i.i.d. observations. In Lemma 2.23 and Example 2.24 below, we will put this example in rigorous terms.

2.23 Lemma. Let $Y_1, Y_2, \ldots$ denote i.i.d. random variables taking values in $\mathbb{R}^k$, with continuous distribution function $F : \mathbb{R}^k \to [0, 1]$. Let $\widehat{F}_n : (t, \omega) \to \widehat{F}_n(t, \omega)$ denote the empirical distribution function based on the first n observations. Consider T compact in $\mathbb{R}^k$, with Borel-σ-field $\mathcal{T}$, and μ a finite measure on $(T, \mathcal{T})$. Write

$$W^n := \sqrt{n}\left(\widehat{F}_n - F\right), \quad n \geq 1$$

Section 2.4 Asymptotic Normality for Minimum Distance Estimator Sequences

and consider the kernel
$$K(s,t) = F(s \wedge t) - F(s)F(t), \quad s,t \in \mathbb{R}^k$$
with minimum taken componentwise in $\mathbb{R}^k$: $s \wedge t = (s_i \wedge t_i)_{1 \leq i \leq k}$. Then all requirements of the asymptotic normality condition $\mathbf{AN}(\vartheta)$ are satisfied, there is a measurable process W which is μ-Gaussian with covariance kernel $K(\cdot,\cdot)$, and we have
$$W_\bullet^n \longrightarrow W_\bullet \quad \text{(weak convergence in } L^2(T,\mathcal{T},\mu)\text{, as } n \to \infty\text{)}.$$
In the case where $d=1$, W is Brownian bridge $B^{0,F}$ time-changed by F.

Proof. (1) For $t, t' \in T$, with minimum $t \wedge t'$ defined componentwise in $\mathbb{R}^k$, we have
$$K(t,t') = F(t \wedge t') - F(t)F(t') = E_F\left(\left[1_{(-\infty,t]}(Y_1) - F(t)\right]\left[1_{(-\infty,t']}(Y_1) - F(t')\right]\right)$$
with intervals in $\mathbb{R}^k$ written according to Assumption 2.1(c), and
$$\sum_{i,i'=1}^l \alpha_i K(t_i,t_{i'})\alpha_{i'} = \operatorname{Var}_F\left(\sum_{i=1}^l \alpha_i \left[1_{(-\infty,t_i]}(Y_1) - F(t_i)\right]\right) \geq 0$$
for arbitrary $l \geq 1$, $\alpha_1,\ldots,\alpha_l$ in $\mathbb{R}$, $t_1,\ldots,t_l$ in T. The distribution function F being continuous, the kernel $K(\cdot,\cdot) : T \times T \to [0,1]$ is thus symmetric, continuous and non-negative definite. By Theorem 2.18, a μ-Gaussian process W with covariance kernel $K(\cdot,\cdot)$ exists.

(2) Let $(Y_i)_{i \geq 1}$ be defined on some $(\Omega, \mathcal{A}, P)$, and put $W^n(t,\omega) := \sqrt{n}(\widehat{F}_n(t,\omega) - F(t))$. We prove that for arbitrary $t_1,\ldots,t_l$ in T, $l \geq 1$, convergence of finite dimensional distributions

(+) $\quad \mathcal{L}\left((W_{t_1}^n,\ldots,W_{t_l}^n) \mid P\right) \longrightarrow \mathcal{N}(0,\Sigma), \quad \Sigma := \left(K(t_i,t_j)\right)_{i,j=1,\ldots,l}$

(weakly in $\mathbb{R}^l$ as $n \to \infty$) holds. Using Cramér–Wold we have to show weak convergence in $\mathbb{R}$
$$\mathcal{L}\left(\sum_{i=1}^l \alpha_i W_{t_i}^n \mid P\right) \longrightarrow \mathcal{N}(0, \alpha^\top \Sigma \alpha), \quad n \to \infty$$
for arbitrary $\alpha = (\alpha_1,\ldots,\alpha_l) \in \mathbb{R}^l$. By definition of W^n and by the central limit theorem we have
$$\sum_{i=1}^l \alpha_i W_{t_i}^n = \frac{1}{\sqrt{n}} \sum_{j=1}^n R_j \longrightarrow \mathcal{N}(0, \operatorname{Var}(R_1))$$
with centred i.i.d. random variables
$$R_j := \sum_{i=1}^l \alpha_i \left[1_{(-\infty,t_i]}(Y_j) - F(t_i)\right], \quad j \in \mathbb{N}.$$

By step (1) we have $\text{Var}(R_1) = \alpha^\top \Sigma \alpha$: this proves (+).

(3) A simple particular case of the above, with $f(t) := K(t,t) = F(t)(1 - F(t))$, is

$$E\left(|W_t^n|^2\right) = f(t) \text{ for all } t, n, \text{ and } E\left(|W_t|^2\right) = f(t) \text{ for all } t\ ;$$

f being continuous and T compact, we have $f \in L^1(T, \mathcal{T}, \mu)$. Hence all requirements of assumption $\mathbf{AN}(\vartheta)$ are satisfied. Thus Theorem 2.4 applies –using condition (2.4") as a sufficient condition for condition (a.ii) in Theorem 2.4 – and gives weak convergence of paths $W_\bullet^n \to W_\bullet$ in $L^2(T, \mathcal{T}, \mu)$. $\qed$

MD estimator sequences based on the empirical distribution function behave as follows.

2.24 Example (Example 2.7 continued). For $\Theta \subset \mathbb{R}^d$ open, consider $(Y_i)_{i \geq 1}$ i.i.d. observations in $\mathbb{R}^k$, with continuous distribution function $F_\vartheta : \mathbb{R}^k \to [0,1]$ under $\vartheta \in \Theta$. Fix T compact in $\mathbb{R}^k$, $\mathcal{T} = \mathcal{B}(T)$, some finite measure μ on $(T, \mathcal{T})$, and assume the following: for all $\vartheta \in \Theta$,

($\diamond$)   the parameterisation $\Theta \ni \vartheta \longrightarrow \Psi_\vartheta := F_\vartheta \in H = L^2(T, \mathcal{T}, \mu)$
satisfies $\mathbf{I}(\vartheta)$ and $\mathbf{D}(\vartheta)$.

Then MD estimator sequences based on the empirical distribution function $(t, \omega) \to \widehat{F}_n(t, \omega)$

(+) $\displaystyle \vartheta_n^* = \operatorname*{arginf}_{\xi \in \Theta} \| \widehat{F}_n - F_\xi(\cdot) \|_{L^2(\mu)}, \quad n \geq 1$

(according to Definition 2.9; a particular construction may be the construction of Proposition 2.10) are strongly consistent and asymptotically normal for all $\vartheta \in \Theta$:

$$\mathcal{L}\left(\sqrt{n}\,(\vartheta_n^* - \vartheta) \mid P_\vartheta\right) \longrightarrow \mathcal{N}\left(0, \Lambda_\vartheta^{-1} \Xi_\vartheta \Lambda_\vartheta^{-1}\right) \quad \text{(weakly in } \mathbb{R}^d, \text{ as } n \to \infty)$$

where Ξ_ϑ and Λ_ϑ take the form

$$(\Lambda_\vartheta)_{i,j} = \int_T \int_T D_i F_\vartheta(t_1)\, D_j F_\vartheta(t_2)\, \mu(dt_1)\mu(dt_2), \quad 1 \leq i,j \leq d,$$

$$(\Xi_\vartheta)_{i,j} = \int_T \int_T D_i F_\vartheta(t_1)\, [F_\vartheta(t_1 \wedge t_2) - F_\vartheta(t_1) F_\vartheta(t_2)]\, D_j F_\vartheta(t_2)\, \mu(dt_1)\mu(dt_2),$$

$$1 \leq i,j \leq d$$

with 'min' taken componentwise in $\mathbb{R}^k$. For the proof of these statements, recall from Example 2.7 that condition $\mathbf{SLLN}(\vartheta)$ holds as a consequence of Glivenko–Cantelli, cf. (∗) in Example 2.7. $\mathbf{I}(\vartheta)$ holds by ($\diamond$) assumed above, hence Proposition 2.11 gives strong consistency of any version of the MD estimator sequence (+). Next, Lemma 2.23 establishes condition $\mathbf{AN}(\vartheta)$ with covariance kernel $K(t_1, t_2) = F_\vartheta(t_1 \wedge$

$t_2) - F_\vartheta(t_1)F_\vartheta(t_2)$. Since $\mathbf{D}(\vartheta)$ holds by assumption ($\diamond$), Theorem 2.22 applies and yields the assertion. □

2.24' Exercise. In dimension $d = 1$, fix a distribution function F on $(\mathbb{R}, \mathcal{B}(\mathbb{R}))$ which admits a continuous and strictly positive Lebesgue density f on $\mathbb{R}$, and consider as a particular case of Example 2.24 a location model

$$(\mathbb{R}, \mathcal{B}(\mathbb{R}), \{F_\vartheta : \vartheta \in \Theta\}), \quad F_\vartheta := F(\cdot - \vartheta), \quad f_\vartheta := f(\cdot - \vartheta)$$

where Θ is open in $\mathbb{R}$. Check that for any choice of a finite measure μ on $(T, \mathcal{T})$, T compact in $\mathbb{R}$, assumptions $\mathbf{I}(\vartheta)$ and $\mathbf{D}(\vartheta)$ are satisfied (with $DF_\vartheta = -f_\vartheta$) for all $\vartheta \in \Theta$. □

MD estimators in i.i.d. models may be defined in many different ways, e.g. based on empirical Laplace transforms when tractable expressions for the Laplace transforms under $\vartheta \in \Theta$ are at hand, from empirical quantile functions, from empirical characteristic functions, and so on. The next example is from Höpfner and Rüschendorf [58].

2.25 Example. Consider real valued i.i.d. symmetric stable random variables $(Y_j)_{j \geq 1}$ with characteristic function

$$u \longrightarrow E_{(\alpha,\xi)}\left(e^{iuY_1}\right) = e^{-\xi|u|^\alpha}, \quad u \in \mathbb{R}$$

and estimate both stability index $\alpha \in (0, 2)$ and weight parameter $\xi \in (0, \infty)$ from the first n observations $Y_1, \ldots, Y_n$.

Put $\vartheta = (\alpha, \xi)$, $\Theta = (0, 2) \times (0, \infty)$. Symmetry of $\mathcal{L}(Y | P_\vartheta)$ – the characteristic functions being real valued – allows to work with the real parts of the empirical characteristic functions $\frac{1}{n} \sum_{j=1}^n e^{iuY_j}$ only, so we put

$$\widehat{\Psi}_n(u) = \frac{1}{n} \sum_{j=1}^n \cos(uY_j), \quad \Psi_\vartheta(u) := e^{-\xi|u|^\alpha}, \quad \vartheta \in \Theta.$$

Fix a sufficiently large compact interval T, symmetric around zero and including open neighbourhoods of the points $u = \pm 1$, $\mathcal{T} = \mathcal{B}(T)$, and a finite measure μ on $(T, \mathcal{T})$, symmetric around zero such that

(+) $$\int_{-\varepsilon}^{\varepsilon} |\log u|^2 \mu(du) < \infty \quad \text{for all } \varepsilon > 0,$$

and assume in addition:

($\diamond$) $\begin{cases} \text{for some open neighbourhood } U \text{ of the point } u = 1, \\ \text{the density of the } \lambda\text{-absolutely continuous part of } \mu \text{ in restriction to } U \\ \text{is strictly positive and bounded away from zero.} \end{cases}$

Then any MD estimator sequence based on the empirical characteristic function

$$\vartheta_n^* := \underset{\vartheta \in \Theta}{\arg\inf} \left\| \widehat{\Psi}_n - \Psi_\vartheta \right\|_{L^2(T, \mathcal{T}, \mu)}$$

according to Definition 2.9 (a particular construction may be the construction in Proposition 2.10) is strongly consistent and asymptotically normal for all $\vartheta \in \Theta$:

$$\mathcal{L}\left(\sqrt{n}\,(\vartheta_n^* - \vartheta) \mid P_\vartheta\right) \longrightarrow \mathcal{N}\left(0,\, \Lambda_\vartheta^{-1} \Xi_\vartheta \Lambda_\vartheta^{-1}\right) \quad \text{(weakly in } \mathbb{R}^2, \text{ as } n \to \infty\text{)}$$

where Ξ_ϑ and Λ_ϑ take the form

$$(\Lambda_\vartheta)_{i,j} = \int_T \int_T D_i \Psi_\vartheta(u)\, D_j \Psi_\vartheta(v)\, \mu(du)\mu(dv),$$

$$(\Xi_\vartheta)_{i,j} = \int_T \int_T D_i \Psi_\vartheta(u) \left[\frac{\Psi_\vartheta(u+v) + \Psi_\vartheta(u-v)}{2} - \Psi_\vartheta(u)\Psi_\vartheta(v)\right]$$
$$\times D_j \Psi_\vartheta(v)\, \mu(du)\mu(dv)$$

for $1 \le i, j \le 2$. The proof for these statements is in several steps.

(1) We show that ($\diamond$) establishes the identifiability condition $\mathbf{I}(\vartheta)$ for all $\vartheta \in \Theta$. Write C_N for the compact $[0,2] \times [\frac{1}{N}, N]$ in $\mathbb{R}^2$. Fix $\vartheta \in \Theta$. Fix some pair (ε_0, N_0) such that the ball $B_{\varepsilon_0}(\vartheta)$ is contained in the interior of C_{N_0}. For $N \ge N_0$ sufficiently large, and for parameter values $\vartheta' = (\alpha', \xi')$ for which $\frac{1}{N} \le \xi' \le N$, we introduce linearisations at $u = 1$

$$\widetilde{\Psi}_{\vartheta'}(t) := e^{-\xi'}[1 - \alpha'\xi'(t-1)] \quad \text{on} \quad \widetilde{U} := B_r(1)$$

of the characteristic functions $\Psi_{\vartheta'}(\cdot)$, where $0 < r < \frac{1}{4N}$ is a radius sufficiently small such that ($\diamond$) holds on $\widetilde{U} = B_r(1)$. For ξ' fixed, the family $\widetilde{\Psi}_{(\alpha', \xi')}$ can be extended to include limits $\alpha' = 0$ or $\alpha' = 2$. We put $\widetilde{\mu} := \lambda_{|\widetilde{U}}$. Then, C_N being compact and the mapping

$$C_N \ni \vartheta' \longrightarrow \widetilde{\Psi}_{\vartheta'} \in L^2(\widetilde{U}, \widetilde{\mu})$$

continuous, we have for every $0 < \varepsilon < \varepsilon_0$

$$\min\left\{\|\widetilde{\Psi}_{\vartheta'} - \widetilde{\Psi}_\vartheta\|_{L^2(\widetilde{U}, \widetilde{\mu})} : \vartheta' \in C_N,\, |\vartheta' - \vartheta| \ge \varepsilon\right\} > 0.$$

This last inequality is identically rewritten in the form

$$\inf\left\{\|\widetilde{\Psi}_{\vartheta'} - \widetilde{\Psi}_\vartheta\|_{L^2(\widetilde{U}, \widetilde{\mu})} : \vartheta' = (\alpha', \xi') \in \Theta,\, \frac{1}{N} \le \xi' \le N,\, |\vartheta' - \vartheta| \ge \varepsilon\right\} > 0.$$

Since $\widetilde{\Psi}_\vartheta$ is uniformly on $\widetilde{U}$ separated from all functions $\widetilde{\Psi}_{\vartheta'}$ with $\xi' < \frac{1}{N}$ or $\xi' > N$ if N is large, the last inequality can be extended to

$$\inf\left\{\|\widetilde{\Psi}_{\vartheta'} - \widetilde{\Psi}_\vartheta\|_{L^2(\widetilde{U}, \widetilde{\mu})} : \vartheta' = (\alpha', \xi') \in \Theta,\, |\vartheta' - \vartheta| \ge \varepsilon\right\} > 0.$$

Now we repeat exactly the same reasoning with the characteristic functions $\Psi_{\vartheta'}$, Ψ_ϑ restricted to $\widetilde{U}$, instead of their linearisations $\widetilde{\Psi}_{\vartheta'}$, $\widetilde{\Psi}_\vartheta$ at $u = 1$ which we considered so far, and obtain

$$\inf\left\{\|\Psi_{\vartheta'} - \Psi_\vartheta\|_{L^2(\widetilde{U}, \widetilde{\mu})} : \vartheta' \in \Theta,\, |\vartheta' - \vartheta| \ge \varepsilon\right\} > 0.$$

Section 2.4 Asymptotic Normality for Minimum Distance Estimator Sequences

Since there is some constant $c > 0$ such that $\mu(dt) \geq (\mu_{|\widetilde{U}})(dt) \geq c\widetilde{\mu}(dt)$, by assumption ($\diamond$), we may pass from $\widetilde{U}$ to T in the last assertion and obtain a fortiori

$$\inf\left\{\|\Psi_{\vartheta'} - \Psi_\vartheta\|_{L^2(T,\mathcal{T},\mu)} : \vartheta' \in \Theta, |\vartheta' - \vartheta| \geq \varepsilon\right\} > 0.$$

Since $\varepsilon > 0$ was arbitrary, this is the identifiability condition $\mathbf{I}(\vartheta)$.

(2) Following Prakasa Rao [107, Proposition 8.3.1], condition $\mathbf{SLLN}(\vartheta)$ holds for all $\vartheta \in \Theta$:

We start again from the distribution functions F_ϑ associated to Ψ_ϑ, and Glivenko–Cantelli. Fix $\vartheta \in \Theta$ and write $A_\vartheta \in \mathcal{A}$ for some set of full P_ϑ-measure such that $\sup_{t \in \mathbb{R}} |\widehat{F}_n(t, \omega) - F_\vartheta(t)| \longrightarrow 0$ as $n \to \infty$ when $\omega \in A_\vartheta$. In particular $\widehat{F}_n(\omega, t) \longrightarrow F_\vartheta(t)$ at continuity points t of F_ϑ, hence empirical measures associated to $(Y_1, \ldots, Y_n)(\omega)$ converge weakly to F_ϑ, for $\omega \in A_\vartheta$. This gives pointwise convergence of the associated characteristic functions $\widehat{\Psi}_n(t, \omega) \longrightarrow \Psi_\vartheta(t)$ for $t \in \mathbb{R}$ when $\omega \in A_\vartheta$. By dominated convergence on T with respect to the finite measure μ this establishes $\mathbf{SLLN}(\vartheta)$.

(3) Condition (+) is a sufficient condition for differentiability $\mathbf{D}(\vartheta)$ at all points $\vartheta \in \Theta$:

By dominated convergence under (+) assumed above

$$\int_{-\varepsilon}^{\varepsilon} |\log u|^2 \mu(du) < \infty \quad \text{for } \varepsilon > 0 \text{ arbitrarily small}$$

the mapping $\Theta \ni \xi \to \Psi_\vartheta \in L^2(T, \mathcal{T}, \mu)$ is differentiable at ϑ, and the derivative $D\Psi_\vartheta$ has components

$$D_\xi \Psi_\vartheta(u) = -|u|^\alpha \Psi_\vartheta(u), \quad D_\alpha \Psi_\vartheta(u) = -(\log|u|) \xi |u|^\alpha \Psi_\vartheta(u), \quad u \in \mathbb{R}$$

which are linearly independent in $L^2(T, \mathcal{T}, \mu)$.

(4) For all $\vartheta \in \Theta$, the asymptotic normality condition $\mathbf{AN}(\vartheta)$ holds with covariance kernel

$$K_\vartheta(u, v) := \frac{1}{2}\left(\Psi_\vartheta(u+v) + \Psi_\vartheta(u-v)\right) - \Psi_\vartheta(u)\Psi_\vartheta(v), \quad u, v \in \mathbb{R}:$$

Fix $\vartheta \in \Theta$. Proceeding as in the proof of Lemma 2.23, for any collection of points $u_1, \ldots, u_l \in T$, we replace the random variable R_j defined there by

$$R_j := \sum_{i=1}^{l} \alpha_i \left[\cos(u_i Y_j) - \Psi_\vartheta(u_i)\right].$$

Calculating the variance of R_1 under ϑ, we arrive at a kernel

$$(u, v) \longrightarrow E_\vartheta\left([\cos(uY_1) - \Psi_\vartheta(u)][\cos(vY_1) - \Psi_\vartheta(v)]\right).$$

As a consequence of $\cos(x)\cos(y) = \frac{1}{2}(\cos(x+y) + \cos(x-y))$, this is the kernel $K(\cdot, \cdot)$ defined above.

(5) Now we conclude the proof: we have conditions **SLLN**(ϑ)+**I**(ϑ)+**D**(ϑ)+**AN**(ϑ) by steps (1)–(4) above, so Theorem 2.22 applies and yields the assertion. □

2.25' Exercise. For the family $\mathcal{P} = \{\Gamma(a, p) : a > 0, p > 0\}$ of Gamma laws on $(\mathbb{R}, \mathcal{B}(\mathbb{R}))$ with densities

$$f_{a,p}(x) = 1_{(0,\infty)}(x) \frac{p^a}{\Gamma(a)} x^{a-1} e^{-px},$$

for i.i.d. observations $Y_1, Y_2, \ldots$ from $\mathcal{P}$, construct MD estimators for (a, p) based on empirical Laplace transforms

$$[0, \infty) \ni \lambda \longrightarrow \frac{1}{n} \sum_{i=1}^{n} e^{-\lambda Y_i} \in [0, 1]$$

and discuss their asymptotics (hint: to satisfy the identifiability condition, work with measures μ on compacts $T = [0, C]$ which satisfy $\mu(dy) \geq c\, dy$ in restriction to small neighbourhoods $[0, \varepsilon)$ of 0^+). □

We conclude this chapter with some remarks. Many classical examples for MD estimator sequences are given in Millar [100, Chap. XIII]. Also, a large number of extensions of the method exposed here are possible. Kutoyants [79] considers MD estimator sequences for i.i.d. realisations of the same point process on a fixed spatial window, in the case where the spatial intensity is parameterised by $\vartheta \in \Theta$. Beyond the world of i.i.d. observations, MD estimator sequences can be considered in ergodic Markov processes when the time of observation tends to infinity, see a large number of diffusion process examples in Kutoyants [80], or in many other stochastic process models.

Chapter 3

Contiguity

Topics for Chapter 3:

3.1 Le Cam's First and Third Lemma
Notations: likelihood ratios in sequences of binary experiments 3.1
Some conventions for sequences of $\overline{\mathbb{R}}^d$-valued random variables 3.1'
Definition of contiguity 3.2
Contiguity and $\mathbb{R}$-tightness of likelihood ratios 3.3–3.4
Le Cam's first lemma: statement and interpretation 3.5–3.5'
Le Cam's third lemma: statement and interpretation 3.6–3.6'
Example: mean shift when limit laws are normal 3.6"

3.2 Proofs for Section 3.1 and some Variants
An ε-δ-characterisation of contiguity 3.7
Proving Proposition 3.3' 3.7'–3.8
Proof of Theorem 3.4(a) 3.9
One-sided contiguity and Le Cam's first lemma 3.10–3.12
Proof of Theorem 3.4(b) 3.13
Proving LeCam's first lemma 3.14–3.15
One-sided contiguity and Le Cam's third lemma 3.16
Proving LeCam's third lemma 3.17
Proof of Proposition 3.6" 3.18

Exercises: 3.5", 3.5"', 3.5"", 3.10', 3.18'

This chapter discusses the notion of contiguity which goes back to Le Cam, see Hájek and Sidák [41], Le Cam [81], Roussas [113], Strasser [121], Liese and Vajda [87], Le Cam and Yang [84], van der Vaart [126], and is of crucial importance in the context of convergence of local models. Mutual contiguity considers sequences of likelihood ratios whose accumulation points in the sense of weak convergence have the interpretation of a likelihood ratio between two equivalent probability laws. Section 3.1 fixes setting and notations, and states two key results on contiguity, termed 'Le Cam's first lemma' and 'Le Cam's third lemma' (3.5 and 3.6 below) since Hájek and Sidák [41, Chap. 6.1]. Section 3.2 contains the proofs together with some variants of the main results.

3.1 Le Cam's First and Third Lemma

For two σ-finite measures P and Q on the same $(\Omega, \mathcal{A})$, P is absolutely continuous with respect to Q (notation $P \ll Q$) if $Q(A) = 0$ implies $P(A) = 0$ for arbitrary $A \in \mathcal{A}$. By the Radon–Nikodym theorem, $P \ll Q$ is equivalent to the following: there is some $\mathcal{A}$-measurable mapping $f : \Omega \to [0, \infty)$ such that $P(A) = \int_A f \, dQ$ for all $A \in \mathcal{A}$; in this case, f is called (a version of) the density of P with respect to Q. P is equivalent to Q (notation $P \sim Q$) if $P \ll Q$ and $Q \ll P$. P and Q are singular (notation $P \perp Q$) if there is some set $A \in \mathcal{A}$ which is of measure zero under P and of full measure under Q. A Lebesgue decomposition of P with respect to Q is any pair (f, N), $N \in \mathcal{A}$ and $f : \Omega \to [0, \infty)$ $\mathcal{A}$-measurable, such that the following holds:

$$Q(N) = 0 \quad \text{and} \quad P(A) = P(A \cap N) + \int_A f \, dQ \quad \text{for all } A \in \mathcal{A}.$$

See e.g. [121, Chap. I] or [127, Satz 1.103].

Throughout this chapter we will consider sequences of probability spaces carrying two probability measures ('binary experiments'), and use the following notations.

3.1 Notations. $(\Omega_n, \mathcal{A}_n)$ denotes a sequence of measurable spaces, $n \geq 1$, every $(\Omega_n, \mathcal{A}_n)$ equipped with a pair P_n, Q_n of probability measures.

(i) Any $\mathcal{A}_n$-measurable mapping $L_n : \Omega_n \to [0, \infty]$ such that the pair $(L_n 1_{\{L_n < \infty\}}, \{L_n = \infty\})$ is a Lebesgue decomposition of P_n with respect to Q_n:

$$Q_n(\{L_n = \infty\}) = 0, \quad P_n(A) = P_n(A \cap \{L_n = \infty\}) + \int_A L_n \, dQ_n, \quad A \in \mathcal{A}_n$$

is a version of the *likelihood ratio of P_n with respect to Q_n*. Two different versions of the likelihood ratio of P_n with respect to Q_n coincide $(P_n + Q_n)$-almost surely; see e.g. [127, Satz 1.110]. The measure $P_n(\cdot \cap \{L_n = \infty\})$ is called the Q_n-singular part of P_n, and the measure $A \to \int_A L_n \, dQ_n$ the Q_n-absolutely continuous part of P_n.

(ii) For any σ-finite measure ν_n on $(\Omega_n, \mathcal{A}_n)$ which dominates P_n and Q_n (take e.g. $\nu_n = P_n + Q_n$), for versions p_n of the ν_n-density of P_n and q_n of the ν_n-density of Q_n,

$$L_n := \frac{p_n}{q_n} 1_{\{q_n > 0\}} + \infty 1_{\{q_n = 0\}}$$

is (a version of) the likelihood ratio of P_n with respect to Q_n.

(iii) For L_n as in (i), $\Lambda_n := \log L_n$ (with conventions $\log(+\infty) = +\infty$ and $\log(0) = -\infty$) mapping Ω_n to $\overline{\mathbb{R}} := [-\infty, +\infty]$ is (a version of) the *log-likelihood ratio of P_n with respect to Q_n*.

(iv) For any version L_n of the likelihood ratio of P_n with respect to Q_n, $\frac{1}{L_n}$ (with conventions $\frac{1}{0} = \infty$ and $\frac{1}{\infty} = 0$) is a version of the likelihood ratio of Q_n with respect to P_n.

Section 3.1 Le Cam's First and Third Lemma

3.1' Notations. For $\overline{\mathbb{R}} := [-\infty, +\infty]$, consider $\overline{\mathbb{R}}^d$-valued random variables X_n living on some $(\Omega_n, \mathcal{A}_n, Q_n)$, $n \geq 1$, and associate to X_n the $\mathbb{R}^d$-valued random variable $\widehat{X}_n := X_n 1_{\{|X_n|<\infty\}}$.

(i) We call the sequence $(X_n)_n$ $\mathbb{R}^d$-*tight under* $(Q_n)_n$ if

(*) $\quad \lim_{n\to\infty} Q_n(X_n \neq \widehat{X}_n) = 0 \quad$ and $\quad \left\{\mathcal{L}\left(\widehat{X}_n \mid Q_n\right) : n \geq 1\right\}$ is tight in $\mathbb{R}^d$.

(ii) For probability measures F on $(\mathbb{R}^d, \mathcal{B}(\mathbb{R}^d))$, we say that $(X_n)_n$ *under* $(Q_n)_n$ *converges $\mathbb{R}^d$-weakly to F* – and write

$$\mathcal{L}(X_n \mid Q_n) \longrightarrow F \quad (\text{weakly in } \mathbb{R}^d \text{ as } n \to \infty)$$

for short – if $(X_n)_n$ under $(Q_n)_n$ is $\mathbb{R}^d$-tight and if the second condition in (*) can be strengthened to

(**) $\quad \mathcal{L}\left(\widehat{X}_n \mid Q_n\right) \longrightarrow F \quad$ (weak convergence in $\mathbb{R}^d$ as $n \to \infty$).

(iii) We call the sequence $(X_n)_n$ *uniformly integrable under* $(Q_n)_n$ if

$$Q_n(X_n \neq \widehat{X}_n) = 0 \text{ for all } n \geq 1, \text{ and } \lim_{K\uparrow\infty} \sup_{n\geq 1} \int_{\{|\widehat{X}_n|>K\}} |\widehat{X}_n|\, dQ_n = 0.$$

3.2 Definition. The sequence $(P_n)_n$ of probability measures is *contiguous to* $(Q_n)_n$ (notation $(P_n)_n \triangleleft (Q_n)_n$) if for arbitrary sequences of events $A_n \in \mathcal{A}_n$ we have as $n \to \infty$

$$Q_n(A_n) \to 0 \quad \Longrightarrow \quad P_n(A_n) \to 0;$$

mutual contiguity (notation $(P_n)_n \triangleleft\triangleright (Q_n)_n$) means that both $(P_n)_n \triangleleft (Q_n)_n$ and $(Q_n)_n \triangleleft (P_n)_n$ hold true.

Hence contiguity $(P_n)_n \triangleleft (Q_n)_n$ is an asymptotic analogue to the classical absolute continuity $P \ll Q$ for probability measures P, Q on a fixed space $(\Omega, \mathcal{A})$. We start with a simple observation.

3.3 Proposition. *Without further conditions, $(L_n)_n$ under $(Q_n)_n$ is $\overline{\mathbb{R}}$-tight.*

Proof. For every $n \geq 1$, the event $\{L_n = \infty\}$ has Q_n-measure zero by Notation 3.1(i); we have to show

for every $\varepsilon > 0$ there is some $K < \infty$ such that $\sup_{n\geq 1} Q_n(\infty > L_n > K) < \varepsilon$

by definition in Notation 3.1'(i). Choosing $\nu_n := P_n + Q_n$ as the dominating measure for P_n and Q_n, we have versions of the densities p_n and q_n such that $p_n + q_n \equiv 1$

on Ω_n. Since L_n is $(P_n + Q_n)$-uniquely determined, we can write

$$Q_n(\infty > L_n > K) = Q_n\left(q_n > 0, \frac{1-q_n}{q_n} > K\right) = Q_n\left(0 < q_n < \frac{1}{K+1}\right)$$
$$= \int_{\Omega_n} 1_{\{q_n < \frac{1}{K+1}\}} q_n \, d\nu_n \leq \frac{1}{K+1} \nu_n(\Omega_n) = \frac{2}{K+1}$$

where the right-hand side does not depend on $n \geq 1$. □

The following results Proposition 3.3', Theorem 3.4, Lemmas 3.5, 3.6 and Proposition 3.6" will be stated in the present section, together with some comments; all proofs – evolving through a series of variants and alternative formulations – will be collected in Section 3.2. The main results are Lemmas 3.5 and 3.6.

3.3' Proposition. For random variables Y_n on $(\Omega_n, \mathcal{A}_n)$ taking values in $(\mathbb{R}^d, \mathcal{B}(\mathbb{R}^d))$, the assertion

$$\{\mathcal{L}(Y_n \mid Q_n) : n \geq 1\} \quad \text{is tight, and} \quad (P_n)_n \triangleleft (Q_n)_n$$

implies

$$\{\mathcal{L}(Y_n \mid P_n) : n \geq 1\} \text{ is tight}.$$

3.4 Theorem. (a) The following assertions (i) and (ii) are equivalent:

(i) $\qquad\qquad\qquad (P_n)_n \triangleleft (Q_n)_n$

(ii) $\qquad$ the sequence $(L_n)_n$ under $(P_n)_n$ is $\mathbb{R}$-tight.

(b) The following assertions (i) and (ii) are equivalent:

(i) $\qquad\qquad\qquad (P_n)_n \triangleleft\triangleright (Q_n)_n$

(ii) $\qquad$ the sequence $(\Lambda_n)_n$ is $\mathbb{R}$-tight under both $(P_n)_n$ and $(Q_n)_n$.

Out of a tight sequence of probability laws on $(\mathbb{R}^d, \mathcal{B}(\mathbb{R}^d))$ we can select weakly convergent subsequences. Hence, considering $\mathbb{R}^d$-tight sequences of $\overline{\mathbb{R}}^d$-valued random variables with the conventions of Notation 3.1'(i), there is no loss of generality – switching to subsequences if necessary – in supposing $\mathbb{R}^d$-weak convergence in the sense of Notation 3.1'(ii).

3.5 Le Cam's First Lemma. Assume that there is a probability law F on $(\mathbb{R}, \mathcal{B}(\mathbb{R}))$ such that
$$\mathcal{L}(\Lambda_n \mid Q_n) \longrightarrow F \quad (\text{weakly in } \mathbb{R} \text{ as } n \to \infty).$$

Then the following holds:

$$(P_n)_n \triangleleft\triangleright (Q_n)_n \iff \int_{\mathbb{R}} e^\lambda F(d\lambda) = 1.$$

3.5' Remark. To any probability measure F on $(\mathbb{R}, \mathcal{B}(\mathbb{R}))$ which may occur in Lemma 3.5, we can associate a real valued random variable Λ on some probability space $(\Omega, \mathcal{A}, Q)$ with distribution F; the easiest choice is $\Lambda := id$ on $(\Omega, \mathcal{A}, Q) := (\mathbb{R}, \mathcal{B}(\mathbb{R}), F)$. Then the equality $\int_{\mathbb{R}} e^\lambda F(d\lambda) = 1$ in Lemma 3.5 signifies that $(\Omega, \mathcal{A})$ carries a second probability measure P defined by

$$dP := e^\Lambda \, dQ$$

which necessarily – by its definition – is equivalent to Q, and gives rise to a binary experiment

(+) $\qquad\qquad (\Omega, \mathcal{A}, \{P, Q\}) \quad$ with $\quad P \sim Q$.

We may view (+) as a 'limit experiment' as $n \to \infty$ for the sequence of binary experiments

$$(\Omega_n, \mathcal{A}_n, \{P_n, Q_n\}), \quad n \geq 1.$$

Thus Le Cam's first lemma relates mutual contiguity $(P_n)_n \triangleleft\triangleright (Q_n)_n$ to convergence of experiments where *weak limits of log-likelihood ratios are log-likelihood ratios between equivalent probability laws*. □

3.5" Exercise. In the case where the weak limit for log-likelihood ratios Λ_n under Q_n is a normal law

$$\mathcal{L}(\Lambda_n \mid Q_n) \longrightarrow \mathcal{N}(\nu, \sigma^2) \quad \text{(weakly in } \mathbb{R} \text{ as } n \to \infty)$$

with strictly positive variance $\sigma^2 > 0$, use Lemma 3.5 to prove the equivalence

$$(P_n)_n \triangleleft\triangleright (Q_n)_n \iff \nu = -\frac{1}{2}\sigma^2.$$

3.5''' Exercise. On $(\Omega_n, \mathcal{A}_n) = (\mathbb{R}^n, \mathcal{B}(\mathbb{R}^n))$, consider probability laws $Q_n = \otimes_{i=1}^n \mathcal{N}(0, 1)$, $P_n = \otimes_{i=1}^n \mathcal{N}(\frac{h}{\sqrt{n}}, 1)$ for some $h \in \mathbb{R}$ which we keep fixed as $n \to \infty$. Write $(X_1, \ldots, X_n)$ for the canonical variable on Ω_n. Check that Λ_n is given by $\frac{h}{\sqrt{n}} \sum_{i=1}^n X_i - \frac{1}{2}h^2$ and think of the Laplace transform of $\mathcal{N}(0, 1)$ to check mutual contiguity $(P_n)_n \triangleleft\triangleright (Q_n)_n$ via Le Cam's First Lemma 3.5. Check also that the 'limit experiment' in Remark 3.5' can be determined as $(\mathbb{R}, \mathcal{B}(\mathbb{R}))$ equipped with $Q = \mathcal{N}(0, 1)$, $P = \mathcal{N}(h, 1)$.

3.5'''' Exercise. On $(\Omega_n, \mathcal{A}_n) = (\mathbb{R}^n, \mathcal{B}(\mathbb{R}^n))$, for some $h > 0$ which we keep fixed as $n \to \infty$, consider probability laws $Q_n = \otimes_{i=1}^n \mathcal{R}(0, 1)$ and $P_n = \otimes_{i=1}^n \mathcal{R}(0+\frac{h}{n}, 1+\frac{h}{n})$. Write $U_n = (0, 1)^n$ for the support of Q_n and $V_n = (0+\frac{h}{n}, 1+\frac{h}{n})^n$ for the support of P_n. Use sequences $A_n = U_n \setminus V_n$ or $\widetilde{A}_n = V_n \setminus U_n$ to check directly from Definition 3.2 that neither $(P_n)_n \triangleleft (Q_n)_n$ nor $(Q_n)_n \triangleleft (P_n)_n$ hold. Show also that instead of a limit law F on $(\mathbb{R}, \mathcal{B}(\mathbb{R}))$ as required in Lemma 3.5 for the sequence of log-likelihood ratios $\mathcal{L}(\Lambda_n \mid Q_n)$ as $n \to \infty$, we have a limit law on $(\overline{\mathbb{R}}, \mathcal{B}(\overline{\mathbb{R}}))$ which takes values 0 with probability e^{-h} and $-\infty$ with probability $1 - e^{-h}$.

The next result is Le Cam's third lemma (the 'second lemma' is related to representation of log-likelihood ratios in smoothly parameterised product experiments, see Chapters 4 and 7).

3.6 Le Cam's Third Lemma. Assume
$$(P_n)_n \triangleleft \triangleright (Q_n)_n .$$
Consider a probability measure $\widetilde{F}$ on $(\mathbb{R}^{1+k}, \mathcal{B}(\mathbb{R}^{1+k}))$ and a sequence of $\mathbb{R}^k$-valued random variables X_n on $(\Omega_n, \mathcal{A}_n)$ such that the following holds:

Then
$$\mathcal{L}((\Lambda_n, X_n) \mid Q_n) \longrightarrow \widetilde{F} \quad \text{(weakly in } \mathbb{R}^{1+k} \text{ as } n \to \infty) .$$
$$\widetilde{G}(d\lambda, dx) := e^\lambda \widetilde{F}(d\lambda, dx), \quad \lambda \in \mathbb{R}, \, x \in \mathbb{R}^k$$
is a probability measure on $(\mathbb{R}^{1+k}, \mathcal{B}(\mathbb{R}^{1+k}))$, and we have
$$\mathcal{L}((\Lambda_n, X_n) \mid P_n) \longrightarrow \widetilde{G} \quad \text{(weakly in } \mathbb{R}^{1+k} \text{ as } n \to \infty) .$$

3.6' Remarks. (a) Under mutual contiguity $(P_n)_n \triangleleft \triangleright (Q_n)_n$, we extend the interpretation in Remark 3.5' and view the law $\widetilde{F}$ in Lemma 3.6 as the distribution of a random variable (Λ, X) defined on some $(\Omega, \mathcal{A}, Q)$. Thus, in addition to Remark 3.5', we have an $\mathbb{R}^k$-valued statistic X in the 'limit experiment' $(\Omega, \mathcal{A}, \{P, Q\})$ where $P \sim Q$ is defined by $dP := e^\Lambda dQ$. In the 'limit experiment' we have
$$\widetilde{F} = \mathcal{L}((\Lambda, X) \mid Q), \quad \widetilde{G} = \mathcal{L}((\Lambda, X) \mid P)$$
where the second statement is a consequence of the definition of P and of $\widetilde{G}$: for suitable $f(\cdot, \cdot)$,
$$E_P(f(\Lambda, X)) = E_Q\left(e^\Lambda f(\Lambda, X)\right) = \int e^\lambda f(\lambda, x) \widetilde{F}(d\lambda, dx)$$
$$= \int f(\lambda, x) \widetilde{G}(d\lambda, dx) .$$

We view $(X_n)_n$ in Lemma 3.6 as a sequence of statistics in the binary experiments $(\Omega_n, \mathcal{A}_n, \{P_n, Q_n\})$ which *under* $(Q_n)_n$ *converges weakly jointly with the log-likelihood ratios* $(\Lambda_n)_n$:

(+) $\qquad \mathcal{L}((\Lambda_n, X_n) \mid Q_n) \longrightarrow \mathcal{L}((\Lambda, X) \mid Q), \quad n \to \infty .$

Now Le Cam's Third Lemma 3.6 can be rephrased as follows: under mutual contiguity $(P_n)_n \triangleleft\triangleright (Q_n)_n$, assertion (+) implies

(++) $\qquad \mathcal{L}((\Lambda_n, X_n) \mid P_n) \longrightarrow \mathcal{L}((\Lambda, X) \mid P), \quad n \to \infty$

with $P \sim Q$ given by $dP := e^\Lambda dQ$.

(b) We sketch a typical application of Le Cam's third lemma. Consider statistical models for i.i.d. observations and $\sqrt{n}$-consistent estimator sequences $(\widetilde{\vartheta}_n)_n$ for an unknown parameter ϑ.

Let $\Theta \subset \mathbb{R}^d$ be open, write $\mathcal{F}_n$ for the σ-field generated by the first n observations, $n \geq 1$. Fix a reference point ϑ and define the sequence of probability laws $(Q_n)_n$ by $(P_\vartheta \mid \mathcal{F}_n)_n$, $n \geq 1$. Define the sequence $(P_n)_n$ by $(P_{\vartheta+h/\sqrt{n}} \mid \mathcal{F}_n)_n$, for some h in $\mathbb{R}^d$ which we keep fixed. Then Λ_n in Lemma 3.6 is the log-likelihood ratio of $P_{\vartheta+h/\sqrt{n}}$ restricted to $\mathcal{F}_n$ with respect to P_ϑ restricted to $\mathcal{F}_n$. In this setting, mutual contiguity $(P_n)_n \triangleleft\triangleright (Q_n)_n$ will hold if the parameterisation is 'smooth enough' (we shall consider this in Chapters 4 and 7 below, see in particular Proposition 7.3 and Lemma 4.11). The reference point ϑ being fixed, we identify the $(X_n)_n$ in Lemma 3.6 with rescaled estimation errors $(\sqrt{n}(\widetilde{\vartheta}_n - \vartheta))_n$ at ϑ. *If we are able to establish weak convergence of pairs* (Λ_n, X_n) *under* ϑ (which requires some tractable relationship between rescaled estimation error $X_n = \sqrt{n}(\widetilde{\vartheta}_n - \vartheta)$ and log-likelihood Λ_n under ϑ), then Le Cam's Third Lemma 3.6 allows us to deduce from convergence at the reference point ϑ

(+) $\qquad \mathcal{L}\left((\Lambda_n, \sqrt{n}(\widetilde{\vartheta}_n - \vartheta)) \mid P_\vartheta\right) = \mathcal{L}((\Lambda_n, X_n) \mid Q_n) \longrightarrow \widetilde{F}$

the convergence under parameter values of the form $\vartheta + h/\sqrt{n}$ which are close to ϑ (in the sense of neighbourhoods shrinking at rate $1/\sqrt{n}$ as $n \to \infty$): first, by Lemma 3.6,

(++) $\qquad \mathcal{L}\left((\Lambda_n, \sqrt{n}(\widetilde{\vartheta}_n - \vartheta)) \mid P_{\vartheta+h/\sqrt{n}}\right) = \mathcal{L}((\Lambda_n, X_n) \mid P_n) \longrightarrow \widetilde{G}$

where the passage from $\widetilde{F}$ to $\widetilde{G}$ is explicit (at least in principle, cf. Lemma 3.6); second, shifting all second components in (++) by h which does not depend on n,

(+++) $\qquad \mathcal{L}\left((\Lambda_n, \sqrt{n}(\widetilde{\vartheta}_n - (\vartheta + h/\sqrt{n}))) \mid P_{\vartheta+h/\sqrt{n}}\right) \longrightarrow \widetilde{H}$

where $\widetilde{H}$ is the image of $\widetilde{G}$ under the mapping $(\lambda, x) \to (\lambda, x - h)$. On the left-hand side of (+++) appear the rescaled estimation errors under $\vartheta + h/\sqrt{n}$ for fixed h as $n \to \infty$.

In this way, our knowledge (+) on limit distributions of rescaled estimation errors at fixed reference points ϑ extends to rescaled estimation errors (+++) on small neighbourhoods of ϑ of radius $O(1/\sqrt{n})$. We will exploit this extensively in Chapter 7. □

There is one important special case where the passage from $\widetilde{F}$ to $\widetilde{G}$ in Le Cam's Third Lemma 3.6 is particularly simple. Normal limit laws $\widetilde{F}$ for $\mathcal{L}((\Lambda_n, X_n) \mid Q_n)$ as $n \to \infty$ in Lemma 3.6 arise in many classical situations. The following proposition shows that in this case $\mathcal{L}(X_n \mid Q_n)$ and $\mathcal{L}(X_n \mid P_n)$ differ only by a mean value shift $\gamma = \lim_{n \to \infty} \mathrm{Cov}_{Q_n}(\Lambda_n, X_n)$:

3.6" Proposition. When $\widetilde{F}$ in Lemma 3.6 is some normal law on $(\mathbb{R}^{1+k}, \mathcal{B}(\mathbb{R}^{1+k}))$ whose first component is not degenerate to a single point, then necessarily

$$\widetilde{F} = \mathcal{N}\left(\begin{pmatrix} -\frac{1}{2}\sigma^2 \\ \mu \end{pmatrix}, \begin{pmatrix} \sigma^2 & \gamma^\top \\ \gamma & \Sigma \end{pmatrix}\right)$$

for some $\mu \in \mathbb{R}^k$, $\gamma \in \mathbb{R}^k$, $\sigma > 0$, and some covariance matrix $\Sigma \in \mathbb{R}^{k \times k}$, symmetric and non-negative definite, and $\widetilde{G}$ associated to $\widetilde{F}$ by Lemma 3.6 takes the form

$$\widetilde{G} = \mathcal{N}\left(\begin{pmatrix} +\frac{1}{2}\sigma^2 \\ \mu + \gamma \end{pmatrix}, \begin{pmatrix} \sigma^2 & \gamma^\top \\ \gamma & \Sigma \end{pmatrix}\right).$$

3.2 Proofs for Section 3.1 and some Variants

We start with an ε-δ-characterisation for contiguity $(P_n)_n \lhd (Q_n)_n$ in analogy to the classical ε-δ-characterisation (e.g. [127, p. 109]) of absolute continuity $P \ll Q$ of probability measures.

3.7 Proposition. The following condition is necessary and sufficient for contiguity $(P_n)_n \lhd (Q_n)_n$: for every $\varepsilon > 0$ there is some $\delta = \delta(\varepsilon) > 0$ such that

$$(+) \qquad \limsup_{n \to \infty} Q_n(A_n) < \delta \quad \Longrightarrow \quad \limsup_{n \to \infty} P_n(A_n) < \varepsilon$$

holds for arbitrary sequences of events A_n in $\mathcal{A}_n$, $n \geq 1$.

Proof. (1) We prove that condition (+) is sufficient for contiguity $(P_n)_n \lhd (Q_n)_n$. To $\varepsilon > 0$ arbitrarily small associate $\delta = \delta(\varepsilon) > 0$ such that (+) holds. Fix any sequence of events A_n in $\mathcal{A}_n$, $n \geq 1$, with the property $\lim_{n \to \infty} Q_n(A_n) = 0$. For this sequence, condition (+) makes sure that we must have $\lim_{n \to \infty} P_n(A_n) = 0$ too: this is contiguity $(P_n)_n \lhd (Q_n)_n$ as in Definition 3.2.

(2) We prove that (+) is a necessary condition. Let us assume that for some $\varepsilon > 0$, irrespectively of the smallness of $\delta > 0$, an implication like (+) never holds true. In this case, considering in particular $\delta = \frac{1}{k}$, $k \in \mathbb{N}$ arbitrarily large, there are sequences of events $(A_n^k)_n$ such that

$$\limsup_{n \to \infty} Q_n(A_n^k) < \frac{1}{k} \text{ holds together with } \limsup_{n \to \infty} P_n(A_n^k) \geq \varepsilon\,.$$

Section 3.2 Proofs for Section 3.1 and some Variants   93

From this we can select a sequence $(n_k)_k$ increasing to ∞ such that for every $k \in \mathbb{N}$ we have
$$P_{n_k}(A_{n_k}^k) > \tfrac{\varepsilon}{2} \text{ together with } Q_{n_k}(A_{n_k}^k) < \tfrac{2}{k}.$$
Using Definition 3.2 along the subsequence $(n_k)_k$, contiguity $(P_{n_k})_k \triangleleft (Q_{n_k})_k$ does not hold. A fortiori, as an easy consequence of Definition 3.2, contiguity $(P_n)_n \triangleleft (Q_n)_n$ does not hold. □

We use the ε-δ-characterisation of contiguity to prove Proposition 3.3'.

3.7' Proof of Proposition 3.3'. We have to show that any sequence of $\mathbb{R}^d$- valued random variables $(Y_n)_n$ which is tight under $(Q_n)_n$ remains tight under $(P_n)_n$ when contiguity $(P_n)_n \triangleleft (Q_n)_n$ holds. Let $\varepsilon > 0$ be arbitrarily small. Assuming contiguity $(P_n)_n \triangleleft (Q_n)_n$, we make use of Proposition 3.7 and select $\delta = \delta(\varepsilon) > 0$ such that the implication (+) in Proposition 3.7 is valid. From tightness of $(Y_n)_n$ under $(Q_n)_n$, there is some large $K = K(\delta) < \infty$ such that
$$\limsup_{n \to \infty} Q_n(|Y_n| > K) < \delta,$$
from which we deduce thanks to (+)
$$\limsup_{n \to \infty} P_n(|Y_n| > K) < \varepsilon.$$
Since $\varepsilon > 0$ was arbitrary, Proposition 3.3' is proved. □

We prepare the next steps by some comments on the conventions in Notations 3.1'.

3.8 Remarks. With Notations 3.1', consider on $(\Omega_n, \mathcal{A}_n, Q_n)$ $\overline{\mathbb{R}}^d$-valued random variables $X_n, n \geq 1$, and their associated $\widehat{X}_n := X_n 1_{\{|X_n| < \infty\}}$ which are $\mathbb{R}^d$-valued. The following assertions (a)–(c) can be checked directly from Notations 3.1'.

(a) The sequence $(X_n)_n$ under $(Q_n)_n$ is $\mathbb{R}^d$-tight if and only if
$$\lim_{K \uparrow \infty} \limsup_{n \to \infty} Q_n(+\infty \geq |X_n| > K) = 0.$$

(b) For probability measures F on $(\mathbb{R}^d, \mathcal{B}(\mathbb{R}^d))$,
$$\mathcal{L}(X_n \mid Q_n) \longrightarrow F \quad (\text{weakly in } \mathbb{R}^d \text{ as } n \to \infty)$$
as defined in Notation 3.1'(ii) is equivalent to the following condition ($\times$):

($\times$)   for all $f \in \mathcal{C}_b(\mathbb{R}^d)$: $\displaystyle \lim_{n \to \infty} \int_{\Omega_n} f(X_n) 1_{\{|X_n| < \infty\}} \, dQ_n = \int_{\mathbb{R}} f \, dF.$

Note that for $f \equiv 1$ we have $\int_{\mathbb{R}} f \, dF = 1$ on the right-hand side of ($\times$), by assumption $F(\mathbb{R}^d) = 1$, hence ($\times$) contains the assertion $\lim_{n \to \infty} Q_n(|X_n| = \infty) = 0$ which was part of the definition of Notation 3.1'(ii).

(c) Uniform integrability of the sequence $(X_n)_n$ under $(Q_n)_n$ as defined in Notation 3.1'(iii) is equivalent to validity of the following assertions (i) and (ii) together:

(i) we have $\sup_{n\geq 1} E_{Q_n}(|X_n|) < \infty$;

(ii) for every $\varepsilon > 0$ there is some $\delta = \delta(\varepsilon) > 0$ such that

$$n \geq 1, \ A_n \in \mathcal{A}_n, \ Q_n(A_n) < \delta \implies \int_{A_n} |X_n|\, dQ_n < \varepsilon.$$

According to the definition in Notation 3.1'(iii), this is proved exactly as in the usual case of real valued random variables living on some fixed probability space. □

By Proposition 3.3 which was proved in Section 3.1, the sequence of likelihood ratios $(L_n)_n$ is $\mathbb{R}$-tight under $(Q_n)_n$, without further assumptions. Now we can prove part (a) of Theorem 3.4: contiguity $(P_n)_n \triangleleft (Q_n)_n$ is equivalent to $\mathbb{R}$-tightness of $(L_n)_n$ under both $(P_n)_n$ and $(Q_n)_n$.

3.9 Proof of Theorem 3.4(a). (1) To prove (i)$\implies$(ii) of Theorem 3.4(a), we assume $(P_n)_n \triangleleft (Q_n)_n$ and have to verify – according to Remark 3.8(a) – that the following holds true:

$$\lim_{K\uparrow\infty} \limsup_{n\to\infty} P_n(L_n \in [K,\infty]) = 0.$$

If this assertion were not true, we could find some $\varepsilon > 0$, a sequence $K_j \uparrow \infty$, and a sequence of natural numbers n_j increasing to ∞ such that

$$P_{n_j}(L_{n_j} \in [K_j,\infty]) > \varepsilon \quad \text{for all } j \geq 1.$$

By Proposition 3.3 we know that $(L_n)_n$ under $(Q_n)_n$ is $\mathbb{R}$-tight, thus we know

$$\lim_{j\to\infty} \limsup_{n\to\infty} Q_n(L_n \in [K_j,\infty]) = 0.$$

Combining the last two formulas, there would be a sequence of events $A_m \in \mathcal{A}_m$

$$A_m := \{L_{n_j} \in [K_j,\infty]\} \text{ in the case where } m = n_j, \text{ and } A_m := \emptyset \text{ else}$$

with the property

$$\lim_{m\to\infty} Q_m(A_m) = 0, \quad \limsup_{m\to\infty} P_m(A_m) \geq \varepsilon$$

in contradiction to the assumption $(P_n)_n \triangleleft (Q_n)_n$. This proves (i)$\implies$(ii) of Theorem 3.4(a).

(2) To prove (ii)$\implies$(i) in Theorem 3.4(a), we assume $\mathbb{R}$-tightness of $(L_n)_n$ under $(P_n)_n$ and have to show that contiguity $(P_n)_n \triangleleft (Q_n)_n$ holds true. To prove this, we

start from a sequence of events $A_n \in \mathcal{A}_n$ with the property $\lim_{n\to\infty} Q_n(A_n) = 0$ and write

$$P_n(A_n) = P_n(A_n \cap \{\infty \geq L_n > K\}) + P_n(A_n \cap \{L_n \leq K\})$$
$$\leq P_n(\infty \geq L_n > K) + \int_{A_n \cap \{L_n \leq K\}} L_n \, dQ_n$$
$$\leq P_n(\infty \geq L_n > K) + K Q_n(A_n).$$

This gives

$$\limsup_{n\to\infty} P_n(A_n) \leq \limsup_{n\to\infty} P_n(\infty \geq L_n > K)$$

where by $\mathbb{R}$-tightness of $(L_n)_n$ under $(P_n)_n$ and by Remark 3.8(i), the righthand side can be made arbitrarily small by suitable choice of K. This gives $\lim_{n\to\infty} P_n(A_n) = 0$ and thus proves (ii)$\Longrightarrow$(i) of Theorem 3.4(a). □

Preparing for the proof of Theorem 3.4(b) which will be completed in Proof 3.13 below, we continue to give characterisations of contiguity $(P_n)_n \triangleleft (Q_n)_n$.

3.10 Proposition. The following assertions are equivalent:

(i) $(P_n)_n \triangleleft (Q_n)_n$;

(ii) the sequence $(L_n)_n$ under $(Q_n)_n$ is uniformly integrable, and we have $\lim_{n\to\infty} P_n(L_n = \infty) = 0$;

(iii) the sequence $(L_n)_n$ under $(Q_n)_n$ is uniformly integrable, and we have $\lim_{n\to\infty} E_{Q_n}(L_n) = 1$.

Proof. (1) Assertions (ii) and (iii) are equivalent since

$$P_n(L_n = \infty) + E_{Q_n}(L_n) = 1, \quad n \geq 1$$

as a direct consequence of the Lebesgue decomposition in Notation 3.1(i) of P_n with respect to Q_n: here $E_{Q_n}(L_n)$ is the total mass of the Q_n-absolutely continuous part of P_n, and $P_n(L_n = \infty)$ the total mass of the Q_n-singular part of P_n.

(2) For $0 < K < \infty$ arbitrarily large we can write

$$P_n(K < L_n \leq \infty) = P_n(L_n = \infty) + \int 1_{\{K < L_n < \infty\}} L_n \, dQ_n$$
$$= (1 - E_{Q_n}(L_n)) + \int 1_{\{K < L_n < \infty\}} L_n \, dQ_n.$$

By Theorem 3.4(a), contiguity $(P_n)_n \triangleleft (Q_n)_n$ is equivalent to $\mathbb{R}$-tightness of $(L_n)_n$ under $(P_n)_n$. The last equation shows that this is again equivalent to uniform integrability of $(L_n)_n$ under $(Q_n)_n$ – combining Notation 3.1'(iii) with $Q_n(L_n = \infty) = 0$

in the Lebesgue decomposition in Notation 3.1(i) – plus any one of the additional conditions considered in step (1). □

3.10' Exercise. On $(\mathbb{R}, \mathcal{B}(\mathbb{R}))$, write $P(\vartheta)$ for the exponential law with parameter 1 shifted by $\vartheta > 0$, i.e. supported by (ϑ, ∞). On $(\Omega_n, \mathcal{A}_n) = (\mathbb{R}^n, \mathcal{B}(\mathbb{R}^n))$ with canonical variable $(X_1, \ldots, X_n)$, consider probability laws $Q_n = \otimes_{i=1}^n P(0)$ and $P_n = \otimes_{i=1}^n P(0+\frac{h}{n})$, for some $h > 0$ which is fixed as $n \to \infty$. Write $U_n = (0, \infty)^n$ for the support of Q_n and $V_n = (0+\frac{h}{n}, \infty)^n$ for the support of P_n. Check that the likelihood ratio L_n is given by

$$L_n = 0 \cdot 1_{U_n \setminus V_n} + e^{+h} 1_{V_n}$$

where

$$Q_n(V_n) = e^{-h} \quad \text{for all } n \geq 1.$$

Deduce contiguity $(P_n)_n \triangleleft (Q_n)_n$ from Proposition 3.10(iii), from Proposition 3.10(ii), from Theorem 3.4(a), from an ε-δ-argument using Proposition 3.7, and finally directly from Definition 3.2.

The sequence of likelihood ratios $(L_n)_n$ under $(Q_n)_n$ being $\mathbb{R}$-tight by Proposition 3.3, for any subsequence $(n_k)_k$ of the natural numbers, there are further subsequences $(n_{k_l})_l$ and probability measures $\check{F}$ on $(\mathbb{R}, \mathcal{B}(\mathbb{R}))$ such that $(L_{n_{k_l}})_l$ under $(Q_{n_{k_l}})_l$ converge to $\check{F}$ weakly in $\mathbb{R}$ as $l \to \infty$. In this sense, the assumption of the following theorem – a one-sided version of Le Cam's First Lemma 3.5 – is (up to selection of subsequences) no restriction at all. Accumulation points $\check{F}$ in this sense are necessarily concentrated on $[0, \infty)$, and may have a point mass $\check{F}(\{0\}) > 0$ at zero.

3.11 Theorem. Assume that there is some probability measure $\check{F}$ on $(\mathbb{R}, \mathcal{B}(\mathbb{R}))$ such that

$$\mathcal{L}(L_n \mid Q_n) \longrightarrow \check{F} \quad \text{(weakly in } \mathbb{R} \text{ as } n \to \infty).$$

Then the following holds:

$$(P_n)_n \triangleleft (Q_n)_n \iff \int_{[0,\infty)} \ell \, \check{F}(d\ell) = 1.$$

Proof. (1) We start with some preliminary remarks. Define truncated identities $g_K(\cdot) \in \mathcal{C}_b(\mathbb{R})$

$$g_K(x) = 0 \vee x \wedge K, \quad x \in \mathbb{R}$$

for $0 < K < \infty$. Then monotone convergence gives

$$\lim_{K \uparrow \infty} \int_0^\infty g_K \, d\check{F} = \int_{[0,\infty)} x \, \check{F}(dx) \in [0, \infty].$$

Section 3.2 Proofs for Section 3.1 and some Variants

Considering $\int_0^\infty (x - g_K(x)) Q_n^{L_n}(dx)$ where $E_{Q_n}(L_n) \le 1$ by definition of a likelihood ratio,

(+) $$\begin{aligned} E_{Q_n}(L_n) - \int g_K \, dQ_n^{L_n} \\ = \int 1_{\{K < L_n < \infty\}} L_n \, dQ_n - K \, Q_n^{L_n}((K, \infty)) \in [0, 1] \, . \end{aligned}$$

By assumption in Theorem 3.11 on weak convergence $\mathcal{L}(L_n | Q_n) \to \check{F}$ we have

(++) $$\lim_{n \to \infty} \int g_K \, dQ_n^{L_n} = \int g_K \, d\check{F}, \quad n \to \infty \, .$$

(2) We show '$\Longrightarrow$'. Let us assume $(P_n)_n \triangleleft (Q_n)_n$ and thus

($\diamond$) $$\lim_{n \to \infty} E_{Q_n}(L_n) = 1, \quad \lim_{K \uparrow \infty} \limsup_{n \to \infty} \int 1_{\{K < L_n < \infty\}} L_n \, dQ_n = 0$$

thanks to Proposition 3.10. In the limit as $n \to \infty$, the second assertion in ($\diamond$) forces right-hand sides in (+) to be small when K is large:

($\star$) $$\lim_{K \uparrow \infty} \limsup_{n \to \infty} \left[\int 1_{\{K < L_n < \infty\}} L_n \, dQ_n - K \, Q_n^{L_n}((K, \infty)) \right] = 0 \, .$$

Then also left-hand sides of (+) will be small when K is large:

($\star\star$) $$\lim_{K \uparrow \infty} \limsup_{n \to \infty} \left[E_{Q_n}(L_n) - \int g_K \, dQ_n^{L_n} \right] = 0 \, .$$

Inserting in this last convergence the first assertion of ($\diamond$), we get via (++)

$$\lim_{K \uparrow \infty} \lim_{n \to \infty} \int g_K \, dQ_n^{L_n} = \lim_{K \uparrow \infty} \int g_K \, d\check{F} = 1$$

which is the desired assertion $\int_{[0,\infty)} x \, \check{F}(dx) = 1$.

(3) We prove '$\Longleftarrow$'. Let us assume $\int_{[0,\infty)} x \, \check{F}(dx) = 1$. A first consequence of this assumption is

($\circ$) $$\lim_{K \uparrow \infty} \lim_{n \to \infty} \int g_K \, dQ_n^{L_n} = \lim_{K \uparrow \infty} \int g_K \, d\check{F} = 1$$

according to (++); a second consequence, considering continuity points K of $\check{F}$, is

($\circ\circ$) $$\lim_{K \uparrow \infty} \lim_{n \to \infty} K \, Q_n^{L_n}((K, \infty)) = \lim_{K \uparrow \infty} K \, \check{F}((K, \infty)) = 0 \, .$$

Inserting ($\circ$) into equations (+) when n tends to ∞, left-hand sides in (+) will be small for large values of K, see ($\star\star$), which again with reference to ($\circ$) gives

($\times$) $$\lim_{n \to \infty} E_{Q_n}(L_n) = 1 \, ,$$

simultaneously, right-hand sides in (+) must be small for large values of K, see ($\star$), which combined with ($\circ\circ$) gives

$$(\times\times) \qquad \lim_{K\uparrow\infty} \limsup_{n\to\infty} \int 1_{\{K<L_n<\infty\}} L_n \, dQ_n = 0 \,.$$

By Proposition 3.10(iii), assertions ($\times$) and ($\times\times$) together establish contiguity $(P_n)_n \triangleleft (Q_n)_n$ as desired. □

3.11' Remark. To any probability measure $\check{F}$ on $(\mathbb{R}, \mathcal{B}(\mathbb{R}))$ which may occur in Theorem 3.11, associate a real valued random variable $L \geq 0$ on some probability space $(\Omega, \mathcal{A}, Q)$ with distribution $\check{F}$. Then the equality $\int_{[0,\infty)} \ell \, \check{F}(d\ell) = 1$ in Theorem 3.11 signifies that $(\Omega, \mathcal{A})$ carries a second probability measure P defined by

$$dP := L \, dQ \,.$$

Since by definition P is absolutely continuous with respect to Q, we have a binary 'limit experiment'

$$(+) \qquad\qquad (\Omega, \mathcal{A}, \{P, Q\}) \quad \text{with} \quad P \ll Q \,;$$

here $\check{F}(\{0\}) = Q(L = 0)$, the Q-weight of the support of the P-singular part of Q, may be strictly positive. In this sense, one-sided contiguity $(P_n)_n \triangleleft (Q_n)_n$ is related to convergence of experiments where *weak limits of likelihood ratios are likelihood ratios between probability measures P, Q such that P is absolutely continuous with respect to Q*.

3.12 Proposition. The following statements are equivalent:

(i) $(P_n)_n \triangleright (Q_n)_n$;

(ii) $\lim_{c\downarrow 0} \limsup_{n\to\infty} Q_n(L_n \leq c) = 0$.

Proof. According to Notation 3.1(iv), the $[0, \infty]$-valued random variable $\widetilde{L}_n := \frac{1}{L_n}$ is a version of the likelihood ratio of Q_n with respect to P_n, and statement (ii) means $\mathbb{R}$-tightness of the sequence $(\frac{1}{L_n})_n$ under $(Q_n)_n$. Changing names $\widetilde{P}_n := Q_n$, $\widetilde{Q}_n := P_n$ such that $\widetilde{L}_n$ is the likelihood ratio of $\widetilde{P}_n$ with respect to $\widetilde{Q}_n$, Theorem 3.4(a) applied to $(\widetilde{P}_n)_n$ and $(\widetilde{Q}_n)_n$ proves Proposition 3.12. □

Now we can prove also the second part of Theorem 3.4:

3.13 Proof of Theorem 3.4(b). We have to show that mutual contiguity $(P_n)_n \triangleleft\triangleright (Q_n)_n$ is equivalent to $\mathbb{R}$-tightness of the sequence $(\Lambda_n)_n$ under both $(P_n)_n$ and $(Q_n)_n$.

Section 3.2 Proofs for Section 3.1 and some Variants

(1) We assume mutual contiguity. Then Propositions 3.3 and 3.12 together show

(+) $$\lim_{K\uparrow\infty} \limsup_{n\to\infty} Q_n \left(L_n \notin [\frac{1}{K}, K]\right) = 0$$

and Theorem 3.4(a) yields

$$\lim_{K\uparrow\infty} \limsup_{n\to\infty} P_n (L_n \notin [0, K]) = 0 .$$

In order to consider small values of L_n under P_n, write for $K > 1$

$$P_n \left(L_n < \frac{1}{K}\right) < \int 1_{\{L_n < \frac{1}{K}\}} L_n \, dQ_n < \frac{1}{K} Q_n \left(L_n < \frac{1}{K}\right) < Q_n \left(L_n < \frac{1}{K}\right)$$

so that the last convergence strengthens to

(++) $$\lim_{K\uparrow\infty} \limsup_{n\to\infty} P_n \left(L_n \notin [\frac{1}{K}, K]\right) = 0 .$$

From (+) and (++), log-likelihood ratios $(\Lambda_n)_n$ are $\mathbb{R}$-tight both under $(P_n)_n$ and under $(Q_n)_n$.

(2) Let us assume $\mathbb{R}$-tightness of $(\Lambda_n)_n$ under both $(P_n)_n$ and $(Q_n)_n$. Then we have $\mathbb{R}$-tightness of

$$(L_n)_n = \left(e^{+\Lambda_n}\right)_n, \quad \left(\frac{1}{L_n}\right)_n = \left(e^{-\Lambda_n}\right)_n, \text{ with conventions of Notation 3.1(iii)}$$

under both $(P_n)_n$ and $(Q_n)_n$. Hence the likelihoods of P_n with respect to Q_n are tight under $(P_n)_n$, and the likelihoods of Q_n with respect to P_n are tight under $(Q_n)_n$. Thus two applications of Theorem 3.4(a) establish the desired mutual contiguity. □

At this stage, Theorem 3.4 being completely proved, we can prove Le Cam's first lemma in the form encountered in Lemma 3.5. An obvious corollary merging Theorem 3.11 and Proposition 3.12 is the following:

3.14 Corollary. Assume that there is a probability measure $\check{F}$ on $(\mathbb{R}, \mathcal{B}(\mathbb{R}))$ such that

$$\mathcal{L}(L_n \mid Q_n) \longrightarrow \check{F} \quad \text{(weakly in } \mathbb{R}, \text{ as } n \to \infty) .$$

Then the following holds:

$$(P_n)_n \triangleleft\triangleright (Q_n)_n \quad \Longleftrightarrow \quad \check{F}(\{0\}) = 0 \text{ and } \int_{[0,\infty)} \ell \, \check{F}(d\ell) = 1 .$$

To deduce Corollary 3.14 from Theorem 3.11 and Proposition 3.12, it is sufficient to consider continuity points $c > 0$ of $\check{F}$ for which $\lim_{n\to\infty} Q_n(L_n \leq c) = \check{F}([0, c])$. Now we deduce Lemma 3.5 from Corollary 3.14:

3.15 Proof of Le Cam's First Lemma 3.5. We show that the setting of Lemma 3.5 specialises the setting of Corollary 3.14; in this sense, Corollary 3.14 is a slightly stronger formulation of Le Cam's first lemma. The difference is in the respective initial conditions on convergence of likelihood ratios. Lemma 3.5 starts from some law F on $(\mathbb{R}, \mathcal{B}(\mathbb{R}))$ such that

$$\mathcal{L}(\Lambda_n \mid Q_n) \longrightarrow F \quad \text{(weakly in } \mathbb{R} \text{ as } n \to \infty)$$

which – writing $\check{F}$ for the image of F under the continuous mapping $\mathbb{R} \ni \lambda \to e^\lambda \in \mathbb{R}$ – translates into

$$\mathcal{L}(L_n \mid Q_n) \longrightarrow \check{F} \quad \text{with } \check{F}((0, \infty)) = 1 \quad \text{(weakly in } \mathbb{R} \text{ as } n \to \infty).$$

In explicit restriction to the subclass of laws $\check{F}$ which are concentrated on $(0, \infty)$, Corollary 3.14 states

$$(P_n)_n \vartriangleleft\vartriangleright (Q_n)_n \iff \int_{(0,\infty)} \ell\, \check{F}(d\ell) = 1$$

where the right-hand side equals $\int_\mathbb{R} e^\lambda F(d\lambda) = 1$ when $\check{F}$ is image of F under the above mapping. $\square$

We turn to the proof of Le Cam's Third Lemma 3.6. Since $(L_n)_n$ under $(Q_n)_n$ is always $\mathbb{R}$-tight by Proposition 3.3, we have $\mathbb{R}^{1+d}$-tightness of pairs $(L_n, X_n)_n$ under $(Q_n)_n$ for any sequence $(X_n)_n$ of $\overline{\mathbb{R}}^d$-valued random variables which is $\mathbb{R}^d$-tight under $(Q_n)_n$, and can select subsequences $(n_l)_l$ and laws $\widetilde{F}$ on $(\mathbb{R}, \mathcal{B}(\mathbb{R}))$ such that $(L_{n_l}, X_{n_l})_l$ under $(Q_{n_l})_l$ converges weakly in $\mathbb{R}^{1+d}$ to $\widetilde{F}$ as $l \to \infty$. All accumulation points $\widetilde{F}$ in this sense are necessarily laws which are concentrated on $[0, \infty) \times \mathbb{R}^d$, and $\widetilde{F}$ may put strictly positive mass to the hyperplane $\{0\} \times \mathbb{R}^d$. Thus, up to selection of subsequences, the only condition in the following one-sided variant of Le Cam's third lemma is contiguity $(P_n)_n \vartriangleleft (Q_n)_n$: in this case, sequences (X_n) converging *jointly with the likelihoods* under $(Q_n)_n$ will also converge under $(P_n)_n$, and the limit law will be 'explicit'.

3.16 Theorem. Assume

$$(P_n)_n \vartriangleleft (Q_n)_n.$$

Consider a probability measure $\widetilde{F}$ on $(\mathbb{R}^{1+d}, \mathcal{B}(\mathbb{R}^{1+d}))$ and a sequence of $\overline{\mathbb{R}}^d$-valued random variables X_n on $(\Omega_n, \mathcal{A}_n)$ such that the following holds:

$$\mathcal{L}((L_n, X_n) \mid Q_n) \longrightarrow \widetilde{F} \quad \text{(weakly in } \mathbb{R}^{1+d} \text{ as } n \to \infty).$$

Then

$$\widetilde{G}(d\ell, dx) := (\ell \vee 0)\, \widetilde{F}(d\ell, dx), \quad \ell \in \mathbb{R},\ x \in \mathbb{R}^d$$

Section 3.2 Proofs for Section 3.1 and some Variants

is a probability measure on $(\mathbb{R}^{1+d}, \mathcal{B}(\mathbb{R}^{1+d}))$, and we have

$$\mathcal{L}((L_n, X_n) \mid P_n) \longrightarrow \widetilde{G} \quad \text{(weakly in } \mathbb{R}^{1+d} \text{ as } n \to \infty).$$

Proof. (I) We prove this result first in the case where all X_n take values in $\mathbb{R}^d$.

(1) We start with some preliminaries. We write $\check{G}$ for the first marginal of the measure $\widetilde{G}$ defined in Theorem 3.16, and $\check{F}$ for the first marginal of $\widetilde{F}$. Then $\check{F}$ is a probability on $(\mathbb{R}, \mathcal{B}(\mathbb{R}))$ such that

$$\mathcal{L}(L_n \mid Q_n) \longrightarrow \check{F} \quad \text{(weakly in } \mathbb{R} \text{ as } n \to \infty).$$

In particular, $\check{F}$ is concentrated on $[0, \infty)$. Now Theorem 3.11 and the assumed contiguity give

(o) $$\int_{[0,\infty)} \ell \, \check{F}(d\ell) = 1,$$

thus $\widetilde{G}$ is a probability. Again by (o), for $\varepsilon > 0$ arbitrarily small, there is $K = K(\varepsilon) < \infty$ such that

(+) $$\int_{\mathbb{R}^{1+d}} l \, 1_{\{l>K\}} \, \widetilde{F}(dl, dx) = \int_{(K,\infty)} l \, \check{F}(dl) < \varepsilon ;$$

increasing $K = K(\varepsilon)$ if necessary, we achieve simultaneously

(++) $$\sup_{n \geq 1} \int L_n \, 1_{\{L_n > K\}} \, dQ_n < \varepsilon$$

from Proposition 3.10 and the assumed contiguity. Contiguity also implies

(+++) $$P_n(L_n = \infty) \longrightarrow 0 \quad \text{as } n \to \infty.$$

(2) First, we prove for functions $f \in \mathcal{C}_b(\mathbb{R}^{1+d})$

$$\lim_{n \to \infty} \int f(L_n, X_n) L_n \, dQ_n = \int f(l, x) \, \widetilde{G}(dl, dx), \quad n \to \infty.$$

Write $M := \sup |f|$ and decompose with truncated identities $g_K(x) = 0 \vee x \wedge K$, $x \in \mathbb{R}$,

$$\int f(L_n, X_n) L_n \, dQ_n = \int f(L_n, X_n) g_K(L_n) \, dQ_n$$
$$+ \int f(L_n, X_n) (L_n - g_K(L_n)) \, dQ_n.$$

Since $(l, x) \to f(l, x) g_K(l)$ is in $\mathcal{C}_b(\mathbb{R}^{1+d})$, weak convergence of pairs $(L_n, X_n)_n$ under $(Q_n)_n$ gives

(◇) $$\lim_{n \to \infty} \int f(L_n, X_n) g_K(L_n) \, dQ_n = \int f(l, x) g_K(l) \, \widetilde{F}(dl, dx).$$

At the same time, we have as a consequence of (++)

$$\sup_{n\geq 1} \int |f(L_n, X_n)| (L_n - g_K(L_n)) \, dQ_n < M\varepsilon$$

and as a consequence of (+)

$$\int |f(l, x)| (l - g_K(l)) \, \widetilde{F}(dl, dx) < M\varepsilon \, .$$

The family of convergences ($\diamond$) for $K = K(\varepsilon)$, $\varepsilon > 0$ arbitrary, in combination with the last lines gives

$$\lim_{n\to\infty} \int f(L_n, X_n) L_n \, dQ_n = \int f(l, x) l \, \widetilde{F}(dl, dx) \, ;$$

$\check{F}$ being concentrated on $[0, \infty)$, the limit is

$$\int f(l, x) (l \vee 0) \widetilde{F}(dl, dx) = \int f(l, x) \, \widetilde{G}(dl, dx) \, .$$

(3) Second, we prove for $f(\cdot, \cdot)$ as above

$$\int f\left(\widehat{(L_n, X_n)}\right) dP_n \longrightarrow \int f(l, x) \widetilde{G}(dl, dx))$$

where

$$\widehat{(L_n, X_n)} = (L_n, X_n) 1_{\{L_n < \infty\}}$$

in accordance with Notations 3.1' since all X_n are $\mathbb{R}^d$-valued, by assumption, in this part of the proof.

This holds since the quantity considered in step (2)

$$\int f(L_n, X_n) L_n \, dQ_n = \int f(L_n, X_n) 1_{\{L_n < \infty\}} \, dP_n$$

(by Lebesgue decomposition of P_n with respect to Q_n) can asymptotically as $n \to \infty$ be replaced by

$$\int f\left((L_n, X_n) 1_{\{L_n < \infty\}}\right) dP_n = \int f\left(\widehat{(L_n, X_n)}\right) dP_n$$

thanks to contiguity (+++) and a bound

$$\left| \int f(L_n, X_n) 1_{\{L_n<\infty\}} \, dP_n - \int f\left((L_n, X_n) 1_{\{L_n<\infty\}}\right) dP_n \right| \leq M \, P_n(L_n = \infty)$$

$$\longrightarrow 0 \, .$$

(4) According to Notation 3.1'(ii), contiguity (+++) combined with step (3) is the desired assertion

$$\mathcal{L}((L_n, X_n) \mid P_n) \longrightarrow \widetilde{G} \quad \text{(weakly in } \mathbb{R}^{1+d}, \text{ as } n \to \infty) \, .$$

This ends the proof of Theorem 3.16 in the case where all X_n take values in $\mathbb{R}^d$.

Section 3.2 Proofs for Section 3.1 and some Variants

(II) Now we extend the result from $\mathbb{R}^d$-valued to $\overline{\mathbb{R}}^d$-valued random variables X_n, $n \geq 1$.

In this case, by Notation 3.1'(ii), the convergence assumption on pairs $(L_n, X_n)_n$ under $(Q_n)_n$ implies
$$\lim_{n \to \infty} Q_n(|X_n| = \infty) = 0,$$
L_n being a.s. finite under Q_n, and thanks to contiguity also
$$\lim_{n \to \infty} P_n(L_n = \infty \text{ or } |X_n| = \infty) = 0.$$
Contiguity and arguments similar to step (3) above allow to replace asymptotically as $n \to \infty$
$$\int f(\widetilde{(L_n, X_n)}) \, dQ_n \text{ by } \int f(L_n, \widehat{X}_n) \, dQ_n$$
$$\int f(\widetilde{(L_n, X_n)}) \, dP_n \text{ by } \int f(L_n, \widehat{X}_n) 1_{\{L_n < \infty\}} \, dP_n$$
where we put $\widehat{X}_n := X_n 1_{\{|X_n| < \infty\}}$, and where $\widetilde{(L_n, X_n)} = (L_n, X_n) 1_{\{L_n < \infty, |X_n| < \infty\}}$ is the notation of 3.1'(ii). The convergence assumption in Theorem 3.16 can thus be rephrased as
$$\int f(L_n, \widehat{X}_n) \, dQ_n \longrightarrow \int f(l, x) \, \widetilde{F}(dl, dx)$$
where we are left to consider pairs $(L_n, \widehat{X}_n)$ for which steps (2) and (3) of the above proof show
$$\int f(L_n, \widehat{X}_n) L_n \, dQ_n = \int f(L_n, \widehat{X}_n) 1_{\{L_n < \infty\}} \, dP_n$$
$$\longrightarrow \int f(l, x) \, \widetilde{G}(dl, dx).$$
Hence the assertion of Theorem 3.16 is proved also for $\overline{\mathbb{R}}^d$-valued random variables X_n, $n \geq 1$. □

Now we can prove Le Cam's Third Lemma 3.6 under mutual contiguity:

3.17 Proof of Le Cam's Third Lemma 3.6. In analogy to the Proof 3.15 which deduces Lemma 3.5 from Corollary 3.14, we shall show that the setting of Lemma 3.6 specialises the setting of Theorem 3.16 (hence, Theorem 3.16 is a slightly stronger formulation of Le Cam's third lemma). Writing d in place of k, Lemma 3.6 starts from some law $\widetilde{F}$ on $(\mathbb{R}^{1+d}, \mathcal{B}(\mathbb{R}^{1+d}))$ such that
$$\mathcal{L}((\Lambda_n, X_n) \mid Q_n) \longrightarrow \widetilde{F} \quad (\text{weakly in } \mathbb{R}^{1+d} \text{ as } n \to \infty).$$

If we define $\widetilde{F}$ as image of $\overline{F}$ under the continuous mapping $\mathbb{R}^{1+d} \ni (\lambda, x) \to (e^\lambda, x) \in \mathbb{R}^{1+d}$, and $\check{F}$ for the first component of $\widetilde{F}$, this translates into

$$\mathcal{L}((L_n, X_n) \mid Q_n) \longrightarrow \widetilde{F} \text{ where } \check{F}((0, \infty)) = 1 \quad \text{(weakly in } \mathbb{R}^{1+d} \text{ as } n \to \infty).$$

By the assumed mutual contiguity $(P_n)_n \triangleleft\triangleright (Q_n)_n$, we have according to Lemma 3.5 or Corollary 3.14

$$1 = \int_{(0,\infty)} \ell \, \check{F}(d\ell) = \int_{(0,\infty) \times \mathbb{R}^d} \ell \, \widetilde{F}(d\ell, dx) = \int_{\mathbb{R}^{1+d}} e^\lambda \, \overline{F}(d\lambda, dx).$$

Hence $\widetilde{G}(d\lambda, dx) = e^\lambda \, \overline{F}(d\lambda, dx)$ defines a probability measure on $(\mathbb{R}^{1+d}, \mathcal{B}(\mathbb{R}^{1+d}))$. The image $\widetilde{G}$ of $\overline{G}$ under the mapping $\mathbb{R}^{1+d} \ni (\lambda, x) \to (e^\lambda, x) \in \mathbb{R}^{1+d}$ being concentrated on $(0, \infty) \times \mathbb{R}^d$, we can write

$$\widetilde{G}(d\ell, dx) = (\ell \vee 0) \, \widetilde{F}(d\ell, dx),$$

thus Theorem 3.16 establishes

$$\mathcal{L}((L_n, X_n) \mid P_n) \longrightarrow \widetilde{G} \quad \text{(weakly in } \mathbb{R}^{1+d} \text{ as } n \to \infty).$$

Again by the same transformation, the last convergence is equivalent to

$$\mathcal{L}((\Lambda_n, X_n) \mid P_n) \longrightarrow \overline{G} \quad \text{(weakly in } \mathbb{R}^{1+d} \text{ as } n \to \infty)$$

which finishes the proof of Lemma 3.6. □

It remains to prove Proposition 3.6" which specialises Le Cam's Third Lemma 3.6 to situations where the limit law $\overline{F}$ in Lemma 3.6 is a normal law with non-degenerate first component.

3.18 Proof of Proposition 3.6". We start from the assumption that $\overline{F}$ in Lemma 3.6 is some normal law on $(\mathbb{R}^{1+k}, \mathcal{B}(\mathbb{R}^{1+k}))$ whose first marginal – the limit law for the log-likelihood ratios $(\Lambda_n)_n$ under $(Q_n)_n$ when mutual contiguity $(P_n)_n \triangleleft\triangleright (Q_n)_n$ holds – is not degenerate at a single point, i.e. has some strictly positive variance $\sigma^2 > 0$. Then we can write $\overline{F}$ in the form

$$\overline{F} = \mathcal{N}\left(\begin{pmatrix} \nu \\ \mu \end{pmatrix}, \begin{pmatrix} \sigma^2 & \gamma^\top \\ \gamma & \Sigma \end{pmatrix} \right)$$

for suitable $\mu, \gamma \in \mathbb{R}^k$, $\nu \in \mathbb{R}$, and $\Sigma \in \mathbb{R}^{k \times k}$ which is symmetric and non-negative definite.

(1) Let us prove that in this representation of $\overline{F}$, we have necessarily

(◇) $$\nu = -\frac{1}{2}\sigma^2.$$

Section 3.2 Proofs for Section 3.1 and some Variants   105

This arises as a consequence of mutual contiguity by which

$$\int e^\lambda \widetilde{F}(d\lambda, dx) = 1$$

according to Le Cam's First Lemma 3.5. For the first marginal,

$$1 = \int e^\lambda \mathcal{N}(\nu, \sigma^2)(d\lambda) = \int \frac{1}{\sqrt{2\pi\sigma^2}} e^{-\frac{1}{2\sigma^2}(\lambda^2 - 2\lambda(\nu + \sigma^2) + \nu^2)} d\lambda$$

amounts – inserting $\pm(\nu + \sigma^2)^2$ in the exponent – to

$$\nu^2 = (\nu + \sigma^2)^2 \iff |\nu| = |\nu + \sigma^2| \iff \nu = -\frac{1}{2}\sigma^2$$

since $\sigma^2 > 0$, by assumption on the first marginal of $\widetilde{F}$. This proves the first assertion of Proposition 3.6".

(2) For the parameters of the normal law

$$\widetilde{F} = \mathcal{N}\left(\begin{pmatrix} -\frac{1}{2}\sigma^2 \\ \mu \end{pmatrix}, \begin{pmatrix} \sigma^2 & \gamma^T \\ \gamma & \Sigma \end{pmatrix}\right)$$

in step (1) above we write

$$\mu = \begin{pmatrix} \mu_1 \\ \vdots \\ \mu_k \end{pmatrix}, \quad \gamma = \begin{pmatrix} \gamma_1 \\ \vdots \\ \gamma_k \end{pmatrix}, \quad m := \begin{pmatrix} -\frac{1}{2}\sigma^2 \\ \mu \end{pmatrix}, \quad \Lambda := \begin{pmatrix} \sigma^2 & \gamma^T \\ \gamma & \Sigma \end{pmatrix}.$$

Then with notations

$$z = \begin{pmatrix} \zeta \\ z_1 \\ \vdots \\ z_k \end{pmatrix} \in \mathbb{R}^{1+k}, \quad e_0 = \begin{pmatrix} 1 \\ 0 \\ \vdots \\ 0 \end{pmatrix} \in \mathbb{R}^{1+k}$$

the Laplace transform of the normal law $\widetilde{F}$ (existing on $\mathbb{R}^{1+k}$) equals

$$z \longrightarrow \int e^{-[\zeta\lambda + \sum_{i=1}^{k} z_i x_i]} \widetilde{F}(d\lambda, dx) = \exp\left\{-z^T m + \frac{1}{2} z^T \Lambda z\right\} =: \widetilde{\phi}(z);$$

as a consequence, the Laplace transform of a law $\widetilde{G}(d\lambda, dx) := e^\lambda \widetilde{F}(d\lambda, dx)$

$$z \longrightarrow \int e^{-[\zeta\lambda + \sum_{i=1}^{k} z_i x_i]} \widetilde{G}(d\lambda, dx) = \int e^{-[(\zeta-1)\lambda + \sum_{i=1}^{k} z_i x_i]} \widetilde{F}(d\lambda, dx)$$

exists on $\mathbb{R}^{1+k}$ and has the form

$$(\diamond) \quad z \longrightarrow \widetilde{\phi}(z - e_0) := \exp\left\{-(z - e_0)^T m + \frac{1}{2}(z - e_0)^T \Lambda (z - e_0)\right\}.$$

We rewrite this function using

$$-(z-e_0)^T m = -(\zeta-1)(-\frac{1}{2}\sigma^2) - \sum_{i=1}^k z_i \mu_i$$

$$\frac{1}{2}(z-e_0)^T \Lambda (z-e_0) = \frac{1}{2} z^T \Lambda z - \zeta \sigma^2 - \sum_{i=1}^k z_i \gamma_i + \frac{1}{2}\sigma^2$$

as

$$z \longrightarrow \exp\left\{-\left[\zeta(+\frac{1}{2}\sigma^2) + \sum_{i=1}^k z_i(\mu_i+\gamma_i)\right] + \frac{1}{2} z^T \Lambda z\right\}$$

or equivalently in compact notation

$$z \longrightarrow \exp\left\{-z^T \widetilde{m} + \frac{1}{2} z^T \Lambda z\right\} \quad \text{where} \quad \widetilde{m} := \begin{pmatrix} +\frac{1}{2}\sigma^2 \\ \mu+\gamma \end{pmatrix}.$$

This is the Laplace transform of $\mathcal{N}(\widetilde{m}, \Lambda)$. By the uniqueness theorem for Laplace transforms (see e.g. [4, Chap. 10.1]), we have identified the law $\widetilde{G}(d\lambda, dx) := e^\lambda \widetilde{F}(d\lambda, dx)$, for $\widetilde{F}$ of step (1), as

$$\widetilde{G} = \mathcal{N}\left(\begin{pmatrix} +\frac{1}{2}\sigma^2 \\ \mu+\gamma \end{pmatrix}, \begin{pmatrix} \sigma^2 & \gamma^T \\ \gamma & \Sigma \end{pmatrix}\right)$$

which proves the second assertion of Proposition 3.6". □

3.18' Exercise. Write M for the space of all piecewise constant right-continuous jump functions $f : [0, \infty) \to \mathbb{N}_0$ such that f has at most finitely many jumps over compact time intervals, all these with jump height $+1$, and starts from $f(0) = 0$. Write η_t for the coordinate projections $\eta_t(f) = f(t), t \geq 0, f \in M$.

Equip M with the σ-field generated by the coordinate projections, consider the filtration $\mathbb{F} = (\mathcal{F}_t)_{t \geq 0}$ where $\mathcal{F}_t := \sigma(\eta_r : 0 \leq r \leq t)$, and call $\eta = (\eta_t)_{t \geq 0}$ the canonical process on $(M, \mathcal{M}, \mathbb{F})$.

For $\lambda > 0$, let $P(\lambda)$ denote the (unique) probability law on $(M, \mathcal{M})$ such that the canonical process $(\eta_t)_{t \geq 0}$ is a Poisson process with parameter λ. Then (e.g. [14, p. 165]) the process

$$L_t^{\lambda'/\lambda} := \left[\frac{\lambda'}{\lambda}\right]^{\eta_t} \exp\{-(\lambda'-\lambda)t\}, \quad t \geq 0$$

is the likelihood ratio process of $P(\lambda')$ relative to $P(\lambda)$ with respect to $\mathbb{F}$, i.e., for all $t \in [0, \infty)$,

$$L_t^{\lambda'/\lambda} \text{ is a version of the likelihood ratio of } P(\lambda')_{|\mathcal{F}_t} \text{ relative to } P(\lambda)_{|\mathcal{F}_t}$$

where $_{|\mathcal{F}_t}$ denotes restriction of a probability measure on $(M, \mathcal{M})$ to the sub-σ-field $\mathcal{F}_t$. Accepting this as background (as in Jacod [63], Kabanov, Liptser and Shiryaev [68], and Brémaud [14]), prove the following.

(1) For fixed reference value $\lambda > 0$, reparameterising the family of laws on $(M, \mathcal{M}, \mathbb{F})$ with respect to λ as

(◇) $\qquad\qquad\qquad \vartheta \longrightarrow P(\lambda e^{\vartheta})\,,\quad \vartheta \in \Theta := \mathbb{R}\,,$

the likelihood ratio process takes the form which has been considered in Exercise 1.16':
$$L_t^{\lambda e^{\vartheta}/\lambda} := \exp\{\vartheta\, \eta_t - \lambda(e^{\vartheta}-1)t\}\,,\quad t \ge 0\,.$$

For t fixed, this is an exponential family in ϑ with canonical statistic η_t. Specify a functional $\gamma : \Theta \to (0, \infty)$ and an estimator $T_t : M \to \mathbb{R}$ for this functional
$$\gamma(\vartheta) := \lambda e^{\vartheta}\,,\quad T_t := \frac{1}{t}\eta_t\,.$$

By classical theory of exponential families, T_t is the best estimator for γ in the sense of uniformly minimum variance within the class of unbiased estimators (e.g. [127, pp. 303 and 157]).

(2) For some $h \in \mathbb{R}$ which we keep fixed as $n \to \infty$, define sequences of probability laws
$$Q_n := P(\lambda)_{|\mathcal{F}_n}\,,\quad P_n := P(\lambda e^{h/\sqrt{n}})_{|\mathcal{F}_n}\,,\quad n \ge 1$$

and prove contiguity $(P_n)_n \triangleleft\triangleright (Q_n)_n$ via Lemma 3.5: for this, use a representation of log-likelihood ratios Λ_n of P_n with respect to Q_n in the form
$$\Lambda_n = h\frac{1}{\sqrt{n}}(\eta_n - \lambda n) - \lambda\left(e^{h/\sqrt{n}} - 1 - \frac{h}{\sqrt{n}}\right)n$$

and weak convergence in $\mathbb{R}$
$$\mathcal{L}(\Lambda_n \mid Q_n) \longrightarrow \mathcal{N}(-\tfrac{1}{2}h^2\lambda, h^2\lambda)\,,\quad n \to \infty\,.$$

(3) Writing $X_n := \sqrt{n}(T_n - \lambda)$, the reference point λ being fixed according to the reparameterisation (◇), extend the last result to joint convergence in $\mathbb{R}^2$
$$\mathcal{L}((\Lambda_n, X_n) \mid Q_n) \longrightarrow \mathcal{N}\left(\begin{pmatrix} -\tfrac{1}{2}h^2\lambda \\ 0 \end{pmatrix}, \begin{pmatrix} h^2\lambda & h\lambda \\ h\lambda & \lambda \end{pmatrix}\right)$$

such that by Le Cam's Third Lemma 3.6
$$\mathcal{L}((\Lambda_n, X_n) \mid P_n) \longrightarrow \mathcal{N}\left(\begin{pmatrix} +\tfrac{1}{2}h^2\lambda \\ h\lambda \end{pmatrix}, \begin{pmatrix} h^2\lambda & h\lambda \\ h\lambda & \lambda \end{pmatrix}\right)\,.$$

In particular, this gives

(◇◇) $\qquad\qquad\qquad \mathcal{L}(X_n - h\lambda \mid P_n) \longrightarrow \mathcal{N}(0, \lambda)\,.$

(4) Deduce from (◇◇) the following 'equivariance' property of the estimator T_n in shrinking neighbourhoods of radius $O(\frac{1}{\sqrt{n}})$ – in the sense of the reparameterisation (◇) above – of the reference point λ, using approximations $e^{h/\sqrt{n}} = 1 + \frac{h}{\sqrt{n}} + O(\frac{1}{n})$ as $n \to \infty$: for every $h \in \mathbb{R}$ fixed, we have weak convergence
$$\mathcal{L}\left(\sqrt{n}(T_n - \lambda e^{h/\sqrt{n}}) \mid P(\lambda e^{h/\sqrt{n}})\right) \longrightarrow \mathcal{N}(0, \lambda) \quad \text{as } n \to \infty$$

where the limit law *does not depend on h*. This means that on small neighbourhoods with radius $O(\frac{1}{\sqrt{n}})$ of a fixed reference point λ, asymptotically as $n \to \infty$, the estimator T_n identifies true parameter values with the same precision. □

Chapter 4

L^2-differentiable Statistical Models

Topics for Chapter 4:

4.1 L^r-differentiability when $r \geq 1$
L^r-differentiability in dominated families 4.1–4.1"
L^r-differentiability in general 4.2–4.2'
Example: one-parametric paths in non-parametric models 4.3
L^r-differentiability implies L^s-differentiability for $1 \leq s \leq r$ 4.4
L^r-derivatives are centred 4.5
Score and information in L^2-differentiable families 4.6
L^2-differentiability and Hellinger distances locally at ϑ 4.7–4.9

4.2 Le Cam's Second Lemma for i.i.d. observations
Assumptions for Section 4.2 4.10
Statement of Le Cam's second lemma 4.11
Some auxiliary results 4.12–4.14
Proof of Le Cam's second lemma 4.15

Exercises: 4.1''', 4.9'

This chapter, closely related to Section 1.1, generalises the notion of 'smoothness' of statistical models. We define it in an L^2-sense which e.g. allows us to consider families of laws which are not pairwise equivalent, or where log-likelihood ratios for ω fixed are not smooth functions of the parameter. To the new notion of smoothness corresponds a new and more general definition of score and information in a statistical model. In the new setting, we will be able to prove rigorously – and under weak assumptions – quadratic expansions of log-likelihood ratios, valid locally in small neighbourhoods of fixed reference points ϑ. Later, such expansions will allow to prove assertions very similar to those intended – with questionable heuristics – in Section 1.3. Frequently called 'second Le Cam lemma', expansions of this type can be proved in large classes of statistical models – for various stochastic phenomena observed over long time intervals or in growing spatial windows – provided the parameterisation is L^2-smooth, and provided we have strong laws of large numbers and corresponding martingale convergence theorems. For i.i.d. models, we present a 'second Le Cam lemma' in Section 4.2 below.

4.1 L^r-differentiable Statistical Models

We start with models which are dominated, and give a preliminary definition of L^r-differentiability of a statistical model at a point ϑ. The domination assumption will be removed in the sequel.

4.1 Motivation. We consider a dominated experiment

$$(\Omega, \mathcal{A}, \mathcal{P} := \{P_\vartheta : \vartheta \in \Theta\}), \quad \Theta \subset \mathbb{R}^d \text{ open}, \quad \mathcal{P} \ll \nu$$

with densities

$$\chi_\vartheta := \frac{dP_\vartheta}{d\nu}, \quad \vartheta \in \Theta.$$

Let $r \geq 1$ be fixed. It is trivial that for all $\vartheta \in \Theta$, $\chi_\vartheta^{1/r}$ belongs to $L^r(\Omega, \mathcal{A}, \nu)$.
(1) In the normed space $L^r(\Omega, \mathcal{A}, \nu)$, the mapping

$$(\diamond) \qquad \Theta \ni \xi \longrightarrow \chi_\xi^{1/r} \in L^r(\Omega, \mathcal{A}, \nu)$$

is (Fréchet) differentiable at $\xi = \vartheta$ with derivative $\widetilde{V}_\vartheta$ if the following (i) and (ii) hold simultaneously:

(i) $\qquad \widetilde{V}_\vartheta$ has components $\widetilde{V}_{\vartheta,1}, \ldots, \widetilde{V}_{\vartheta,d} \in L^r(\Omega, \mathcal{A}, \nu)$,

(ii) $\qquad \dfrac{1}{|\xi - \vartheta|} \left\| \chi_\xi^{1/r} - \chi_\vartheta^{1/r} - (\xi - \vartheta)^\top \widetilde{V}_\vartheta \right\|_{L^r(\nu)} \longrightarrow 0 \quad \text{as} \quad |\xi - \vartheta| \longrightarrow 0.$

(2) We prove: in a statistical model $\mathcal{P}$, (i) and (ii) imply the assertion

$$(*) \qquad \widetilde{V}_{\vartheta,i} = 0 \quad \nu\text{-almost surely on } \{\chi_\vartheta = 0\}, \quad 1 \leq i \leq d.$$

To see this, consider sequences $(\xi_n)_n \subset \Theta$ of type $\xi_n = \vartheta \pm \delta_n e_i$ where e_i is the i-th unit vector in $\mathbb{R}^d$, and $(\delta_n)_n$ any sequence of strictly positive real numbers tending to 0. Then (ii) with $\xi = \xi_n$ restricted to the event $\{\chi_\vartheta = 0\}$ gives

$$\left\| 1_{\{\chi_\vartheta = 0\}} \left[\frac{1}{\delta_n} \chi_{\vartheta + \delta_n e_i}^{1/r} - \widetilde{V}_{\vartheta,i} \right] \right\|_{L^r(\nu)} \longrightarrow 0,$$

$$\left\| 1_{\{\chi_\vartheta = 0\}} \left[\frac{1}{\delta_n} \chi_{\vartheta - \delta_n e_i}^{1/r} + \widetilde{V}_{\vartheta,i} \right] \right\|_{L^r(\nu)} \longrightarrow 0$$

as $n \to \infty$. Selecting subsequences $(n_k)_k$ along which ν-almost sure convergence holds, we have

$$\text{on } \{\chi_\vartheta = 0\}: \quad \frac{1}{\delta_{n_k}} \chi_{\vartheta + \delta_{n_k} e_i}^{1/r} \longrightarrow \widetilde{V}_{\vartheta,i} \quad \text{and} \quad \frac{1}{\delta_{n_k}} \chi_{\vartheta - \delta_{n_k} e_i}^{1/r} \longrightarrow -\widetilde{V}_{\vartheta,i}$$

ν-almost surely as $k \to \infty$. The densities of the left-hand sides being non-negative, $\widetilde{V}_{\vartheta,i}$ necessarily equals 0 ν-almost surely on the event $\{\chi_\vartheta = 0\}$. Thus any version of the derivative $\widetilde{V}_\vartheta$ in (i) and (ii) has the property $(*)$.

(3) By step (2) combined with the definition (ii) of the Fréchet derivative, we can modify $\widetilde{V}_\vartheta$ on some ν-null set in $\mathcal{A}$ in order to achieve

$$\widetilde{V}_\vartheta = 0 \quad \text{on } \{\chi_\vartheta = 0\} .$$

This allows to transform $\widetilde{V}_\vartheta$ into a new object

$$V_{\vartheta,i} := 1_{\{\chi_\vartheta > 0\}} \, r \, \chi_\vartheta^{-1/r} \, \widetilde{V}_{\vartheta,i} , \quad 1 \le i \le d$$

which gives (note that with respect to (i) we change the integrating measure from ν to P_ϑ)

$$V_\vartheta \text{ with components } V_{\vartheta,1}, \ldots, V_{\vartheta,d} \in L^r(\Omega, \mathcal{A}, P_\vartheta) ,$$

$$\widetilde{V}_{\vartheta,i} = \frac{1}{r} \chi_\vartheta^{1/r} \, V_{\vartheta,i} , \quad 1 \le i \le d .$$

With V_ϑ thus associated to $\widetilde{V}_\vartheta$, Fréchet differentiability of the mapping $(\diamond)$ takes the following form:

4.1′ Definition (preliminary). For $\Theta \in \mathbb{R}^d$ open, consider a dominated family $\mathcal{P} = \{P_\xi : \xi \in \Theta\}$ of probability measures on $(\Omega, \mathcal{A})$ with ν-densities χ_ξ, $\xi \in \Theta$. Fix $r \ge 1$ and $\vartheta \in \Theta$. The model $\mathcal{P}$ is called L^r-differentiable in ϑ with derivative V_ϑ if the following (i) and (ii) hold simultaneously:

(i) $\qquad V_\vartheta$ has components $V_{\vartheta,1}, \ldots V_{\vartheta,d} \in L^r(\Omega, \mathcal{A}, P_\vartheta)$

(ii) $\qquad \frac{1}{|\xi - \vartheta|^r} \int_\Omega \left| \chi_\xi^{1/r} - \chi_\vartheta^{1/r} - \frac{1}{r} \chi_\vartheta^{1/r} (\xi - \vartheta)^\top V_\vartheta \right|^r d\nu \longrightarrow 0$
as $|\xi - \vartheta| \to 0$.

We remark that the integral in Definition 4.1′(ii) does not depend on the choice of the dominating measure ν. Considering for $\mathcal{P}$ different dominating measures $\nu \ll \widetilde{\nu}$, and densities χ_ϑ (with respect to ν) and $\widetilde{\chi}_\vartheta$ (with respect to $\widetilde{\nu}$), we have $\widetilde{\chi}_\vartheta = \chi_\vartheta \frac{d\nu}{d\widetilde{\nu}}$ $\widetilde{\nu}$-almost surely. Hence, in a representation analogous to Definition 4.1′(ii) with $\widetilde{\chi}_\xi, \widetilde{\chi}_\vartheta$ and $\widetilde{\nu}$, the factor $\frac{d\nu}{d\widetilde{\nu}}$ cancels out. Thus (any version of) the derivative V_ϑ in Definition 4.1′(i) depends only on the law P_ϑ in the family $\mathcal{P}$, and not on the choice of a dominating measure for $\mathcal{P}$.

4.1″ Remark. Fix $r \ge 1$. If a statistical model $\mathcal{P} := \{P_\xi : \xi \in \mathbb{R}^d\}$ is L^r-differentiable in all points $\vartheta \in \Theta$ in the sense of Definition 4.1′, and if in particular

($\circ$) $\quad \Theta \ni \xi \longrightarrow \chi_\xi(\omega) \in [0, \infty) \quad$ is continuous and admits partial derivatives

Section 4.1 L^r-differentiable Statistical Models

for every $\omega \in \Omega$ fixed, we shall show that the components $V_{\vartheta,i}$ of the L^r-derivatives V_ϑ coincide P_ϑ-almost surely with

(oo) $$\omega \longrightarrow 1_{\{\chi_\vartheta > 0\}}(\omega) \left(\frac{\partial}{\partial \vartheta_i} \log \chi\right)(\vartheta, \omega)$$

which was considered in Definition 1.2, for all $1 \leq i \leq d$. We do have $V_{\vartheta,i} \in L^r(P_\vartheta)$ by Definition 4.1'(i), and shall see later in Corollary 4.5 that $V_{\vartheta,i}$ is necessarily centred under P_ϑ. When $r = 2$, (oo) being the classical definition of the score at ϑ (Definition 1.2), L^2-differentiability is a notion of smoothness of parameterisation which extends the classical setting of Chapter 1. We will return to this in Definition 4.6 below.

To check that components $V_{\vartheta,i}$ of L^r-derivatives V_ϑ coincide with (oo) P_ϑ-almost surely, consider in Definition 4.1' sequences $\xi_n = \vartheta + \delta_n e_i$ (with e_i the i-th unit vector in $\mathbb{R}^d$, and $\delta_n \downarrow 0$). From $L^r(\nu)$-convergence in Definition 4.1'(ii), we can select subsequences $(n_k)_k$ along which ν-almost sure convergence holds

$$\left| \frac{\chi_{\vartheta+\delta_{n_k} e_i}^{1/r} - \chi_\vartheta^{1/r}}{\delta_{n_k}} - \frac{1}{r} \chi_\vartheta^{1/r} V_{\vartheta,i} \right| \longrightarrow 0 \quad \text{as} \quad k \to \infty.$$

Thus for the mapping (o), the gradient $\nabla(\chi^{1/r})(\vartheta, \cdot)$ coincides ν-almost surely with $\frac{1}{r} \chi_\vartheta^{1/r} V_\vartheta = \widetilde{V}_\vartheta$. On $\{\chi_\vartheta > 0\}$ we thus have

$$\nabla(\log \chi)(\vartheta, \cdot) = r \nabla(\log(\chi^{1/r}))(\vartheta, \cdot) = r \frac{\nabla(\chi^{1/r})(\vartheta, \cdot)}{\chi^{1/r}(\vartheta, \cdot)} = V_\vartheta \quad \nu\text{-almost surely}$$

whereas on $\{\chi_\vartheta = 0\}$ we put $V_\vartheta \equiv 0 \equiv \widetilde{V}_\vartheta$ according to Assertion 4.1($*$). This gives the representation (oo). □

4.1''' Exercise. Consider the location model $\mathcal{P} := \{F(\cdot - \xi) : \xi \in \mathbb{R}\}$ generated by the doubly exponential distribution $F(dx) = \frac{1}{2} e^{-|x|} dx$ on $(\mathbb{R}, \mathcal{B}(\mathbb{R}))$. Prove that $\mathcal{P}$ is L^2-differentiable in $\xi = \vartheta$, for every $\vartheta \in \mathbb{R}$, with derivative $V_\vartheta = \text{sgn}(\cdot - \vartheta)$.

The next step is to remove the domination assumption from the preliminary Definition 4.1'.

4.2 Definition. For $\Theta \subset \mathbb{R}^d$ open, consider a (not necessarily dominated) family $\mathcal{P} = \{P_\xi : \xi \in \Theta\}$ of probability measures on $(\Omega, \mathcal{A})$. For $\xi', \xi \in \Theta$ let $L^{\xi'/\xi}$ denote a version of the likelihood ratio of $P_{\xi'}$ with respect to P_ξ, as defined in Notation 3.1(i). Fix $r \geq 1$. The model $\mathcal{P}$ is called L^r-*differentiable in* ϑ *with derivative* V_ϑ if the following (i), (iia), (iib) hold simultaneously:

(i) V_ϑ has components $V_{\vartheta,1}, \ldots V_{\vartheta,d} \in L^r(\Omega, \mathcal{A}, P_\vartheta)$,

(iia) $$\frac{1}{|\xi - \vartheta|^r} P_\xi(L^{\xi/\vartheta} = \infty) \longrightarrow 0, \quad |\xi - \vartheta| \to 0,$$

(iib) $\quad \dfrac{1}{|\xi-\vartheta|^r} \displaystyle\int_\Omega \left| (L^{\xi/\vartheta})^{1/r} - 1 - \dfrac{1}{r}(\xi-\vartheta)^\top V_\vartheta \right|^r dP_\vartheta \longrightarrow 0, \quad |\xi-\vartheta| \to 0.$

4.2' Remarks. (a) For arbitrary statistical models $\mathcal{P} = \{P_\xi : \xi \in \Theta\}$ and reference points $\vartheta \in \Theta$, it is sufficient to check (iia) and (iib) along sequences $(\xi_n)_n \subset \Theta$ which converge to ϑ as $n \to \infty$.

(b) Fix any sequence $(\xi_n)_n \subset \Theta$ which converges to ϑ as $n \to \infty$. Then the countable subfamily $\{P_{\xi_n}, n \geq 1, P_\vartheta\}$ is dominated by some σ-finite measure ν (e.g. take $\nu := P_\vartheta + \sum_{n\geq 1} 2^{-n} P_{\xi_n}$) which depends on the sequence under consideration. Let $\chi_{\xi_n}, n \geq 1, \chi_\vartheta$ denote densities of $P_{\xi_n}, n \geq 1, P_\vartheta$ with respect to ν. Along this subfamily, the integral in Definition 4.1'(ii)

$$\int_\Omega \left| \chi_{\xi_n}^{1/r} - \chi_\vartheta^{1/r} - \dfrac{1}{r}\chi_\vartheta^{1/r}(\xi_n-\vartheta)^\top V_\vartheta \right|^r d\nu = \int_{\{\chi_\vartheta=0\}} \cdots d\nu + \int_{\{\chi_\vartheta>0\}} \cdots d\nu$$

splits into the two expressions which appear in (iia) and (iib) of Definition 4.2:

$$\int_{\{\chi_\vartheta=0\}} \chi_{\xi_n} d\nu + \int_{\{\chi_\vartheta>0\}} \left| \left(\dfrac{\chi_{\xi_n}}{\chi_\vartheta}\right)^{1/r} - 1 - \dfrac{1}{r}(\xi_n-\vartheta)^\top V_\vartheta \right|^r \chi_\vartheta d\nu$$

$$= P_{\xi_n}\left(L^{\xi_n/\vartheta} = \infty\right) + \int_\Omega \left| (L^{\xi_n/\vartheta})^{1/r} - 1 - \dfrac{1}{r}(\xi_n-\vartheta)^\top V_\vartheta \right|^r dP_\vartheta.$$

Thus the preliminary Definition 4.1' and the final Definition 4.2 are equivalent in restriction to the subfamily $\{P_{\xi_n}, n \geq 1, P_\vartheta\}$. Since we can consider arbitrary sequences $(\xi_n)_n$ converging to ϑ, we have proved that Definition 4.2 extends Definition 4.1' consistently. $\square$

4.3 Example. On an arbitrary space $(\Omega, \mathcal{A})$ consider $\mathcal{P} := \mathcal{M}^1(\Omega, \mathcal{A})$, the set of all probability measures on $(\Omega, \mathcal{A})$. Fix $r \geq 1$ and $P \in \mathcal{P}$. In the non-parametric model $\mathcal{E} = (\Omega, \mathcal{A}, \mathcal{P})$, we shall show that one-parametric paths $\{Q_\vartheta : |\vartheta| < \varepsilon\}$ through P in directions

$$g \in L^r(P) \quad \text{such that} \quad \int_\Omega g\, dP = 0$$

(defined in analogy to Example 1.3) are L^r-differentiable at all points of the parameter interval, and have L^r-derivative $V_0 = g$ at the origin $Q_0 = P$.

(a) Consider first the case of directions g which are bounded, and write $M := \sup |g|$. Then the one-dimensional path through P in $\mathcal{P}$

(*) $\quad \mathcal{E}_P^g := \left(\Omega, \mathcal{A}, \left\{Q_\vartheta : |\vartheta| < \dfrac{1}{M}\right\}\right), \quad dQ_\vartheta := (1 + \vartheta g) dP$

is L^r-differentiable at every parameter value ϑ, $|\vartheta| < \dfrac{1}{M}$, and the derivative V_ϑ at ϑ is given by

$$V_\vartheta = \dfrac{g}{1 + \vartheta g} \in L^r(\Omega, \mathcal{A}, Q_\vartheta)$$

Section 4.1 L^r-differentiable Statistical Models

(in analogy to expression (1.3'), and in agreement with Remark 4.1"). At $\vartheta = 0$ we have simply $V_0 = g$.

We check this as follows. From $|g| \leq M$ and $|\vartheta| < \frac{1}{M}$, V_ϑ is bounded. Thus $V_\vartheta \in L^r(\Omega, \mathcal{A}, Q_\vartheta)$ holds for all ϑ in (*) and all $r \geq 1$. This is condition (i) in Definition 4.2. Since $\mathcal{S}_P^g$ is dominated by $\nu := P$, conditions (iia) and (iib) of Definition 4.2 are equivalent to Condition 4.1'(ii) which we shall check now. For all $r \geq 1$,

$$\frac{(1+\xi g)^{1/r} - (1+\vartheta g)^{1/r}}{\xi - \vartheta} = \frac{1}{r}(1+\zeta g)^{\frac{1}{r}-1} g = \frac{1}{r}(1+\zeta g)^{1/r} \frac{g}{1+\zeta g}$$

for every ω fixed, with some ζ (depending on ω) between ξ and ϑ. For ε small enough and $\zeta \in B_\varepsilon(\vartheta)$, $|\zeta|$ remains separated from $\frac{1}{M}$: thus dominated convergence under P_ϑ as $\xi \to \vartheta$ in the above line gives

$$\int_\Omega \left| \frac{(1+\xi g)^{1/r} - (1+\vartheta g)^{1/r}}{\xi - \vartheta} - \frac{1}{r}(1+\vartheta g)^{1/r} \frac{g}{1+\vartheta g} \right|^r d\nu \to 0$$

as $|\xi - \vartheta| \to 0$

which is Condition 4.1'(ii).

(b) Consider now arbitrary directions $g \in L^r(P)$ with $E_P(g) = 0$. As in the second part of Example 1.3, use of truncation avoids boundedness assumptions. With $\psi \in \mathcal{C}_0^1(\mathbb{R})$ as there

$$\psi(x) = x \quad \text{on } \{|x| < \frac{1}{3}\}, \quad \psi(x) = 0 \quad \text{on } \{|x| > 1\}, \quad \max_{x \in \mathbb{R}} |\psi| < \frac{1}{2}$$

we define

(**) $\mathcal{E} = (\Omega, \mathcal{A}, \{Q_\vartheta : |\vartheta| < 1\})$,
    $Q_\vartheta(d\omega) := \left(1 + [\psi(\vartheta g(\omega)) - \int \psi(\vartheta g) dP]\right) P(d\omega)$.

In the special case of bounded g as in (a), one-parametric paths (**) and (*) coincide for parameter values ϑ close to 0. Uniformly in (ϑ, ω), densities

$$f(\vartheta, \omega) := 1 + [\psi(\vartheta g(\omega)) - \int \psi(\vartheta g) dP]$$

in (**) are bounded away from both 0 and 2, by choice of ψ. Since ψ is Lipschitz, dominated convergence gives

$$\frac{d}{d\vartheta} \int \psi(\vartheta g) dP = \int g \psi'(\vartheta g) dP.$$

First, with $\widetilde{V}_\vartheta$ defined from the mapping ($\diamond$) in 4.1, our assumptions on ψ and g imply that

$$\widetilde{V}_\vartheta = \frac{d}{d\vartheta} f_\vartheta^{1/r} = \frac{1}{r} f_\vartheta^{(\frac{1}{r}-1)} \frac{d}{d\vartheta} f_\vartheta = \frac{1}{r} f_\vartheta^{(\frac{1}{r}-1)} \left[g\psi'(\vartheta g) - \int g\psi'(\vartheta g) dP \right]$$

belongs to $L^r(\Omega, \mathcal{A}, P)$ (since f_ϑ is bounded away from 0; ψ' is bounded), or equivalently that

$$V_\vartheta = r f_\vartheta^{-1/r} \widetilde{V}_\vartheta = \frac{g\psi'(\vartheta g) - \int g\psi'(\vartheta g)\,dP}{1 + [\psi(\vartheta g(\omega)) - \int \psi(\vartheta g)dP]} = \frac{\partial}{\partial \vartheta} \log f(\vartheta, \cdot)$$

belongs to $L^r(\Omega, \mathcal{A}, Q_\vartheta)$, in analogy to Example 1.3(b). Thus condition (i) in Definition 4.1' is checked. Second, exploiting again our assumptions on ψ and g to work with dominated convergence, we obtain

$$\frac{1}{|\xi - \vartheta|^r} \int_\Omega \left| f_\xi^{1/r} - f_\vartheta^{1/r} - (\xi - \vartheta) \widetilde{V}_\vartheta \right|^r dP \longrightarrow 0 \text{ as } |\xi - \vartheta| \to 0.$$

This establishes condition (ii) in Definition 4.1'. According to Definition 4.1', we have proved L^r-differentiability in the one-parametric path $\mathcal{E}$ at every parameter value $|\vartheta| < 1$. In particular, the derivative at $\vartheta = 0$ is

(o) $\qquad\qquad\qquad V_0(\omega) = g(\omega), \quad \omega \in \Omega$

which does not depend on the particular construction – through choice of ψ – of the path $\mathcal{E}$ through P. Note that the last assertion (o) holds for arbitrary $P \in \mathcal{P}$ and for arbitrary directions $g \in L^r(P)$ such that $E_P(g) = 0$. $\square$

4.4 Theorem. Consider a family $\mathcal{P} = \{P_\xi : \xi \in \Theta\}$ where $\Theta \subset \mathbb{R}^d$ is open. If $\mathcal{P}$ is L^r-differentiable at $\xi = \vartheta$ with derivative V_ϑ for some $r > 1$, then $\mathcal{P}$ is L^s-differentiable at $\xi = \vartheta$ with same derivative V_ϑ for all $1 \le s \le r$.

Proof. Our proof follows Witting [127, pp. 175–176]
(1) Fix $1 \le s < r$, put $t := \frac{r}{r-s}$ (such that $t > 1$, and $(\frac{1}{r/s}) + \frac{1}{t} = 1$ or $\frac{1}{ts} = \frac{1}{s} - \frac{1}{r}$) and define functions φ, ψ from $[0, \infty)$ to $(0, \infty)$ by

$$\varphi(y) := \frac{s}{r} \frac{y^{1/s} - 1}{y^{1/r} - 1}, \quad \psi(y) := \varphi^{\frac{rs}{r-s}}(y) = \varphi^{ts}(y), \quad y \ge 0.$$

Check that φ is continuous at $y = 1$ (with $\varphi(1) = 1$), and thus continuous on $[0, \infty)$. For

$$y > \max\left\{\left(\frac{r}{r-s}\right)^r, 1\right\}$$

we have the inclusions

$$y^{1/r} > \frac{r}{r-s} \implies \left(\frac{r}{s} - 1\right) y^{1/r} > \frac{r}{s} \implies \frac{r}{s}(y^{1/r} - 1) > y^{1/r}$$

and thus by definition of t

$$\psi(y) = \varphi^{ts}(y) = \left(\frac{y^{1/s} - 1}{\frac{r}{s}(y^{1/r} - 1)}\right)^{ts} < \left(\frac{y^{1/s} - 1}{y^{1/r}}\right)^{ts} < \left(y^{\frac{1}{s} - \frac{1}{r}}\right)^{ts} = y.$$

Section 4.1 L^r-differentiable Statistical Models

From this and continuity of ψ on $[0, \infty)$, the function ψ satisfies a linear growth condition

(○) $\qquad 0 \leq \psi(y) = \varphi^{ts}(y) \leq M(1+y) \quad \text{for all } y \geq 0$

for some constant M.

(2) We exploit the assumption of L^r-differentiability at ϑ. For sequences $(\xi_m)_m$ converging to ϑ in Θ, consider likelihood ratios $L^{\xi_m/\vartheta}$ of P_{ξ_m} with respect to P_ϑ. As a consequence of (i) in Definition 4.2, the mapping $u \to \|u^\top V_\vartheta\|_{L^r(P_\vartheta)}$ is continuous on $\mathbb{R}^d$, hence bounded on the unit sphere. Thus $(\xi_m - \vartheta)^\top V_\vartheta$ vanishes in $L^r(P_\vartheta)$. From (iib) of Definition 4.2 and inverse triangular inequality

$$\left| \left\| (L^{\xi_m/\vartheta})^{1/r} - 1 \right\|_{L^r(P_\vartheta)} - \left\| \frac{1}{r} (\xi_m - \vartheta)^\top V_\vartheta \right\|_{L^r(P_\vartheta)} \right| \leq o(|\xi_m - \vartheta|)$$

and both last assertions together establish

(◇) $\qquad (L^{\xi_m/\vartheta})^{1/r} \longrightarrow 1 \quad \text{in } L^r(P_\vartheta) \text{ as } m \to \infty.$

First, (◇) implying convergence in probability, we have

(+) $\qquad g(L^{\xi_m/\vartheta}) \longrightarrow 1 \quad \text{in } P_\vartheta\text{-probability}$

for any continuous function $g : [0, \infty) \to \mathbb{R}$ with $g(1) = 1$; second, (◇) gives

(++) $\qquad E_\vartheta(L^{\xi_m/\vartheta}) = E_\vartheta\left([(L^{\xi_m/\vartheta})^{1/r}]^r \right) \longrightarrow 1 \quad \text{as } m \to \infty.$

(3) We wish to prove

(◇◇) $\qquad \varphi(L^{\xi_m/\vartheta}) \longrightarrow 1 \quad \text{in } L^{st}(P_\vartheta) \text{ as } m \to \infty.$

Fix $q \geq 1$. Recall that for variables $f_m \in L^q(P_\vartheta)$ converging to some random variable f in P_ϑ-probability as $m \to \infty$, we have

$$\{|f_m|^q : m \geq 1\} \text{ uniformly integrable under } P_\vartheta$$
$$\iff f \in L^q(P_\vartheta) \text{ and } f_m \longrightarrow f \text{ in } L^q(P_\vartheta);$$

in the special case where $q = 1$ and $f_m \geq 0$ for all m, we can extend this to a chain of equivalences

$$\iff f \in L^1(P_\vartheta) \text{ and } E_\vartheta(f_m) \longrightarrow E_\vartheta(f) \text{ as } m \to \infty;$$

for the last equivalence, see [20, Nr. 21 in Chapter II]). In the case where $q = 1$, based on (+) and (++) in step (2), the above equivalences applied to $f \equiv 1$ and $f_m = L^{\xi_m/\vartheta}$ establish

$$\{L^{\xi_m/\vartheta} : m \geq 1\} \text{ uniformly integrable under } P_\vartheta$$

and thus, thanks to the linear growth condition (○) for the function $\psi = \varphi^{st} \geq 0$ in step (1),

$$\{\varphi^{st}(L^{\xi_m/\vartheta}) : m \geq 1\} \text{ uniformly integrable under } P_\vartheta.$$

Finally, based on convergence in probability (+), we put $q = st$ and apply the first of the equivalences above to $f \equiv 1$ and $f_m = \varphi(L^{\xi_m/\vartheta})$: this establishes (⋄⋄).

(4) With these preparations, we come to the core of the proof. Regarding first (i) and (iia) in Definition 4.2, it is clear that these conditions are valid for $1 \le s \le r$ when they are valid for $r > 1$. We turn to condition (iib) in Definition 4.2. By definition of the function φ in step (1), we can upper-bound the expression

(×) $$\left\| s\left\{(L^{\xi/\vartheta})^{1/s} - 1\right\} - (\xi - \vartheta)^\top V_\vartheta \right\|_{L^s(P_\vartheta)}$$

in a first step (using triangular inequality) by

$$\left\| \left[r\left\{(L^{\xi/\vartheta})^{1/r} - 1\right\} - (\xi - \vartheta)^\top V_\vartheta\right] \varphi(L^{\xi/\vartheta}) \right\|_{L^s(P_\vartheta)}$$
$$+ \left\| (\xi - \vartheta)^\top V_\vartheta \left[\varphi(L^{\xi/\vartheta}) - 1\right] \right\|_{L^s(P_\vartheta)}$$

and in a second step using Hölder inequality (we have $(\frac{1}{r/s}) + \frac{1}{t} = 1$) by

$$\left\| r\left\{(L^{\xi/\vartheta})^{1/r} - 1\right\} - (\xi - \vartheta)^\top V_\vartheta \right\|_{L^r(P_\vartheta)} \cdot \left\| \varphi(L^{\xi/\vartheta}) \right\|_{L^{ts}(P_\vartheta)}$$
$$+ \left\| (\xi - \vartheta)^\top V_\vartheta \right\|_{L^r(P_\vartheta)} \cdot \left\| \varphi(L^{\xi/\vartheta}) - 1 \right\|_{L^{ts}(P_\vartheta)}.$$

In this last right-hand side, as ξ tends to ϑ, the first product is of order

$$o(|\xi - \vartheta|) \cdot \left\| \varphi(L^{\xi/\vartheta}) \right\|_{L^{ts}(P_\vartheta)} = o(|\xi - \vartheta|)$$

and the second of order

$$|\xi - \vartheta| \cdot \left\| \varphi(L^{\xi/\vartheta}) - 1 \right\|_{L^{ts}(P_\vartheta)} = o(|\xi - \vartheta|)$$

by L^r-differentiability at ϑ – we use (iib) of Definition 4.2 in the first case, and $\max_{|u|=1} \|u^\top V_\vartheta\|_{L^r(P_\vartheta)} < \infty$ in the second case – in combination with L^{ts}-convergence (⋄⋄). Summarising, we have proved that (×) is of order $o(|\xi - \vartheta|)$ as $\xi \to \vartheta$: thus condition (iib) in Definition 4.2 holds for $1 \le s < r$ if it holds for $r > 1$. □

As a consequence, we now prove that derivatives V_ϑ arising in Definitions 4.1' or 4.2 are always centred:

4.5 Corollary. Consider a family $\mathcal{P} = \{P_\xi : \xi \in \Theta\}$ where $\Theta \subset \mathbb{R}^d$ is open, fix $r \ge 1$. If $\mathcal{P}$ is L^r-differentiable at $\xi = \vartheta$ with derivative V_ϑ, then $E_\vartheta(V_\vartheta) = 0$.

Proof. From Theorem 4.4, $\mathcal{P}$ is L^1-differentiable at $\xi = \vartheta$ with same derivative V_ϑ. Consider a unit vector e_i in $\mathbb{R}^d$, a sequence $\delta_n \downarrow 0$, a dominating measure ν for the countable family $\{P_{\vartheta + \delta_n e_i}, n \ge 1, P_\vartheta\}$, and associated densities. Then Definition 4.1'

Section 4.1 L^r-differentiable Statistical Models

with $r = 1$ gives

$$\frac{\chi_{\vartheta+\delta_n e_i} - \chi_\vartheta}{\delta_n} - \chi_\vartheta V_{\vartheta,i} \longrightarrow 0 \quad \text{in } L^1(\nu) \text{ as } n \to \infty$$

and thus $E_{P_\vartheta}(V_{\vartheta,i}) = \int \chi_\vartheta V_{\vartheta,i} \, d\nu = 0$. This holds for $1 \le i \le d$. □

From now on we focus on the case $r = 2$ which is of particular importance. Assuming L^2-differentiability at all points $\vartheta \in \Theta$, the derivatives have components

(∗) $\qquad V_{\vartheta,i} \in L^2(P_\vartheta) \quad \text{such that} \quad E_{P_\vartheta}(V_{\vartheta,i}) = 0, \quad 1 \le i \le d$

by Definition 4.2(i) and Corollary 4.5. In the special case of dominated models admitting continuous densities for which partial derivatives exist, V_ϑ necessarily has all properties of the score considered in Definition 1.2, cf. Remark 4.1". In this sense, the present setting generalises Definition 1.2 and allows to transfer notions such as score or information to L^2-differentiable statistical models.

4.6 Definition. Consider $\mathcal{P} = \{P_\xi : \xi \in \Theta\}$ where $\Theta \subset \mathbb{R}^d$ is open. If a point $\vartheta \in \Theta$ is such that

(⋄) $\qquad$ the family $\mathcal{P}$ is L^2-differentiable at ϑ with derivative V_ϑ ,

we call V_ϑ the *score* and

$$J_\vartheta := E_\vartheta(V_\vartheta V_\vartheta^\top)$$

the *Fisher information* at ϑ. The family $\mathcal{P}$ is called L^2-differentiable if (⋄) holds for all $\vartheta \in \Theta$.

In the light of Definition 4.2, Hellinger distance $H(\cdot, \cdot)$ between probability measures

$$H^2(Q_1, Q_2) = \frac{1}{2} \int \left| \chi_1^{1/2} - \chi_2^{1/2} \right|^2 d\nu \in [0, 1]$$

on $(\Omega, \mathcal{A})$ (as in Definition 1.18, not depending on the choice of the σ-finite measure ν which dominates Q_1 and Q_2, and not on versions of the densities $\chi_i = \frac{dQ_i}{d\nu}, i = 1, 2$) gives the geometry of experiments which are L^2-differentiable. This will be seen in Proposition 4.8 below.

4.7 Proposition. Without further assumptions on a statistical model $\mathcal{P} = \{P_\xi : \xi \in \Theta\}$, we have

$$2 H^2(P_{\xi'}, P_\xi) = E_\xi \left(\left(\sqrt{L^{\xi'/\xi}} - 1 \right)^2 \right) + P_{\xi'}(L^{\xi'/\xi} = \infty)$$

for all ξ, ξ' in Θ, and

$$E_\xi(\sqrt{L^{\xi'/\xi}} - 1) = -H^2(P_{\xi'}, P_\xi) \,.$$

Proof. By Notations 3.1, for any σ-finite measure ν on $(\Omega, \mathcal{A})$ which dominates $P_{\xi'}$ and P_ξ and for any choice $\chi_{\xi'}$, χ_ξ of ν-densities, $L^{\xi'/\xi}$ coincides $(P_\xi + P_{\xi'})$-almost surely with $\frac{\chi_{\xi'}}{\chi_\xi} 1_{\{\chi_\xi > 0\}} + \infty 1_{\{\chi_\xi = 0\}}$. As in step (1) of the proof of Proposition 3.10,

$$E_\xi(1 - L^{\xi'/\xi}) = P_{\xi'}(\{\chi_\xi = 0\}) = P_{\xi'}(L^{\xi'/\xi} = \infty)$$

is the total mass of the P_ξ-singular part of $P_{\xi'}$. Now the decomposition

$$\int_\Omega \left(\sqrt{\chi_{\xi'}} - \sqrt{\chi_\xi}\right)^2 d\nu = \int_{\{\chi_\xi > 0\}} \left(\sqrt{\chi_{\xi'}} - \sqrt{\chi_\xi}\right)^2 d\nu + P_{\xi'}(\chi_\xi = 0)$$

yields the first assertion. Since $\sqrt{L^{\xi'/\xi}} \in L^2(P_\xi)$, all expectations in

$(+) \qquad E_\xi\left(\sqrt{L^{\xi'/\xi}} - 1\right) = \frac{1}{2} E_\xi(L^{\xi'/\xi} - 1) - \frac{1}{2} E_\xi\left(\left(\sqrt{L^{\xi'/\xi}} - 1\right)^2\right)$

are well defined, where equality in $(+)$ is the elementary $\sqrt{a} - 1 = \frac{1}{2}(a - 1) - \frac{1}{2}(\sqrt{a} - 1)^2$ for $a \geq 0$. Taking together the preceding three lines yields the second assertion. $\square$

4.8 Proposition. In a model $\mathcal{P} = \{P_\xi : \xi \in \Theta\}$ where $\Theta \subset \mathbb{R}^d$ is open, consider sequences $(\xi_m)_m$ which approach ϑ from direction $u \in S_{d-1}$ (S_{d-1} the unit sphere in $\mathbb{R}^d$) at arbitrary rate $(\delta_m)_m$:

$$\delta_m := |\xi_m - \vartheta| \to 0, \qquad u_m := \frac{\xi_m - \vartheta}{|\xi_m - \vartheta|} \longrightarrow u, \qquad m \to \infty.$$

If $\mathcal{P}$ is L^2-differentiable in $\xi = \vartheta$ with derivative V_ϑ, then (with notation $a_n \sim b_n$ for $\lim_{n \to \infty} \frac{a_n}{b_n} = 1$)

$$2 H^2\left(P_{\xi_m}, P_\vartheta\right) \sim E_\vartheta\left(\left(\sqrt{L^{\xi_m/\vartheta}} - 1\right)^2\right) \sim \delta_m^2 \frac{1}{4} u^\top J_\vartheta u$$

as $m \to \infty$ whenever $u^\top J_\vartheta u$ is strictly positive.

Proof. With respect to a given sequence $\xi_m \to \vartheta$ satisfying the assumptions of the proposition, define a dominating measure ν and densities χ_{ξ_m}, χ_ϑ as in Remark 4.2'(b). By L^2-differentiability at ϑ we approximate first

$$\frac{\sqrt{\chi_{\xi_m}} - \sqrt{\chi_\vartheta}}{|\xi_m - \vartheta|} \quad \text{by} \quad \frac{1}{2} \sqrt{\chi_\vartheta}\, u_m^\top V_\vartheta \quad \text{in } L^2(\nu)$$

as $m \to \infty$, using Definition 4.2(iib), and then

$$\frac{1}{2} \sqrt{\chi_\vartheta}\, u_m^\top V_\vartheta \quad \text{by} \quad \frac{1}{2} \sqrt{\chi_\vartheta}\, u^\top V_\vartheta \quad \text{in } L^2(\nu)$$

since $(\xi_m)_m$ approaches ϑ from direction u. This gives

$$\frac{2}{\delta_m^2} H^2\left(P_{\xi_m}, P_\vartheta\right) = \int_\Omega \left(\frac{\sqrt{\chi_{\xi_m}} - \sqrt{\chi_\vartheta}}{|\xi_m - \vartheta|}\right)^2 dv$$
$$\longrightarrow \frac{1}{4} E_\vartheta\left([u^\top V_\vartheta]^2\right) = \frac{1}{4} u^\top J_\vartheta u$$

as $m \to \infty$, by Definition 4.6 of the Fisher information. L^2-differentiability at ϑ also guarantees

$$P_{\xi_m}\left(L^{\xi_m/\vartheta} = \infty\right) = o\left(\delta_m^2\right)$$

as $m \to \infty$, from Definition 4.2(iia), thus the first assertion of Proposition 4.7 completes the proof. □

The sphere S_{d-1} being compact in $\mathbb{R}^d$, an arbitrary sequence $(\xi_n)_n \subset \Theta$ converging to ϑ contains subsequences $(\xi_{n_m})_m$ which approach ϑ from some direction $u \in S_{d-1}$, as required in Proposition 4.8.

4.9 Remark. By Definition 4.6, in an L^2-differentiable experiment, the Fisher information J_ϑ at ϑ is symmetric and non-negative definite. If J_ϑ admits an eigenvalue 0, for a corresponding eigenvector u and for sequences $(\xi_m)_m$ approaching ϑ from direction u, the Hellinger distance will be unable to distinguish ξ_m from ϑ at rate $|\xi_m - \vartheta|$ as $m \to \infty$, and the proof of the last proposition only gives

$$H^2\left(P_{\xi_m}, P_\vartheta\right) = o\left(|\xi_m - \vartheta|^2\right), \quad E_\vartheta\left((\sqrt{L^{\xi_m/\vartheta}} - 1)^2\right) = o\left(|\xi_m - \vartheta|^2\right)$$

as $m \to \infty$. This explains the crucial role of assumptions on strictly positive definite Fisher information in an experiment. Only in this last case, the geometry of the experiment in terms of Hellinger distance is locally at ϑ equivalent – up to some deformation expressed by J_ϑ – to Euclidean geometry on the parameter space Θ as a subset of $\mathbb{R}^d$. □

4.2 Le Cam's Second Lemma for i.i.d. Observations

Under independent replication of experiments, L^2-differentiability of a statistical model $\mathcal{E}$ at a point ϑ makes log-likelihood ratios in product models $\mathcal{E}_n$ resemble – locally in small neighbourhoods of ϑ which we reparameterise via $\vartheta + h/\sqrt{n}$ in terms of a local parameter h – the log-likelihoods in the Gaussian shift model

$$\{\mathcal{N}(J_\vartheta h, J_\vartheta) : h \in \mathbb{R}^d\}$$

where J_ϑ is the Fisher information at ϑ in $\mathcal{E}$. This goes back to Le Cam, see [81]; since Hájek and Sidák [41], results which establish an approximation of this type are called a 'second Le Cam lemma'. From the next chapter on, we will exploit similar approximations of log-likelihood ratios. Beyond the i.i.d. setting, such approximations do exist in a large variety of contexts where a statistical model is smoothly parameterised and where strong laws of large numbers and central limit theorems are at hand (e.g. autoregressive processes, ergodic diffusions, ergodic Markov processes, ...).

4.9' Exercise. For $J \in \mathbb{R}^{d \times d}$ symmetric and strictly positive definite, calculate log-likelihood ratios in the experiment $\{\mathcal{N}(Jh, J) : h \in \mathbb{R}^d\}$ and compare to the expansion given in 4.11 below.

Throughout this section the following assumptions will be in force.

4.10 Assumptions and Notations for Section 4.2. (a) We work with an experiment

$$\mathcal{E} = (\Omega, \mathcal{A}, \mathcal{P} = \{P_\xi : \xi \in \Theta\}), \quad \Theta \subset \mathbb{R}^d \text{ open}$$

with likelihood ratios $L^{\xi'/\xi}$ of $P_{\xi'}$ with respect to P_ξ. Notation ϑ for a point in the parameter space implies that the following is satisfied:

$$\mathcal{P} = \{P_\xi : \xi \in \Theta\} \text{ is } L^2\text{-differentiable at } \xi = \vartheta \text{ with derivative } V_\vartheta.$$

In this case, we write J_ϑ for the Fisher information at ϑ in $\mathcal{E}$, cf. Definition 4.6:

$$J_\vartheta = E_\vartheta \left(V_\vartheta V_\vartheta^\top \right).$$

(b) In n-fold product experiments

$$\mathcal{E}_n = \left(\Omega_n := \bigtimes_{i=1}^n \Omega, \; \mathcal{A}_n := \bigotimes_{i=1}^n \mathcal{A}, \; \left\{ P_\xi^n := \bigotimes_{i=1}^n P_\xi : \xi \in \Theta \right\} \right),$$

$L_n^{\xi'/\xi}$ and $\Lambda_n^{\xi'/\xi}$ denote the likelihood ratio and the log-likelihood ratio of $P_{\xi'}^n$ with respect to P_ξ^n. At points $\vartheta \in \Theta$ where $\mathcal{E}$ is L^2-differentiable, cf. (a), we write

$$S_n(\vartheta)(\omega_1, \ldots, \omega_n) := \frac{1}{\sqrt{n}} \sum_{j=1}^n V_\vartheta(\omega_j), \quad \omega = (\omega_1, \ldots, \omega_n) \in \Omega_n$$

and have from the Definition 4.6 (which includes Corollary 4.5) and the central limit theorem

$$\mathcal{L}\left(S_n(\vartheta) \mid P_\vartheta^n \right) \longrightarrow \mathcal{N}(0, J_\vartheta) \quad \text{(weak convergence in } \mathbb{R}^d\text{) as } n \to \infty.$$

Here is the main result of this section:

4.11 Le Cam's Second Lemma for i.i.d. Observations. At points $\vartheta \in \Theta$ which satisfy the Assumptions 4.10, for bounded sequences $(h_n)_n$ in $\mathbb{R}^d$ (such that $\vartheta + h_n/\sqrt{n}$ is in Θ), we have

$$\Lambda_n^{(\vartheta + h_n/\sqrt{n})/\vartheta} = h_n^\top S_n(\vartheta) - \frac{1}{2} h_n^\top J_\vartheta h_n + o_{(P_\vartheta^n)}(1), \quad n \to \infty$$

where $\mathcal{L}(S_n(\vartheta) \mid P_\vartheta^n)$ converges weakly in $\mathbb{R}^d$ as $n \to \infty$ to $\mathcal{N}(0, J_\vartheta)$.

The proof of Lemma 4.11 will be given in 4.15, after a series of auxiliary results.

4.12 Proposition. For bounded sequences $(h_n)_n$ in $\mathbb{R}^d$ and points $\xi_n := \vartheta + h_n/\sqrt{n}$ in Θ, we have

$$\sum_{j=1}^n \left(\sqrt{L^{\xi_n/\vartheta}}(\omega_j) - 1 \right) = \frac{1}{2} \frac{1}{\sqrt{n}} \sum_{j=1}^n h_n^\top V_\vartheta(\omega_j) - \frac{1}{8} h_n^\top J_\vartheta h_n + o_{(P_\vartheta^n)}(1), \quad n \to \infty.$$

Proof. (1) Fix ϑ satisfying Assumptions 4.10, and define for $0 \neq h \in \mathbb{R}^d$

$$r_\vartheta(n, h) := \frac{1}{|h|/\sqrt{n}} \left[\sqrt{L^{(\vartheta + h/\sqrt{n})/\vartheta}} - 1 - \frac{1}{2} \frac{1}{\sqrt{n}} h^\top V_\vartheta \right]$$

in the experiment $\mathcal{E}$. Then for any choice of a constant $C < \infty$, L^2-differentiability at ϑ implies

(⋄) $$\sup_{0 < |h| \leq C} E_\vartheta \left([r_\vartheta(n, h)]^2 \right) \longrightarrow 0 \quad \text{as } n \to \infty:$$

to see this, select in $\{|h| \leq C\}$ at stage n some h_n such that $E_\vartheta([r_\vartheta(n, h_n)]^2)$ is sufficiently close to the left-hand side of (⋄), and apply part (iib) of Definition 4.2 to the sequence $\xi_n := \vartheta + h_n/\sqrt{n}$ as $n \to \infty$.

(2) For $n \geq 1$, consider in $\mathcal{E}_n$

(∗)
$$A_{n,h}(\omega) := \sum_{j=1}^n \left[\sqrt{L^{(\vartheta + h/\sqrt{n})/\vartheta}} - 1 - \frac{1}{2} \frac{1}{\sqrt{n}} h^\top V_\vartheta \right](\omega_j)$$

$$= \frac{|h|}{\sqrt{n}} \sum_{j=1}^n r_\vartheta(n, h)(\omega_j)$$

for $0 \neq h \in \mathbb{R}^d$. From the second assertion in Proposition 4.7 combined with Corollary 4.5 we know

$$E_\vartheta(A_{n,h}) = n \, E_\vartheta \left[\sqrt{L^{(\vartheta + h/\sqrt{n})/\vartheta}} - 1 - \frac{1}{2} \frac{1}{\sqrt{n}} h^\top V_\vartheta \right] = -n \, H^2(P_{\vartheta + h/\sqrt{n}}, P_\vartheta).$$

Consider now a bounded sequence $(h_n)_n$ in $\mathbb{R}^d$. The unit sphere $\mathcal{S}^{d-1}$ being compact, we can select subsequences $(h_{n_k})_k$ whose directions $u_k := \frac{h_{n_k}}{|h_{n_k}|}$ approach limits $u \in \mathcal{S}^{d-1}$. Thus we deduce from Proposition 4.8

$(**)$ $\qquad E_\vartheta(A_{n,h_n}) = -n\, H^2(P_{\vartheta+h_n/\sqrt{n}}, P_\vartheta) = -\frac{1}{8} h_n^\top J_\vartheta\, h_n + o(1)$

as $n \to \infty$. On the other hand, calculating variances for $(*)$ we find from $(\diamond)$

$$\sup_{|h|\leq C} \mathrm{Var}_\vartheta(A_{n,h}) \leq C^2 \sup_{|h|\leq C} \mathrm{Var}_\vartheta(r_\vartheta(n,h))$$
$$\leq C^2 \sup_{|h|\leq C} E_\vartheta\big([r_\vartheta(n,h)]^2\big) \longrightarrow 0$$

as $n \to \infty$. Thus the sequence $(*)$ behaves as the sequence of its expectations $(**)$:

$$A_{n,h_n} = E_\vartheta(A_{n,h_n}) + o_{(P_\vartheta^n)}(1) = -\frac{1}{8} h_n^\top J_\vartheta\, h_n + o_{(P_\vartheta^n)}(1)$$

as $n \to \infty$ which is the assertion. $\qquad\square$

4.13 Proposition. For bounded sequences $(h_n)_n$ in $\mathbb{R}^d$ and points $\xi_n := \vartheta + h_n/\sqrt{n}$, we have

$$\sum_{j=1}^n \left(\sqrt{L^{\xi_n/\vartheta}(\omega_j)} - 1\right)^2 = \frac{1}{4} h_n^\top J_\vartheta\, h_n + o_{(P_\vartheta^n)}(1), \quad n \to \infty.$$

Proof. We write $\rho_n(\omega), \widetilde{\rho}_n(\omega), \ldots$ for remainder terms $o_{(P_\vartheta^n)}(1)$ defined in $\mathcal{E}_n$. By Definition 4.6 of the Fisher information and the strong law of large numbers,

$$\frac{1}{n}\sum_{j=1}^n (V_\vartheta\, V_\vartheta^\top)(\omega_j) = E_\vartheta(V_\vartheta\, V_\vartheta^\top) + \rho_n(\omega) = J_\vartheta + \rho_n(\omega)$$

as $n \to \infty$. Thus we have for bounded sequences $(h_n)_n$

$(+)$ $\qquad \displaystyle\sum_{j=1}^n \left(\frac{1}{\sqrt{n}} h_n^\top V_\vartheta(\omega_j)\right)^2 = h_n^\top J_\vartheta\, h_n + o_{(P_\vartheta^n)}(1), \quad n \to \infty.$

With notation $r_\vartheta(n,h)$ introduced at the start of the proof of Proposition 4.12 we write

$(++)$ $\qquad \sqrt{L^{\xi_n/\vartheta}(\omega_j)} - 1 = \dfrac{1}{2}\dfrac{1}{\sqrt{n}} h_n^\top V_\vartheta(\omega_j) + \dfrac{|h_n|}{\sqrt{n}} r_\vartheta(n,h_n)(\omega_j).$

As $n \to \infty$, we take squares of both the right-hand sides and the left-hand sides in $(++)$, and sum over $1 \leq j \leq n$. Thanks to $(\diamond)$ in the proof of Proposition 4.12

$$\sup_{|h|\leq C} E_\vartheta\big(r_\vartheta(n,h)^2\big) \longrightarrow 0 \quad \text{as } n \to \infty$$

Section 4.2 Le Cam's Second Lemma for i.i.d. Observations

and Cauchy–Schwarz inequality, the terms

$$\frac{1}{n}\sum_{j=1}^{n}[r_\vartheta(n,h_n)]^2(\omega_j) \quad \text{and} \quad \frac{1}{n}\sum_{j=1}^{n}\left|\frac{h_n^\top}{|h_n|}V_\vartheta(\omega_j)\,r_\vartheta(n,h_n)(\omega_j)\right|$$

then vanish as $n \to \infty$ on the right-hand sides, and we obtain

$$\sum_{j=1}^{n}\left(\sqrt{L^{\xi_n/\vartheta}}(\omega_j) - 1\right)^2 = \sum_{j=1}^{n}\left(\frac{1}{2}\frac{1}{\sqrt{n}}h_n^\top V_\vartheta(\omega_j)\right)^2 + \tilde{p}_n(\omega)$$

$$= \frac{1}{4}h_n^\top J_\vartheta h_n + o_{(P_\vartheta^n)}(1)$$

as a consequence of (++), ($\diamond$) in the proof of Proposition 4.12, and (+).   □

4.14 Proposition. For bounded sequences $(h_n)_n$ in $\mathbb{R}^d$ and $\xi_n := \vartheta + h_n/\sqrt{n}$,

$$\Lambda_n^{\xi_n/\vartheta}(\omega) = 2\sum_{j=1}^{n}\left(\sqrt{L^{\xi_n/\vartheta}}(\omega_j) - 1\right) - \sum_{j=1}^{n}\left(\sqrt{L^{\xi_n/\vartheta}}(\omega_j) - 1\right)^2 + o_{(P_\vartheta^n)}(1)$$

as $n \to \infty$.

Proof. (1) The idea of the proof is a logarithmic expansion

$$\log(1+z) = z - \frac{1}{2}z^2 + o(z^2) \quad \text{as} \quad z \to 0$$

which for bounded sequences $(h_n)_n$ and $\xi_n = \vartheta + h_n/\sqrt{n}$ should give

$$\Lambda_n^{\xi_n/\vartheta}(\omega) = \sum_{j=1}^{n}\log L^{\xi_n/\vartheta}(\omega_j) = 2\sum_{j=1}^{n}\log\left(1 + \left(\sqrt{L^{\xi_n/\vartheta}} - 1\right)(\omega_j)\right)$$

$$= 2\sum_{j=1}^{n}\left(\sqrt{L^{\xi_n/\vartheta}} - 1\right)(\omega_j) - \sum_{j=1}^{n}\left(\sqrt{L^{\xi_n/\vartheta}} - 1\right)^2(\omega_j) + \cdots$$

where we have to consider carefully the different remainder terms which arise as $n \to \infty$.

(2) In $\mathcal{E}_n$, we do have

($\circ$) $\quad \Lambda_n^{\xi_n/\vartheta}(\omega) = \sum_{j=1}^{n}\log L^{\xi_n/\vartheta}(\omega_j) \quad$ for P_ϑ^n-almost all $\omega = (\omega_1,\ldots,\omega_n) \in \Omega_n$

which justifies the first '=' in the chain of equalities in (1) above. To see this, fix the sequence $(\xi_n)_n$, choose on $(\Omega, \mathcal{A})$ some dominating measure ν for the restricted experiment $\{P_{\xi_n}, n \geq 1, P_\vartheta\}$, and select densities χ_{ξ_n}, $n \geq 1$, χ_ϑ. In the restricted product experiment $\{P_{\xi_n}^n, n \geq 1, P_\vartheta^n\}$, the set

$$A_n := \{\omega \in \Omega_n : \chi_\vartheta(\omega_j) > 0 \text{ for all } 1 \leq j \leq n\} \in \mathcal{A}_n$$

has full measure under P_ϑ^n, the likelihood ratio $L_n^{\xi_n/\vartheta}$ coincides $(P_{\xi_n}^n + P_\vartheta^n)$-almost surely with

$$\omega \longrightarrow 1_{A_n}(\omega) \prod_{j=1}^n \frac{\chi_{\xi_n}(\omega_j)}{\chi_\vartheta(\omega_j)} + \infty \, 1_{A_n^c}(\omega),$$

and the expressions

$$\Lambda_n^{\xi_n/\vartheta}(\omega) = \log(L_n^{\xi_n/\vartheta}(\omega)) \quad \text{and} \quad \sum_{j=1}^n \log L^{\xi_n/\vartheta}(\omega_j)$$

are well defined and $[-\infty, +\infty)$-valued in restriction to A_n, and coincide on A_n. This is ($\circ$).

(3) We exploit L^2-differentiability at ϑ in the experiment $\mathcal{E}$. Fix $\delta > 0$, write $Z_n = \sqrt{L^{\xi_n/\vartheta}} - 1$ where $\xi_n = \vartheta + h_n/\sqrt{n}$. Then we have

$$P_\vartheta(|Z_n| > \delta) = P_\vartheta(Z_n > \delta) + P_\vartheta(Z_n < -\delta)$$

$$\leq P_\vartheta\left(Z_n > \delta, \frac{1}{2\sqrt{n}} h_n^\top V_\vartheta \leq \frac{\delta}{2}\right) + P_\vartheta\left(\frac{1}{2\sqrt{n}} h_n^\top V_\vartheta > \frac{\delta}{2}\right)$$

$$+ P_\vartheta\left(Z_n < -\delta, \frac{1}{2\sqrt{n}} h_n^\top V_\vartheta \geq -\frac{\delta}{2}\right) + P_\vartheta\left(\frac{1}{2\sqrt{n}} h_n^\top V_\vartheta < -\frac{\delta}{2}\right)$$

$$\leq P_\vartheta\left(\left|\sqrt{L^{\xi_n/\vartheta}} - 1 - \frac{1}{2}(\xi_n - \vartheta)^\top V_\vartheta\right| > \frac{\delta}{2}\right) + P_\vartheta\left(\left|h_n^\top V_\vartheta\right| > \delta\sqrt{n}\right).$$

Since $(h_n)_n$ is a bounded sequence, we have for suitable C as $n \to \infty$

$$P_\vartheta\left(|h_n^\top V_\vartheta| > \delta\sqrt{n}\right) \leq P_\vartheta\left(|V_\vartheta| > \frac{\delta\sqrt{n}}{C}\right)$$

$$\leq \frac{C^2}{\delta^2 n} E_\vartheta\left(|V_\vartheta|^2 1_{\{|V_\vartheta| > \frac{\delta\sqrt{n}}{C}\}}\right) = o\left(\frac{1}{n}\right)$$

since V_ϑ is in $L^2(P_\vartheta)$. A simpler argument works for the first term on the right-hand side above since

$$E_\vartheta\left(\left|\sqrt{L^{\xi_n/\vartheta}} - 1 - \frac{1}{2}(\xi_n - \vartheta)^\top V_\vartheta\right|^2\right) = o(|\xi_n - \vartheta|^2) = o\left(\frac{1}{n}\right)$$

by part (iib) of Definition 4.2. Taking all this together, we have

$$P_\vartheta(|Z_n| > \delta) = o\left(\frac{1}{n}\right) \quad \text{as} \quad n \to \infty$$

for $\delta > 0$ fixed.

(4) In the logarithmic expansion of step (1), we consider remainder terms

$$R(z) := \left|\log(1+z) - z + \frac{1}{2}z^2\right|, \quad z \in (-1, \infty)$$

Section 4.2 Le Cam's Second Lemma for i.i.d. Observations

together with random variables in the product experiment $\mathcal{E}_n$

$$Z_{n,j}(\omega) := \left(\sqrt{L^{\xi_n/\vartheta}} - 1\right)(\omega_j), \quad \omega = (\omega_1, \ldots, \omega_n) \in \Omega_n,$$

and shall prove in step (5) below

(+) $$\sum_{j=1}^{n} R(Z_{n,j}) = o_{(P_\vartheta^n)}(1) \quad \text{as } n \to \infty.$$

Then (+) will justify the last '=' in the chain of heuristic equalities in step (1), and thus will finish the proof of Proposition 4.14.

(5) To prove (+), we will consider 'small' and 'large' absolute values of $Z_{n,j}$ separately, 'large' meaning

$$\sum_{j=1}^{n} R(Z_{n,j}) 1_{\{|Z_{n,j}|>\delta\}} = \sum_{j=1}^{n} R(Z_{n,j}) 1_{\{Z_{n,j}>\delta\}} + \sum_{j=1}^{n} R(Z_{n,j}) 1_{\{Z_{n,j}<-\delta\}}$$

for any $\delta > 0$ fixed. For $Z_{n,j}$ positive, we associate to $\delta > 0$ the quantity

$$\gamma = \gamma(\delta) := \inf\{R(z) : z > \delta\} > 0$$

which allows to write

$$\left\{ \sum_{j=1}^{n} R(Z_{n,j}) 1_{\{Z_{n,j}>\delta\}} > \gamma \right\} = \left\{ Z_{n,j} > \delta \text{ for at least one } 1 \le j \le n \right\}.$$

Using step (3) above, the probability of the last event under P_ϑ^n is smaller than

$$n \, P_\vartheta(Z_n > \delta) = o(1) \quad \text{as} \quad n \to \infty.$$

Negative values of $Z_{n,j}$ can be treated analogously. We thus find that the contribution of 'large' absolute values of $Z_{n,j}$ to the sum (+) is negligible:

(++) for any $\delta > 0$ fixed: $$\sum_{j=1}^{n} R(Z_{n,j}) = \sum_{j=1}^{n} R(Z_{n,j}) 1_{\{|Z_{n,j}|\le\delta\}} + o_{(P_\vartheta^n)}(1)$$

as $n \to \infty$. It remains to consider 'small' absolute values of $Z_{n,j}$. Fix $\varepsilon > 0$ arbitrarily small. As a consequence of $R(z) = o(z^2)$ for $z \to 0$, we can associate

$$\delta = \delta(\varepsilon) > 0 \text{ such that } |R(z)| < \varepsilon z^2 \text{ on } \{|z| \le \delta\}$$

to every value of $\varepsilon > 0$. Since $(h_n)_n$ is a bounded sequence, we have as $n \to \infty$

(×) $$\sum_{j=1}^{n} Z_{n,j}^2(\omega) = \sum_{j=1}^{n} \left(\sqrt{L^{\xi_n/\vartheta}}(\omega_j) - 1\right)^2 = \frac{1}{4} h_n^T J_\vartheta h_n + o_{(P_\vartheta^n)}(1)$$

from Proposition 4.13, and at the same time

$$(+++) \qquad \sum_{j=1}^{n} R(Z_{n,j}) 1_{\{|Z_{n,j}| \leq \delta\}} \leq \varepsilon \cdot \sum_{j=1}^{n} Z_{n,j}^2$$

for $\delta = \delta(\varepsilon)$. Now, (×) implies tightness of $\mathcal{L}\left(\sum_{j=1}^{n} Z_{n,j}^2 \mid P_\vartheta^n\right)$ as $n \to \infty$, and we have the possibility to choose $\varepsilon > 0$ arbitrarily small. Combining (+++) and (++) we thus obtain

$$\sum_{j=1}^{n} R(Z_{n,j}) = o_{(P_\vartheta^n)}(1) \quad \text{as } n \to \infty.$$

This is (+), and the proof of Proposition 4.14 is finished. $\square$

Now we can conclude this section:

4.15 Proof of Le Cam's Second Lemma 4.11. Under Assumptions 4.10, we put together Propositions 4.14, 4.12 and 4.13

$$\Lambda_n^{\xi_n/\vartheta}(\omega) = 2 \sum_{j=1}^{n} \left(\sqrt{L^{\xi_n/\vartheta}}(\omega_j) - 1\right) - \sum_{j=1}^{n} \left(\sqrt{L^{\xi_n/\vartheta}}(\omega_j) - 1\right)^2 + o_{(P_\vartheta^n)}(1)$$

$$= 2 \left[\frac{1}{2} \frac{1}{\sqrt{n}} \sum_{j=1}^{n} h_n^T V_\vartheta(\omega_j) - \frac{1}{8} h_n^T J_\vartheta \, h_n \right] - \left[\frac{1}{4} h_n^T J_\vartheta \, h_n \right] + o_{(P_\vartheta^n)}(1)$$

$$= \frac{1}{\sqrt{n}} \sum_{j=1}^{n} h_n^T V_\vartheta(\omega_j) - \frac{1}{2} h_n^T J_\vartheta \, h_n + o_{(P_\vartheta^n)}(1)$$

and have proved the representation of log-likelihoods in Lemma 4.11. Weak convergence of the scores

$$S_n(\vartheta, \omega) = \frac{1}{\sqrt{n}} \sum_{j=1}^{n} V_\vartheta(\omega_j) \quad \text{under } P_\vartheta^n$$

has already been stated in Assumption 4.10(b). Le Cam's Second Lemma 4.11 is now proved. $\square$

Chapter 5

Gaussian Shift Models

Topics for Chapter 5:

5.1 Gaussian Shift Experiments
A classical normal distribution model 5.1
Gaussian shift experiment $\mathcal{E}(J)$ 5.2–5.3
Equivariant estimators 5.4
Boll's convolution theorem 5.5
Subconvex loss functions 5.6
Anderson's lemma, and some consequences 5.6'–5.8
Total variation distance 5.8'
Under 'very diffuse' a priori, arbitrary estimators are approximately equivariant 5.9
Main result: the minimax theorem 5.10

5.2 *Brownian Motion with Unknown Drift as a Gaussian Shift Experiment
Local dominatedness/equivalence of probability measures 5.11
Filtered spaces and density processes 5.12
Canonical path space $(C, \mathcal{C}, \mathbb{G})$ for continuous processes 5.13
Example: Brownian motion with unknown drift, special case $d = 1$ 5.14
The statistical model 'scaled BM with unknown drift' in dimension $d \geq 1$ 5.15
Statistical consequence: optimal estimation of the drift parameter 5.16

Exercises: 5.4', 5.4'', 5.4''', 5.5', 5.10'

This chapter considers the Gaussian shift model and its statistical properties. The main stages of Section 5.1 are Boll's Convolution Theorem 5.5 for equivariant estimators, the proof that arbitrary estimators are approximately equivariant under a very diffuse prior, and – as a consequence of both – the Minimax Theorem 5.10 which establishes a lower bound for the maximal risk of arbitrary estimators, in terms of the central statistic. We will see a stochastic process example in Section 5.2.

5.1 Gaussian Shift Experiments

Up to a particular representation of the location parameter, the following experiment is well known:

5.1 Example. Fix $J \in \mathbb{R}^{d \times d}$ symmetric and strictly positive definite, and consider the normal distribution model

$$\left(\mathbb{R}^d, \mathcal{B}(\mathbb{R}^d), \{ P_h := \mathcal{N}(Jh, J) : h \in \mathbb{R}^d \} \right).$$

Here densities $f_h = \frac{dP_h}{d\lambda}$ with respect to Lebesgue measure λ on $\mathbb{R}^d$ are given by

$$f(h, x) = (2\pi)^{-d/2} (\det J)^{-1/2} \exp\left(-\frac{1}{2}(x - Jh)^\top J^{-1}(x - Jh) \right),$$

$$x \in \mathbb{R}^d, \quad h \in \mathbb{R}^d$$

and all laws P_h, $h \in \mathbb{R}^d$, are pairwise equivalent.

(a) It follows that likelihood ratios of P_h with respect to P_0 have the form

(+) $\qquad L^{h/0} := \dfrac{dP_h}{dP_0} = \exp\left(h^\top S - \dfrac{1}{2} h^\top J h \right), \quad h \in \mathbb{R}^d$

where $S(x) := x$ denotes the canonical variable on $\mathbb{R}^d$ for which, as a trivial assertion,

(++) $\qquad \mathcal{L}(S \mid P_0) = \mathcal{N}(0, J).$

(b) Taking an arbitrary $h_0 \in \mathbb{R}^d$ as reference point, we obtain in the same way

$$L^{h/h_0} := \dfrac{dP_h}{dP_{h_0}} = \exp\left((h - h_0)^\top (S - Jh_0) - \dfrac{1}{2}(h - h_0)^\top J (h - h_0) \right),$$

$$h \in \mathbb{R}^d$$

and have, again as a trivial assertion,

$$\mathcal{L}(S - Jh_0 \mid P_{h_0}) = \mathcal{N}(0, J).$$

(c) Thus a structure analogous to (+) and (++) persists if we reparameterise around arbitrary reference points $h_0 \in \mathbb{R}^d$: we always have

(×) $\qquad L^{(h_0 + \widetilde{h})/h_0} = \exp\left(\widetilde{h}^\top (S - Jh_0) - \dfrac{1}{2} \widetilde{h}^\top J \widetilde{h} \right), \quad \widetilde{h} \in \mathbb{R}^d$

together with

(××) $\qquad \mathcal{L}(S - Jh_0 \mid P_{h_0}) = \mathcal{N}(0, J).$

(d) The above quadratic shape (+) of likelihoods combined with a distributional property (++) was seen to appear as a limiting structure – by Le Cam's Second Lemma 4.10 and 4.11 – over shrinking neighbourhoods of fixed reference points ϑ in L^2-differentiable experiments. $\square$

Section 5.1 Gaussian Shift Experiments

5.2 Definition. A model $(\Omega, \mathcal{A}, \{P_h : h \in \mathbb{R}^d\})$ is called a *Gaussian shift experiment* $\mathcal{E}(J)$ if there exists a statistic

$$S : (\Omega, \mathcal{A}) \longrightarrow (\mathbb{R}^d, \mathcal{B}(\mathbb{R}^d))$$

and a deterministic matrix

$$J \in \mathbb{R}^{d \times d} \text{ symmetric and strictly positive definite}$$

such that for every $h \in \mathbb{R}^d$,

(+) $\qquad \omega \longrightarrow \exp\left(h^\top S(\omega) - \frac{1}{2} h^\top J\, h\right) =: L^{h/0}(\omega)$

is a version of the likelihood ratio of P_h with respect to P_0. We call $Z := J^{-1} S$ *central statistic* in the Gaussian shift experiment $\mathcal{E}(J)$.

In a Gaussian shift experiment, the name 'central statistic' indicates a benchmark for good estimation, simultaneously under a broad class of loss functions: this will be seen in Corollary 5.8 and Theorem 5.10 below.

For every given matrix $J \in \mathbb{R}^{d \times d}$ symmetric and strictly positive definite, a Gaussian shift experiment $\mathcal{E}(J)$ exists by Example 5.1. The following proposition shows that $\mathcal{E}(J)$ as a statistical experiment is completely determined from the matrix $J \in \mathbb{R}^{d \times d}$ symmetric and strictly positive definite.

5.3 Proposition. In an experiment $\mathcal{E}(J)$ with the properties of Definition 5.2, the following holds:

(a) all laws P_h, $h \in \mathbb{R}^d$, are equivalent probability measures;

(b) we have $\mathcal{L}(Z - h \mid P_h) = \mathcal{N}(0, J^{-1})$ for all $h \in \mathbb{R}^d$;

(c) we have $\mathcal{L}(S - Jh \mid P_h) = \mathcal{N}(0, J)$ for all $h \in \mathbb{R}^d$;

(d) we have for all $h \in \mathbb{R}^d$

$$L^{(h+\widetilde{h})/h} = \exp\left(\widetilde{h}^\top (S - Jh) - \frac{1}{2} \widetilde{h}^\top J\, \widetilde{h}\right), \quad \widetilde{h} \in \mathbb{R}^d.$$

It follows that in the classical sense of Definition 1.2, $M_h := S - Jh$ is the score in $h \in \mathbb{R}^d$ and

$$J = E_h(M_h\, M_h^\top)$$

the Fisher information in h. The Fisher information does not depend on the parameter $h \in \mathbb{R}^d$.

Proof. (1) For any $h \in \mathbb{R}^d$, the likelihood ratio $L^{h/0}$ in Definition 5.2 is strictly positive and finite on Ω: hence neither a singular part of P_h with respect to P_0 nor a singular part of P_0 with respect to P_h exists, and we have $P_0 \sim P_h$.

(2) Recall that the Laplace transform of a normal law $\mathcal{N}(0, \Lambda)$ on $(\mathbb{R}^d, \mathcal{B}(\mathbb{R}^d))$ is

$$\mathbb{R}^d \ni u \longrightarrow \int e^{-u^\top x} \mathcal{N}(0, \Lambda)(dx) = e^{+\frac{1}{2} u^\top \Lambda u}$$

($\Lambda \in \mathbb{R}^{d \times d}$ symmetric and non-negative definite); the characteristic function of $\mathcal{N}(0, \Lambda)$ is

$$\mathbb{R}^d \ni u \longrightarrow \int e^{i u^\top x} \mathcal{N}(0, \Lambda)(dx) = e^{-\frac{1}{2} u^\top \Lambda u} .$$

(3) In $\mathcal{E}(J)$, J being deterministic, (1) and Definition 5.2 give

$$1 = E_0 \left(L^{(-h)/0} \right) = E_0 \left(e^{-h^\top S} \right) e^{-\frac{1}{2} h^\top J h} , \quad h \in \mathbb{R}^d$$

which specifies the Laplace transform of the law of S under P_0 and establishes

$$\mathcal{L}(S \mid P_0) = \mathcal{N}(0, J) .$$

For the central statistic $Z = J^{-1} S$, the law of $Z - h$ under P_h is determined – via scaling and change of measure, by (1) and Definition 5.2 – from the Laplace transform of the law of S under P_0: we have

$$E_h \left(e^{-\lambda^\top (Z-h)} \right) = E_0 \left(e^{-\lambda^\top (Z-h)} L^{h/0} \right) = e^{+\lambda^\top h} e^{-\frac{1}{2} h^\top J h} E_0 \left(e^{-(J^{-1}\lambda - h)^\top S} \right)$$

$$= e^{+\lambda^\top h} e^{-\frac{1}{2} h^\top J h} e^{+\frac{1}{2}(J^{-1}\lambda - h)^\top J (J^{-1}\lambda - h)} = e^{+\frac{1}{2} \lambda^\top J^{-1} \lambda}$$

for all $\lambda \in \mathbb{R}^d$ and all $h \in \mathbb{R}^d$. This shows

$$\mathcal{L}(Z - h \mid P_h) = \mathcal{N}(0, J^{-1})$$

for arbitrary $h \in \mathbb{R}^d$. This is (b), and (c) follows via standard transformations of normal laws.

(4) From (1) and Definition 5.2 we obtain the representation (d) in the same way as ($\times$) in Example 5.1(b). Then (d) and (c) together show that the experiment $\mathcal{E}(J)$ admits score and Fisher information as indicated. □

For the statistical properties of a parametric experiment, the space $(\Omega, \mathcal{A})$ supporting the family of laws $\{P_\zeta : \zeta \in \Theta\}$ is of no importance: the structure of likelihoods $L^{\zeta/\vartheta}$ matters when ζ and ϑ range over Θ. Hence, one may encounter the Gaussian shift experiment $\mathcal{E}(J)$ in quite different contexts.

In a Gaussian shift experiment $\mathcal{E}(J)$, the problem of estimation of the unknown parameter $h \in \mathbb{R}^d$ seems completely settled in a very classical way. The central statistic

Section 5.1 Gaussian Shift Experiments

Z is a maximum likelihood estimator. The theorems by Rao–Blackwell and Lehmann–Scheffé (e.g. [127, pp. 349 and 354], using the notions of a canonical statistic in a d-parametric exponential family, of sufficiency and of completeness, assert that in the class of all unbiased and square integrable estimators for the unknown parameter in $\mathcal{E}(J)$, Z is the unique estimator which uniformly on Θ minimises the variance.

However, a famous example by Stein (see e.g. [60, pp. 25–27] or [59, p. 93]) shows that for normal distribution models in dimension $d \geq 3$, there are estimators admitting bias which improve quadratic risk strictly beyond the best unbiased estimator. Hence it is undesirable to restrict from the start the class of estimators under consideration by imposing unbiasedness. The following definition does not involve any such restrictions.

5.4 Definition. In a statistical model $(\Omega', \mathcal{A}', \{P'_h : h \in \Theta'\})$, $\Theta' \subset \mathbb{R}^d$, an estimator κ for the unknown parameter $h \in \Theta'$ is called *equivariant* if

$$\mathcal{L}\left(\kappa - h \mid P'_h\right) = \mathcal{L}\left(\kappa \mid P'_0\right) \quad \text{for all } h \in \Theta'.$$

An equivariant estimator simply 'works equally well' at all points of the statistical model. By Proposition 5.3(b), the central statistics Z is equivariant in the Gaussian shift model $\mathcal{E}(J)$.

5.4' Exercise. With C the space of continuous functions $\mathbb{R}^d \to \mathbb{R}$, write $C_p \subset C$ for the cone of strictly positive f vanishing at ∞ faster than any polynomial: for every $\ell \in \mathbb{N}$, $\sup_{\{|x|>R\}}\{|x|^\ell f(x)\} \longrightarrow 0$ as $R \to \infty$. Consider an experiment $\mathcal{E}' = (\Omega', \mathcal{A}', \{P'_h : h \in \mathbb{R}^d\})$ of mutually equivalent probability laws for which paths

$$\mathbb{R}^d \ni \widetilde{h} \longrightarrow L^{(h_0+\widetilde{h})/h_0}(\omega) \in (0, \infty)$$

belong to C_p almost surely, for arbitrary $h_0 \in \mathbb{R}^d$ fixed, and such that the laws

$$\mathcal{L}\left(\left(L^{(h_0+\widetilde{h})/h_0}\right)_{\widetilde{h} \in \mathbb{R}^d} \mid P'_{h_0} \right)$$

on $(C, \mathcal{C})$ do not depend on $h_0 \in \mathbb{R}^d$ (as an example, all this is satisfied in Gaussian shift experiments $\mathcal{E}(J)$ in virtue of Proposition 5.3(c) and (d) for arbitrary dimension $d \geq 1$). As a consequence, laws of pairs

$$\mathcal{L}\left(\left(\int_{\mathbb{R}^d} \widetilde{h}\, L^{(h_0+\widetilde{h})/h_0}\, d\widetilde{h},\ \int_{\mathbb{R}^d} L^{(h_0+\widetilde{h})/h_0}\, d\widetilde{h} \right) \mid P'_{h_0} \right)$$

are well defined and do not depend on h_0. Prove the following (a) and (b):

(a) In $\mathcal{E}'$, a Bayesian estimator h^* with 'uniform over $\mathbb{R}^d$ prior' for the unknown parameter $h \in \mathbb{R}^d$

$$h^*(\omega) := \frac{\int_{\mathbb{R}^d} h'\, L^{h'/0}(\omega)\, dh'}{\int_{\mathbb{R}^d} L^{h'/0}(\omega)\, dh'}$$

(sometimes called Pitman estimator) is well defined, and is an equivariant estimator.

(b) In $\mathcal{E}'$, a maximum likelihood estimator $\widehat{h}$ for the unknown parameter $h \in \mathbb{R}^d$

$$\widehat{h}(\omega) = \mathrm{argmax}\{L^{h'/0}(\omega) : h' \in \mathbb{R}^d\}$$

(with measurable selection of an argmax – if necessary – similar to Proposition 2.10) is well defined, and is equivariant.

(c) In general statistical models, estimators according to (a) or to (b) will have different properties. Beyond the scope of the chapter, we add the following remark, for further reading: in the statistical model with likelihoods

$$L^{u/0} = \exp\left(W_u - \frac{1}{2}|u|\right), \quad u \in \mathbb{R}$$

where $(W_u)_{u \in \mathbb{R}}$ is a two-sided Brownian motion and dimension is $d = 1$ (the parameter is 'time': this is a two-sided variant of Example 1.16', in the case where X in Example 1.16' is Brownian motion), all assumptions above are satisfied, (a) and (b) hold true, and the Bayesian estimator h^* outperforms the maximum likelihood estimator $\widehat{h}$ under squared loss. See the references quoted at the end of Example 1.16'. For the third point, see [114]; for the second point, see [56, Lemma 5.2]. □

5.4'' Exercise. Prove the following: in a Gaussian shift model $\mathcal{E}(J)$, the Bayesian estimator with 'uniform over $\mathbb{R}^d$ prior' (from Exercise 5.4') for the unknown parameter $h \in \mathbb{R}^d$

$$h^*(\omega) := \frac{\int_{\mathbb{R}^d} h'\, L^{h'/0}(\omega)\, dh'}{\int_{\mathbb{R}^d} L^{h'/0}(\omega)\, dh'}, \quad \omega \in \Omega$$

coincides with the central statistic Z, and thus with the maximum likelihood estimator in $\mathcal{E}(J)$. Hint: varying the i-th component of h', start from one-dimensional integration

$$0 = \int_{-\infty}^{\infty} \frac{\partial}{\partial h'_i} L^{h'/0}(\omega)\, dh'_i = \int_{-\infty}^{\infty} (S(\omega) - Jh')_i\, L^{h'/0}(\omega)\, dh'_i, \quad 1 \le i \le d$$

and prove

$$Z(\omega) \int_{\mathbb{R}^d} L^{h'/0}(\omega)\, dh' = \int_{\mathbb{R}^d} h'\, L^{h'/0}(\omega)\, dh'.$$

Then Z and h^* coincide on Ω, for the Gaussian shift model $\mathcal{E}(J)$. □

5.4''' Exercise. In a Gaussian shift model $\mathcal{E}(J)$ with unknown parameter $h \in \mathbb{R}^d$ and central statistic $Z = J^{-1}S$, fix some point $h_0 \in \Theta$ and some $0 < \alpha < 1$, and use a statistic

$$T := \alpha Z + (1 - \alpha) h_0$$

as an estimator for the unknown parameter. Prove that T is not an equivariant estimator, and specify the law $\mathcal{L}(T - h | P_h)$ for all $h \in \mathbb{R}^d$ from Proposition 5.3. □

The following implies a criterion for optimality within the class of all equivariant estimators in $\mathcal{E}(J)$. This is the first main result of this section.

Section 5.1 Gaussian Shift Experiments

5.5 Convolution Theorem (Boll [13]). Consider a Gaussian shift experiment $\mathcal{E}(J)$. If κ is an equivariant estimator for the unknown parameter $h \in \mathbb{R}^d$, there is some probability measure Q on $(\mathbb{R}^d, \mathcal{B}(\mathbb{R}^d))$ such that

$$\mathcal{L}(\kappa - h \mid P_h) = \mathcal{N}(0, J^{-1}) \star Q \quad \text{for all} \quad h \in \mathbb{R}^d.$$

In addition, the law Q coincides with $\mathcal{L}(\kappa - Z \mid P_0)$.

Proof. κ being equivariant for the unknown parameter $h \in \mathbb{R}^d$, it is sufficient to prove

$(\diamond)$ $\qquad \mathcal{L}(\kappa \mid P_0) = \mathcal{N}(0, J^{-1}) \star \mathcal{L}(\kappa - Z \mid P_0).$

(1) Using characteristic functions for laws on $\mathbb{R}^d$, we fix $t \in \mathbb{R}^d$. Equivariance of κ gives

$(+)$ $\qquad E_0\bigl(e^{it^\top \kappa}\bigr) = E_h\bigl(e^{it^\top(\kappa-h)}\bigr) = E_0\bigl(e^{it^\top(\kappa-h)} L^{h/0}\bigr), \quad h \in \mathbb{R}^d.$

Let us replace $h \in \mathbb{R}^d$ by $z \in \mathbb{C}^d$ in $(+)$ to obtain an analytic function

$$f: \mathbb{C}^d \ni z \longrightarrow E_0\bigl(e^{it^\top(\kappa-z)} e^{z^\top S - \frac{1}{2} z^\top J z}\bigr) \in \mathbb{C}.$$

By $(+)$, the restriction of $f(\cdot)$ to $\mathbb{R}^d \subset \mathbb{C}^d$ being constant, $f(\cdot)$ is constant on $\mathbb{C}^d$, thus in particular

$$E_0\bigl(e^{it^\top \kappa}\bigr) = f(0) = f(-iJ^{-1}t).$$

Calculating the value of f at $-iJ^{-1}t$ we find

$(*)$ $\qquad E_0\bigl(e^{it^\top \kappa}\bigr) = e^{-\frac{1}{2} t^\top J^{-1} t} E_0\bigl(e^{it^\top(\kappa-Z)}\bigr).$

(2) We have $(*)$ for every $t \in \mathbb{R}^d$. We thus have equality of characteristic functions on $\mathbb{R}^d$. The right-hand side of $(+)$ admits an interpretation as characteristic function of

$$\mathcal{L}\bigl(\widetilde{\xi} + (\kappa - Z) \mid P_0\bigr)$$

for *some* random variable $\widetilde{\xi}$ independent of $\kappa - Z$ under P_0 and such that $\mathcal{L}(\widetilde{\xi} \mid P_0) = \mathcal{N}(0, J^{-1})$. Hence, writing $Q := \mathcal{L}(\kappa - Z \mid P_0)$, the right-hand side of $(*)$ is the characteristic function of a convolution $\mathcal{N}(0, J^{-1}) \star Q$.

(3) It is important to note the following: the above steps (1) and (2) did *not* establish (and the assertion of the theorem did *not* claim) that Z and $(\kappa - Z)$ should be independent under P_0. $\square$

Combined with Proposition 5.3(b), Boll's Convolution Theorem 5.5 states that in a Gaussian shift experiment $\mathcal{E}(J)$, estimation errors of equivariant estimators are 'more spread out' than the estimation error of the central statistic Z, as an estimator for the unknown parameter. By Theorem 5.5, the best possible concentration of estimation errors (within the class of all equivariant estimators) is attained for $\kappa = Z$: then

$\mathcal{L}(Z-h \mid P_h)$ equals $\mathcal{N}(0, J^{-1})$, and $Q = \epsilon_0$ is a point mass at 0. However, a broad class of estimators with possibly interesting properties is not equivariant.

5.5' Exercise. For $R < \infty$ arbitrarily large, consider closed balls C centred at 0 with radius R.
 (a) In a Gaussian shift model $\mathcal{E}(J)$, check that a Bayesian with uniform prior over the compact C

$$h_C^*(\omega) := \frac{\int_C h' \, L^{h'/0}(\omega) \, dh'}{\int_C L^{h'/0}(\omega) \, dh'}, \quad \omega \in \Omega$$

is not an equivariant estimator for the unknown parameter.
 (b) For arbitrary estimators $T : (\Omega, \mathcal{A}) \to (\mathbb{R}^d, \mathcal{B}(\mathbb{R}^d))$ for the unknown parameter which may exist in $\mathcal{E}(J)$, consider quadratic loss and Bayes risks

$$R(T, C) := \frac{1}{\lambda(C)} \int_C E_h(|T - h|^2) \, dh \le \infty.$$

In the class of all $\mathcal{A}$-measurable estimators T for the unknown parameter, h_C^* minimises the squared risk for h chosen randomly from C, and provides a lower bound for the maximal squared risk over C:

$$(+) \qquad \sup_{h \in C} E_h(|T - h|^2) \ge R(T, C) \ge R(h_C^*, C).$$

This is seen as follows: the first inequality in (+) being a trivial one (we replace an integrand by its upper bound), it is sufficient to prove the second. Put $\widetilde{\Omega} := \Omega \times C$ and $\widetilde{\mathcal{A}} := \mathcal{A} \otimes \mathcal{B}(C)$; on the extended space $(\widetilde{\Omega}, \widetilde{\mathcal{A}})$, let $id : (\omega, h) \to (\omega, h)$ denote the canonical statistic. Define a probability measure

$$\widetilde{P}(d\omega, dh') := 1_C(h') \frac{dh'}{\lambda(C)} P_{h'}(d\omega) = P_0(d\omega) \, 1_C(h') \frac{L^{h'/0}(\omega)}{\lambda(C)} \, dh'$$

on $(\widetilde{\Omega}, \widetilde{\mathcal{A}})$, and write $\widetilde{E}$ for expectation under $\widetilde{P}$. On the space $(\widetilde{\Omega}, \widetilde{\mathcal{A}}, \widetilde{P})$, identify h_C^* as conditional expectation of the second component of id given the first component of id. Note that the random variable h – the second component of id – belongs to $L^2(\widetilde{P})$ since C is compact. Then the projection property of conditional expectations in $L^2(\widetilde{P})$ gives

$$\widetilde{E}(|h - T|^2) \ge \widetilde{E}(|h - h_C^*|^2)$$

for all random variables $T \in L^2(\widetilde{P})$ which depend only on the first component of id. But this is the class of all $T : \Omega \to \mathbb{R}^d$ which are $\mathcal{A}$-measurable, thus the class of all possible estimators for the unknown parameter in the experiment $\mathcal{E}(J)$, and we obtain the second inequality in (+). □

5.6 Definition. In an arbitrary experiment $(\Omega', \mathcal{A}', \{P_h' : h \in \Theta'\})$, $\Theta' \subset \mathbb{R}^d$,
 (i) we call a loss function

$$\ell : \mathbb{R}^d \longrightarrow [0, \infty) \quad \mathcal{B}(\mathbb{R}^d)\text{-measurable}$$

subconvex or *bowl-shaped* if all level sets

$$A_c := \{x \in \mathbb{R}^d : \ell(x) \le c\}, \quad c \ge 0$$

Section 5.1 Gaussian Shift Experiments

are convex and symmetric with respect to the origin (i.e. $x \in A_c \iff -x \in A_c$, for $x \in \mathbb{R}^d$);

(ii) we associate – with respect to a loss function which we keep fixed – a risk function

$$R(T,\cdot) : \quad \Theta' \ni h \longrightarrow R(T,h) := \int_{\Omega'} \ell(T-h) \, dP_h \in [0,\infty]$$

to every estimator $T : (\Omega', \mathcal{A}') \to (\mathbb{R}^d, \mathcal{B}(\mathbb{R}^d))$ for the unknown parameter $h \in \Theta'$.

Note that risk functions according to Definition 5.6(ii) are well defined, but not necessarily finite-valued. When A is a subset of Θ of finite Lebesgue measure, we also write

$$R(T,A) := \int \pi_A(dh) \, R(T,h) \in [0,\infty]$$

for the Bayes risk where π_A is the uniform distribution on A. For short, we write π_n for the uniform distribution on the closed ball $B_n := \{|x| \leq n\}$. The following lemma is based on an inequality for volumes of convex combinations of convex sets in $\mathbb{R}^d$, see [1, p. 170–171]) for the proof.

5.6' Lemma (Anderson [1]). Consider $C \subset \mathbb{R}^d$ convex and symmetric with respect to the origin. Consider $f : \mathbb{R}^d \to [0,\infty)$ integrable, symmetric with respect to the origin, and such that all sets $\{x \in \mathbb{R}^d : f(x) \geq c\}$ are convex, $0 < c < \infty$. Then

$$\int_C f(x) \, dx \geq \int_{C+y} f(x) \, dx$$

for all $y \in \mathbb{R}^d$, where $C + y$ is the set C shifted by y.

We will use Anderson's lemma in the following form:

5.7 Corollary. For matrices $\Lambda \in \mathbb{R}^{d \times d}$ which are symmetric and strictly positive definite, sets $C \in \mathcal{B}(\mathbb{R}^d)$ convex and symmetric with respect to the origin, subconvex loss functions $\ell : \mathbb{R}^d \to [0,\infty)$, points $a \in \mathbb{R}^d \setminus \{0\}$ and probability measures Q on $(\mathbb{R}^d, \mathcal{B}(\mathbb{R}^d))$, we have

$$\mathcal{N}(0,\Lambda)(C) \geq \mathcal{N}(0,\Lambda)(C-a) ;$$
$$\int \ell(x) \, \mathcal{N}(0,\Lambda)(dx) \leq \int \ell(x-a) \, \mathcal{N}(0,\Lambda)(dx) ;$$
$$\int \ell(x) \, \mathcal{N}(0,\Lambda)(dx) \leq \int \ell(x) \, [\mathcal{N}(0,\Lambda) \star Q](dx) .$$

Proof. Writing $f(\cdot)$ for the Lebesgue density of the normal law $\mathcal{N}(0, \Lambda)$ on $\mathbb{R}^d$, the first assertion rephrases Anderson's Lemma 5.6'. An immediate consequence is

$$\int \widetilde{\ell}(x) \, \mathcal{N}(0, \Lambda)(dx) \leq \int \widetilde{\ell}(x - a) \, \mathcal{N}(0, \Lambda)(dx)$$

for 'elementary' subconvex loss functions which are finite sums

$$\widetilde{\ell} = \sum_{i=1}^{m} \alpha_i \, 1_{\mathbb{R}^d \setminus C_i},$$

$\alpha_i > 0$, $C_i \in \mathcal{B}(\mathbb{R}^d)$ convex and symmetric with respect to the origin, $m \geq 1$.

The second assertion follows if we pass to general subconvex loss functions $\ell(\cdot)$ by monotone convergence $\ell_n \uparrow \ell$ where all ℓ_n are 'elementary': writing $C_{n,j} := \{\ell \leq \frac{j}{2^n}\}$ we can take

$$\ell_n := \sum_{j=1}^{n2^n} \frac{1}{2^n} 1_{\mathbb{R}^d \setminus C_{n,j}} = \sum_{j'=0}^{n2^n - 1} \frac{j'}{2^n} 1_{\{\frac{j'}{2^n} < \ell \leq \frac{j'+1}{2^n}\}} + n \, 1_{\{\ell > n\}}, \quad n \geq 1.$$

The second assertion now allows us to compare

$$\int \ell(x) \left[\mathcal{N}(0, \Lambda) \star Q \right] (dx) = \int \left[\int \ell(y + b) \, \mathcal{N}(0, \Lambda)(dy) \right] Q(db)$$

$$\geq \int \left[\int \ell(y) \, \mathcal{N}(0, \Lambda)(dy) \right] Q(db)$$

$$= \int \ell(y) \, \mathcal{N}(0, \Lambda)(dy)$$

which is the third assertion. □

The last inequalities of Corollary 5.7 allow to rephrase Boll's Convolution Theorem 5.5. This yields a powerful way to compare (equivariant) estimators, in the sense that 'optimality' appears decoupled from any particular choice of a loss function which we might invent to penalise estimation errors.

5.8 Corollary. In the Gaussian shift experiment $\mathcal{E}(J)$, with respect to any subconvex loss function, the central statistic Z minimises the risk

$$R(\kappa, h) \geq R(Z, h), \quad h \in \mathbb{R}^d$$

in the class of all equivariant estimators κ for the unknown parameter. □

Proof. Both κ and Z being equivariant, their risk functions are constant over $\mathbb{R}^d$, thus it is sufficient to prove the assertion for the parameter value $h = 0$. We have

$$\mathcal{L}(Z \mid P_0) = \mathcal{N}(0, J^{-1})$$

Section 5.1 Gaussian Shift Experiments

from Proposition 5.3. Theorem 5.5 associates to κ a probability measure Q such that
$$\mathcal{L}(\kappa \mid P_0) = \mathcal{N}(0, J^{-1}) \star Q .$$
The loss function $\ell(\cdot)$ being subconvex, the third assertion in Corollary 5.7 shows
$$R(Z, 0) = E_0(\ell(Z - 0)) = \int \ell(x) \, \mathcal{N}(0, J^{-1})(dx)$$
$$\leq \int \ell(x) \left[\mathcal{N}(0, J^{-1}) \star Q \right](dx) = E_0(\ell(\kappa - 0)) = R(\kappa, 0)$$
which is the assertion. □

Given Theorem 5.5 and Corollary 5.8, we are able to compare equivariant estimators; the next aim is the comparison of arbitrary estimators for the unknown parameter. We quote the following from [121, Chap. I.2] or [124, Chap. 2.4]:

5.8' Remark. *Total variation distance* $d_1(P, Q)$ *between probability measures* P, Q *living on the same space* $(\Omega', \mathcal{A}')$ *is defined by*
$$d_1(P, Q) := \sup_{A \in \mathcal{A}'} |P(A) - Q(A)| = \sup_{A \in \mathcal{A}'} \left| \int_A (p - q) \, d\mu \right| \in [0, 1],$$
with μ some dominating measure and $p = \frac{dP}{d\mu}, q = \frac{dQ}{d\mu}$ versions of the densities, and one has
$$\sup_{A \in \mathcal{A}'} |P(A) - Q(A)| =$$
(+)
$$\sup \left\{ \left| \int \phi \, dP - \int \phi \, dQ \right| : \phi \text{ an } \mathcal{A}'\text{-measurable function } \Omega' \to [0, 1] \right\}.$$

The following lemma represent a key tool: in a Gaussian shift experiment $\mathcal{E}(J)$, arbitrary estimators η for the unknown parameter can be viewed as 'approximately equivariant' in the absence of any a priori knowledge except that the unknown parameter should range over large balls centred at the origin.

5.9 Lemma. *In a Gaussian shift experiment $\mathcal{E}(J)$, every estimator η for the unknown parameter $h \in \mathbb{R}^d$ is associated to a sequence of probability measures $(Q_n)_n$ on $(\mathbb{R}^d, \mathcal{B}(\mathbb{R}^d))$ such that*

(i) $\quad d_1 \left(\int \pi_n(dh) \, \mathcal{L}(\eta - h \mid P_h), \, \mathcal{N}(0, J^{-1}) \star Q_n \right) \longrightarrow 0 \quad \text{as } n \to \infty.$

As a consequence, for any choice of a loss function $\ell(\cdot)$ which is subconvex and bounded, we have

(ii)
$$R(\eta, B_n) = \int \pi_n(dh) \, E_h(\ell(\eta - h))$$
$$= \int \ell(x) \left[\mathcal{N}(0, J^{-1}) \star Q_n \right](dx) + \|\ell\|_\infty \cdot o(1)$$

as $n \to \infty$ (with B_n the closed ball $\{|h| \leq n\}$ in $\mathbb{R}^d$, and π_n the uniform distribution on B_n). In the last line, $\|\ell\|_\infty$ is the sup-norm of the loss function, and the $o(1)$-terms do not depend on $\ell(\cdot)$.

Proof. Assertion (ii) is a consequence of (i), via (+) in Remark 5.8'. We prove (i) in several steps.

(1) The central statistic $Z = J^{-1}S$ is a sufficient statistic in the Gaussian shift experiment $\mathcal{E}(J)$ (e.g. [4, Chap. II.1 and II.2], or [127, Chap. 3.1]). Thus, for any random variable η taking values in $(\mathbb{R}^d, \mathcal{B}(\mathbb{R}^d))$, there is a regular version of the conditional law of η given $Z = \cdot$ which does not depend on the parameter $h \in \mathbb{R}^d$: there is a transition kernel $K(\cdot, \cdot)$ on $(\mathbb{R}^d, \mathcal{B}(\mathbb{R}^d))$ such that

$K(\cdot, \cdot)$ is a regular version of $(z, A) \to P_h^{\eta|Z=z}(A)$ for every value of $h \in \mathbb{R}^d$.

We write this as $K(z, A) = P_\bullet^{\eta|Z=z}(A)$. In the same sense, the conditional law of $\eta - Z$ given $Z = \cdot$ does not depend on the parameter, and we define a sequence of probability measures $(Q_n)_n$ on $(\mathbb{R}^d, \mathcal{B}(\mathbb{R}^d))$:

$$Q_n(A) := \int \pi_n(dz)\, P_\bullet^{(\eta-Z)|Z=z}(A), \quad A \in \mathcal{B}(\mathbb{R}^d),\ n \geq 1.$$

Note that the sequence $(Q_n)_n$ is defined only in terms of the pair (η, Z). Comparing with the first expression in (i), this definition signifies that the observed value z of the central statistic starts to take over the role of the parameter h.

(2) For every fixed value of $x \in \mathbb{R}^d$, we have uniformly in $A \in \mathcal{B}(\mathbb{R}^d)$ the bound

(*) $$\left| \int \pi_n(dh)\, K(x+h, A+h) - Q_n(A-x) \right| \leq \frac{\lambda(B_n \triangle (B_n + x))}{\lambda(B_n)}$$

where $A - x$ is the set A shifted by x, and $\triangle$ denotes symmetric difference. To see this, write

$$\int \pi_n(dh)\, K(x+h, A+h) = \frac{1}{\lambda(B_n)} \int \lambda(dh)\, 1_{B_n}(h)\, K(x+h, A+h)$$
$$= \frac{1}{\lambda(B_n)} \int \lambda(dh')\, 1_{B_n+x}(h')\, K(h', A - x + h')$$
$$= \frac{1}{\lambda(B_n)} \int \lambda(dh')\, 1_{B_n+x}(h')\, P_\bullet^{\eta|Z=h'}(A - x + h')$$
$$= \frac{1}{\lambda(B_n)} \int \lambda(dh')\, 1_{B_n+x}(h')\, P_\bullet^{\eta-Z|Z=h'}(A - x)$$

and compare the last right-hand side to the definition of Q_n in step (1)

$$Q_n(A - x) = \frac{1}{\lambda(B_n)} \int \lambda(dh')\, 1_{B_n}(h')\, P_\bullet^{\eta-Z|Z=h'}(A - x).$$

Section 5.1 Gaussian Shift Experiments

It follows that the difference on the left-hand side of (∗) is uniformly in A smaller than

$$\frac{1}{\lambda(B_n)} \int \lambda(dh') \left|1_{B_n}(h') - 1_{B_n+x}(h')\right| = \frac{\lambda(B_n \triangle (B_n + x))}{\lambda(B_n)}.$$

(3) To conclude the proof of (i), we condition the first law in (i) with respect to the central statistic Z. For $A \in \mathcal{B}(\mathbb{R}^d)$ we obtain from Proposition 5.3(b), definition of $K(\cdot, \cdot)$ and substitution $z = x + h$

$$\int \pi_n(dh)\, P_h(\eta - h \in A) = \int \pi_n(dh) \left[\int P_h^Z(dz)\, P_h^{\eta|Z=z}(A+h)\right]$$
$$= \int\int \pi_n(dh)\, \mathcal{N}(h, J^{-1})(dz)\, K(z, A+h)$$
$$= \int \mathcal{N}(0, J^{-1})(dx) \left[\int \pi_n(dh)\, K(x+h, A+h)\right]$$

whereas the second law in (i) charges A with mass

$$\left(\mathcal{N}(0, J^{-1}) \star Q_n\right)(A) = \int \mathcal{N}(0, J^{-1})(dx)\, Q_n(A - x).$$

By the bounds (∗) obtained in step (2) which are uniform in A for fixed value of x, we can compare the last right-hand sides. Thus the definition of total variation distance gives

$$d_1\left(\int \pi_n(dh)\, \mathcal{L}(\eta - h \mid P_h),\, \mathcal{N}(0, J^{-1}) \star Q_n\right)$$
$$\leq \int \mathcal{N}(0, J^{-1})(dx)\, \frac{\lambda(B_n \triangle (B_n + x))}{\lambda(B_n)}$$

where the integrand on the right-hand side, trivially bounded by 2, converges to 0 pointwise in $x \in \mathbb{R}^d$ as $n \to \infty$. Assertion (i) now follows from dominated convergence. □

Lemma 5.9 is the key to the minimax theorem in Gaussian shift experiments $\mathcal{E}(J)$. It allows to compare all possible estimators for the unknown parameter $h \in \mathbb{R}^d$, with respect to any subconvex loss function: it turns out that for all choices of $\ell(\cdot)$, the maximal risk on $\mathbb{R}^d$ is minimised by the central statistic. This is – following Theorem 5.5 – the second main result of this section.

5.10 Minimax Theorem. In the Gaussian shift experiment $\mathcal{E}(J)$, the central statistic Z is a minimax estimator for the unknown parameter with respect to any subconvex loss function $\ell(\cdot)$; we have

$$\sup_{h \in \mathbb{R}^d} R(\eta, h) \geq \int \ell(z)\, \mathcal{N}(0, J^{-1})(dz) = R(Z, 0) = \sup_{h \in \mathbb{R}^d} R(Z, h).$$

Proof. Consider risk with respect to any subconvex loss function $\ell(\cdot)$. The last equality is merely equivariance of Z as an estimator for the unknown parameter $h \in \mathbb{R}^d$, and we have to prove the first sign '$\geq$'. Consider any estimator η for the unknown parameter, define π_n, B_n, Q_n as in Lemma 5.9, and recall that the sequence $(Q_n)_n$ depends only on the pair (η, Z). A trivial chain of inequalities is

(+)
$$\infty \geq \sup_{h \in \mathbb{R}^d} R(\eta, h) \geq \sup_{h \in B_n} R(\eta, h)$$
$$\geq \int \pi_n(dh) R(\eta, h) =: R(\eta, B_n)$$

for arbitrary $n \in \mathbb{N}$. We shall show that

(++)
$$\liminf_{n \to \infty} R(\eta, B_n) \geq \int (\ell \wedge N)(x) \, \mathcal{N}(0, J^{-1})(dx)$$
for every constant $N < \infty$.

Given (+) and (++), monotone convergence as $N \to \infty$ will finish the proof of the theorem.

In order to prove (++), observe first that it is sufficient to work with loss functions $\ell(\cdot)$ which are subconvex and bounded. For such $\ell(\cdot)$, Lemma 5.9(ii) shows

$$R(\eta, B_n) = \int \ell(x) \left[\mathcal{N}(0, J^{-1}) \star Q_n \right](dx) + \|\ell\|_\infty \cdot o(1)$$

as $n \to \infty$, where Anderson's inequality 5.7 gives

$$\int \ell(x) \left[\mathcal{N}(0, J^{-1}) \star Q_n \right](dx) \geq \int \ell(x) \, \mathcal{N}(0, J^{-1})(dx) \quad \text{for every } n.$$

Both assertions together establish (++); the proof is complete. □

To resume, a Gaussian shift experiment $\mathcal{E}(J)$ allows – thanks to the properties of its central statistic – for two remarkable results: first, Boll's Convolution Theorem 5.5 for equivariant estimators; second, Lemma 5.9 for arbitrary estimators η, according to which the structure of risks of equivariant estimators is mimicked by Bayes risks over large balls B_n. All this is independent of the choice of a subconvex loss function. The Minimax Theorem 5.10 is then an easy consequence of both results. So neither a notion of 'maximum likelihood' nor any version of a 'Bayes property' turns out to be intrinsic for good estimation: it is the existence of a central statistic which allows for good estimation in the Gaussian shift experiment $\mathcal{E}(J)$.

5.10' Exercise. In a Gaussian shift model $\mathcal{E}(J)$ with unknown parameter $h \in \mathbb{R}^d$ and central statistic $Z = J^{-1}S$, fix some point $h_0 \in \Theta$, and consider for $0 < \alpha < 1$ estimators

$$T := \alpha Z + (1-\alpha) h_0$$

according to exercise 5.4''' which are not equivariant. Under squared loss, calculate the risk of T as a function of $h \in \mathbb{R}^d$. Evaluate what T 'achieves' at the particular point h_0 in comparison to the central statistic, and the price to be paid for this at parameter points $h \in \mathbb{R}^d$ which are distant from h_0. Historically, estimators of similar structure have been called 'superefficient at h_0' and have caused some trouble; in the light of Theorem 5.10 it is clear that any denomination of this type is misleading. □

5.2 *Brownian Motion with Unknown Drift as a Gaussian Shift Experiment

In the following, all sections or subsections preceded by an asterisk * will require techniques related to continuous-time martingales, semi-martingales and stochastic analysis, and a reader not interested in stochastic processes may skip these and keep on with the statistical theory. Some typical references for sections marked by * are e.g. Liptser and Shiryaev [88], Metivier [98], Jacod and Shiryaev [64], Ikeda and Watanabe [61], Chung and Williams [15], Karatzas and Shreve [69], Revuz and Yor [112]. In the present section, we introduce the notion of a density process (or likelihood ratio process), and then look in particular to statistical models for Brownian motion with unknown drift as an illustration to Section 5.1.

5.11 Definition. Consider a probability space $(\Omega, \mathcal{A})$ equipped with a right-continuous filtration $\mathbb{F} = (\mathcal{F}_t)_{t \geq 0}$. For any probability measure P on $(\Omega, \mathcal{A})$ and any $0 \leq t < \infty$, we write P_t for the restriction of P to the σ-field $\mathcal{F}_t$. For pairs P', P of probability measures on $(\Omega, \mathcal{A})$, we write

$$P' \stackrel{\text{loc}}{\ll} P \text{ relative to } \mathbb{F}$$

if $P'_t \ll P_t$ for all $0 \leq t < \infty$, and

$$P' \stackrel{\text{loc}}{\sim} P \text{ relative to } \mathbb{F}$$

if $P'_t \sim P_t$ for all $0 \leq t < \infty$.

Typical situations will combine local equivalence $P' \stackrel{\text{loc}}{\sim} P$ relative to $\mathbb{F}$ with singularity $P' \perp P$ on $\mathcal{A}$. As an example, think of a process X on $(\Omega, \mathcal{A})$ which is Brownian motion under P, and Brownian motion with drift $\mu \neq 0$ under P', and of the strong law of large numbers: then $\lim_{t \to \infty} \frac{1}{t} X_t$ equals μ almost surely under P', and equals 0 almost surely under P. For the following result, see [64, Chap. III], for background, see [98].

5.12 Theorem. Consider a probability space $(\Omega, \mathcal{A})$, a right-continuous filtration $\mathbb{F} = (\mathcal{F}_t)_{t \geq 0}$, and probability measures $P' \stackrel{\text{loc}}{\ll} P$ relative to $\mathbb{F}$.

(a) Then there is a $(P, \mathbb{F})$-martingale $M = (M_t)_{t \geq 0}$ with the properties

$$M \geq 0, \quad E_P(M_t) = 1 \text{ for all } t < \infty, \quad P\text{-almost all paths of } M \text{ are càdlàg}$$

such that the following holds:

for all $t < \infty$: M_t is a version of the density $\frac{d P'_t}{d P_t}$.

The P-martingale M is uniquely determined up to $(P' + P)$-indistinguishability, and is called the *density process of P' with respect to P relative to* $\mathbb{F}$.

(b) For $\mathbb{F}$-stopping times τ consider the σ-field of events up to time τ

$$\mathcal{F}_\tau := \{ F \in \mathcal{F}_\infty : F \cap \{\tau \leq t\} \in \mathcal{F}_t \text{ for all } 0 \leq t < \infty \}$$

where $\mathcal{F}_\infty := \sigma\left(\bigcup_{0 \leq t < \infty} \mathcal{F}_t\right)$. The two mappings

$$F \longrightarrow P'(F \cap \{\tau < \infty\}) \quad \text{and} \quad F \longrightarrow P(F \cap \{\tau < \infty\})$$

define measures on $\mathcal{F}_\tau$ of total mass ≤ 1, and we have

(+) $\qquad P'(F \cap \{\tau < \infty\}) = E_P \left(1_{F \cap \{\tau < \infty\}} M_\tau \right) .$

(c) For $\mathbb{F}$-stopping times τ satisfying the condition

(∗) $\qquad \tau < \infty$ P'-almost surely and P-almost surely

the restrictions $P'_\tau := P'|_{\mathcal{F}_\tau}$ and $P_\tau := P|_{\mathcal{F}_\tau}$ of P' and P to $\mathcal{F}_\tau$ are probability measures such that

$$P'_\tau \ll P_\tau, \quad M_\tau 1_{\{\tau < \infty\}} \text{ is a version of } \frac{d P'_\tau}{d P_\tau}.$$

(d) For $\mathbb{F}$-stopping times τ which are P-almost surely finite, we write M_τ for $M_\tau 1_{\{\tau < \infty\}}$ under P and have equivalence of the following three assertions:

(α) $\quad P'(\{\tau < \infty\}) = 1$,

(β) $\quad E_P(M_\tau) = 1$,

(γ) $\quad (M_{\tau \wedge N})_{N \in \mathbb{N}_0}$ is a uniformly integrable P-martingale.

Proof. (1) For $T \in \mathbb{N}$ fixed, we have $P'_T \ll P_T$ on $\mathcal{F}_T$, hence there is a density

$$f_T := \frac{d P'_T}{d P_T}, \quad \mathcal{F}_T\text{-measurable, } [0, \infty)\text{-valued, unique up to } (P'_T + P_T)\text{-null sets}.$$

Define $M = (M_t)_{0 \leq t \leq T}$ to be the càdlàg modification of the martingale $t \to E_P(f_T \mid \mathcal{F}_t), 0 \leq t \leq T$. Then for every $0 \leq t \leq T$ and $F \in \mathcal{F}_t$, we can write

$$P'_t(F) = P'_T(F) = E_P(1_F f_T) = E_P(1_F M_T) = E_P(1_F M_t)$$

$$= \int_F M_t \, dP = \int_F M_t \, dP_t$$

which shows that in restriction to $\mathcal{F}_t$, M_t is a version of the density $\frac{d P'_t}{d P_t}$.

(2) For $T \in \mathbb{N}$ we can paste together the processes $(M_t)_{0 \le t \le T}$ constructed so far into one process $M = (M_t)_{t \ge 0}$ with the desired properties. This is (a).

(3) Consider $\mathbb{F}$-stopping times τ. For $F \in \mathcal{F}_\tau$ and $N \in \mathbb{N}$, consider first the subset $F \cap \{\tau \le N\}$ which belongs to $\mathcal{F}_N$ and to $\mathcal{F}_{\tau \wedge N}$. Combining (a) above with the stopping theorem for bounded stopping times we get

$$P'(F \cap \{\tau \le N\}) = E_P\left(1_{F \cap \{\tau \le N\}} M_N\right) = E_P\left(1_{F \cap \{\tau \le N\}} M_{\tau \wedge N}\right)$$
$$= E_P\left(1_{F \cap \{\tau \le N\}} M_\tau\right)$$

where $N \in \mathbb{N}$. Since M is non-negative, monotone convergence as $N \uparrow \infty$ yields (+). This is (b).

(4) Under the additional condition (∗), (+) in (b) implies (c).

(5) Consider a stopping time τ such that $P(\tau < \infty) = 1$. Then equivalence of (α) and (β) in (d) is a direct consequence of (+) in (b), taking $F = \Omega$. If $E_P(M_\tau) = 1$, taking conditional expectations under P of M_τ given $\mathcal{F}_{\tau \wedge N}$ yields a uniformly integrable P-martingale which P-almost surely coincides with $(M_{\tau \wedge N})_{N \in \mathbb{N}_0}$; conversely, a uniformly integrable P-martingale $(M_{\tau \wedge N})_{N \in \mathbb{N}_0}$ converges P-almost surely and in $L^1(P)$ to M_τ as $N \to \infty$. This is the equivalence of (γ) and (β) in (d). $\square$

For statistical purposes, the filtered spaces $(\Omega, \mathcal{A}, \mathbb{F})$ in Definition 5.11 and Theorem 5.12 are frequently path spaces for certain classes of stochastic processes.

5.13 Notations. In dimension $d \ge 1$, write C for the space of continuous functions $f : [0, \infty) \to \mathbb{R}^d$ equipped with the topology of locally uniform convergence, and $\mathcal{C}$ for the Borel σ-field. At the same time, $\mathcal{C}$ is the σ-field generated by the coordinate projections $f \to f(t), t \ge 0$ (cf. [11], [64, Chap. VI.1]). On $(C, \mathcal{C})$, we write $(\eta_t)_{t \ge 0}$ for the canonical process. i.e. the process of coordinate projections $\eta_t(f) := f(t)$, $t \ge 0$, $f \in C$. Let $\mathbb{G}$ denote the (right-continuous) filtration generated by η

$$\mathbb{G} = (\mathcal{G}_t)_{t \ge 0}, \quad \mathcal{G}_t := \bigcap_{r > t} \sigma\left(\eta_s : 0 \le s \le r\right),$$

and call $\mathcal{G}_t$ (it contains all events in the process η up to time t and infinitesimally beyond) for short the σ-field of events up to time t.

We start with Brownian motion with unknown drift in dimension $d = 1$, without scaling constant.

5.14 Example. On the path space $(C, \mathcal{C}, \mathbb{G})$ of Notations 5.13 with $d = 1$, consider probability measures

$$Q_\mu := \mathcal{L}\left((B_t + \mu t)_{t \ge 0}\right), \quad \mu \in \mathbb{R}$$

where B is one-dimensional standard Brownian motion starting in 0. For $\mu = 0$, $Q := Q_0$ is Wiener measure on $(C, \mathcal{C})$; the canonical process $(t, \omega) \to \eta(t, \omega)$ is Brownian motion under Q, and Brownian motion with drift μ under Q_μ. We shall show that for all $\mu \in \mathbb{R}$

$$Q_\mu \stackrel{\text{loc}}{\sim} Q \quad \text{relative to } \mathbb{G}$$

and that

(*) $$M_\mu = (M_{\mu,t})_{t \geq 0}, \quad M_{\mu,t} := \exp\left\{\mu \eta_t - \frac{1}{2}\mu^2 t\right\}$$

is the density process of Q_μ with respect to Q relative to $\mathbb{G}$. For pairs $\mu' \neq \mu$ in $\mathbb{R}$, this implies that

(**) $$L^{\mu'/\mu} = \left(L_t^{\mu'/\mu}\right)_{t \geq 0}, \quad L_t^{\mu'/\mu} := \exp\left\{(\mu' - \mu) m_t^{(\mu)} - \frac{1}{2}(\mu' - \mu)^2 t\right\}$$

is the density process of $Q_{\mu'}$ with respect to Q_μ relative to $\mathbb{G}$, with notation $m^{(\mu)}$ for the local martingale part of the observation η under Q_μ (in particular, $L^{\mu/0}$ is the Q-martingale M_μ).

Thus for every fixed time $0 < t < \infty$, with pair (S, J) in Definition 5.2 defined by (η_t, t), we have the structure of a Gaussian shift experiment $\mathcal{E}(t) = (C, \mathcal{G}_t, \{Q_\mu : \mu \in \mathbb{R}\})$ which corresponds to time-continuous observation of a trajectory under unknown μ up to time t.

Proof. Fix $\mu \in \mathbb{R}$ and consider the process M_μ on $(C, \mathcal{C}, Q)$ defined by (*). The process takes values in $(0, \infty)$ and has $M_{\mu,0} = 1$ Q-almost surely. By Ito formula, M_μ is a solution to $dM_{\mu,t} = \mu M_{\mu,t} d\eta_t$ under Q. Since η is a $(Q, \mathbb{G})$-martingale, M_μ is a local $(Q, \mathbb{G})$-martingale and a nonnegative supermartingale. By the classical formula for Laplace transforms of normal laws $\mathcal{N}(0, t)$

$$E_Q(e^{\mu \eta_t}) = E(e^{\mu B_t}) = e^{+\frac{1}{2}\mu^2 t} \quad \text{for } 0 \leq t < \infty$$

M_μ satisfies $E(M_{\mu,t}) = 1$ for all $t < \infty$ and thus is a martingale. Fix a time horizon $T < \infty$ and define from M_μ a probability measure $\widetilde{Q}_{\mu,T}$ on $(C, \mathcal{C})$ with the property

$$\widetilde{Q}_{\mu,T}(A) := E_Q(M_{\mu,T} 1_A) \quad \text{for all } A \in \mathcal{C}$$
$$= E_Q(M_{\mu,s} 1_A) \quad \text{whenever } 0 \leq s \leq T \text{ and } A \in \mathcal{G}_s.$$

M_μ being given by (*), Girsanov theorem (e.g. [14, App. A.3.3], [69, Sect. 3.5], [64, Chap. III.3]) establishes that $(\eta_t)_{t \geq 0}$ remains a semi-martingale under $\widetilde{Q}_{\mu,T}$, with angle bracket under $\widetilde{Q}_{\mu,T}$ identical to the angle bracket $\langle \eta \rangle_t \equiv t$ under Q, and that

$$(\eta_t - \mu(t \wedge T))_{t \geq 0} \quad \text{is a local martingale under } \widetilde{Q}_{\mu,T}.$$

From this, by P. Lévy's characterisation of Brownian motion (e.g. [61, Chap. II.7]):

$$(\eta_t - \mu(t \wedge T))_{t \geq 0} \quad \text{is a Brownian motion under } \widetilde{Q}_{\mu,T}.$$

Section 5.2 *Brownian Motion with Unknown Drift as a Gaussian Shift Experiment

Thus under $\widetilde{Q}_{\mu,T}$ and up to time T, the canonical $(\eta_t)_{0\le t\le T}$ is Brownian motion with drift μ. The σ-field $\mathcal{G}_T$ being generated by the coordinate projections, there is at most one such probability on $\mathcal{G}_T$. Hence the restrictions of the laws $\widetilde{Q}_{\mu,T}$ and Q_μ to $\mathcal{G}_T$ coincide: thus $M_{\mu,T}$ is a version of the density $\frac{dQ_{\mu,T}}{dQ_T}$ which gives $Q_{\mu,T} \sim Q_T$ since M is strictly positive.

As a consequence, the density process of Q_μ with respect to Q coincides with M_μ up to time T; $T < \infty$ being arbitrary, we have identified M_μ as the density process of Q_μ with respect to Q relative to $\mathbb{G}$.

The last part of the assertion is proved as follows: since $Q_{\mu'} \overset{\text{loc}}{\sim} Q_\mu$ for all pairs $\mu' \ne \mu$ in $\mathbb{R}$, the density process $L^{\mu'/\mu}$ of $Q_{\mu'}$ with respect to Q_μ relative to $\mathbb{G}$ is obtained from ratios

$$M_{\mu',t}/M_{\mu,t} = \exp\left((\mu'-\mu)\eta_t - \frac{1}{2}(\mu'^2 - \mu^2)t\right) \quad \text{under } Q$$

where we have to rewrite η in terms of its semi-martingale decomposition under Q_μ: but

$$(\mu'-\mu)\eta_t - \frac{1}{2}(\mu'^2 - \mu^2)t = (\mu'-\mu)(\eta_t - \mu t) - \frac{1}{2}(\mu'-\mu)^2 t$$
$$= (\mu'-\mu)m_t^{(\mu)} - \frac{1}{2}(\mu'-\mu)^2 t$$

where $(m_t^{(\mu)})_{t\ge 0}$ is the martingale part of the canonical process $(\eta_t)_{t\ge 0}$ under Q_μ. $\square$

The general case (dimension $d \ge 1$, scaling matrix for d-dimensional Brownian motion) is as follows.

5.15 The Experiment 'Scaled Brownian Motion with Unknown Drift'. Fix a matrix $\Lambda \in \mathbb{R}^{d\times d}$ which is symmetric and strictly positive definite, and let $\Lambda^{1/2}$ denote its square root. On the path space $(C, \mathcal{C}, \mathbb{G})$ of Notation 5.13 with d-dimensional canonical process $(\eta_t)_{t\ge 0}$, consider probability measures

$$Q_h := \mathcal{L}\left((\Lambda^{1/2} B_t + (\Lambda h)t)_{t\ge 0}\right), \quad h \in \mathbb{R}^d$$

where $B = (B_t)_{t\ge 0}$ is d-dimensional standard Brownian motion starting from $B_0 \equiv 0$. Then we have

$$Q_h \overset{\text{loc}}{\sim} Q \quad \text{relative to } \mathbb{G}$$

for all $h \in \mathbb{R}^d$, with density process

(*) $\qquad L^{h/0} = (L_t^{h/0})_{t\ge 0}, \quad L_t^{h/0} := \exp\left\{h^\top \eta_t - \frac{1}{2} h^\top \Lambda t\, h\right\}$

of Q_h with respect to Q_0 relative to $\mathbb{G}$. For pairs $h' \neq h$ in $\mathbb{R}^d$, this implies that

(**)
$$L^{h'/h} = \left(L_t^{h'/h}\right)_{t \geq 0},$$
$$L_t^{h'/h} := \exp\left\{(h'-h)^\top m_t^{(h)} - \frac{1}{2}(h'-h)^\top \Lambda t\, (h'-h)\right\}$$

is the density process of $Q_{h'}$ with respect to Q_h relative to $\mathbb{G}$, with notation $m^{(h)}$ for the local martingale part of the canonical η under Q_h; note that

$$\mathcal{L}(m^{(h)} \mid Q_h) = \mathcal{L}(\Lambda^{1/2} B) \quad \text{does not depend on } h \in \mathbb{R}^d.$$

For $0 < t < \infty$ fixed and for (S, J) in Definition 5.2 defined by $(\eta_t, \Lambda t)$, we have the structure of a Gaussian shift experiment $\mathcal{E}(\Lambda t) = (C, \mathcal{G}_t, \{Q_h : h \in \mathbb{R}^d\})$ which corresponds to time-continuous observation of the trajectory η under unknown h up to time t.

Proof. Note that in the definition of Q_h, the drift term of the form Λh, $h \in \mathbb{R}^d$, involves the matrix Λ, i.e. the covariance matrix of η_1 under Q_0. Fix $h \in \mathbb{R}^d$. The proof is similar to the proof in Example 5.14, and we mention only the main steps. Again we start from a martingale M_h on $(C, \mathcal{C}, Q_0)$ defined as in (*):

$$M_h = (M_{h,t})_{t \geq 0}, \quad M_{h,t} := \exp\left\{h^\top \eta_t - \frac{1}{2} h^\top \Lambda t\, h\right\}$$

and define for $0 < T < \infty$ probability laws $\widetilde{Q}_{h,T}$ on $(C, \mathcal{C})$ from M_h

$$\widetilde{Q}_{h,T}(A) := E_Q(M_{h,T} 1_A) \quad \text{for all } A \in \mathcal{C}$$

such that by Girsanov's theorem and the definition of M_h

$$(\eta_t - (t \wedge T) \Lambda h)_{t \geq 0} \quad \text{is a local martingale under } \widetilde{Q}_{h,T}.$$

By P. Lévy's characterisation of Brownian motion, this implies that

$$(\eta_t - (t \wedge T) \Lambda h)_{t \geq 0} \quad \text{is a } d\text{-dimensional Brownian motion under } \widetilde{Q}_{h,T}$$

from which we deduce that $\widetilde{Q}_{h,T}$ coincides with the restriction of Q_h to $\mathcal{G}_T$: $M_{h,T}$ being strictly positive, we have $Q_{h,T} \sim Q_{0,T}$, and $M_{h,T}$ is a version of the density $\frac{d\,Q_{h,T}}{d\,Q_{0,T}}$. The remaining parts of the proof are along the lines of Example 5.14. □

5.16 Remark. As a consequence of Example 5.15, we have optimal estimators for the drift parameter in the sense of Boll's Convolution Theorem 5.5, its reformulation in terms of subconvex loss functions (Corollary 5.8), and in the sense of the Minimax Theorem 5.10: when we observe d-dimensional scaled Brownian motion with

Section 5.2 *Brownian Motion with Unknown Drift as a Gaussian Shift Experiment

unknown parameter $h \in \mathbb{R}^d$ up to (deterministic) time $0 < t < \infty$, the central statistic

$$Z = \Lambda^{-1} \begin{pmatrix} \eta_t^{(1)}/t \\ \vdots \\ \eta_t^{(d)}/t \end{pmatrix}$$

(with $\eta^{(1)}, \ldots, \eta^{(d)}$ the components of the canonical process η) is the best equivariant estimator and the minimax estimator for the unknown parameter, by the properties of the Gaussian shift experiment $\mathcal{E}(\Lambda t) = (C, \mathcal{G}_t, \{Q_h : h \in \mathbb{R}^d\})$. Note that this estimator only makes use of the last observation η_t.

Chapter 6

Quadratic Experiments and Mixed Normal Experiments

Topics for Chapter 6:

6.1 Quadratic and Mixed Normal Experiments
Quadratic experiments 6.1–6.1"
Mixed normal experiments 6.2–6.3
Score and observed information in quadratic experiments 6.4–6.4'
Strongly equivariant estimators in mixed normal experiments 6.5–6.5'
Convolution theorem in mixed normal experiments 6.6–6.6"
On approximate equivariance of arbitrary estimators 6.7
Minimax theorem for mixed normal experiments 6.8

6.2 *Likelihood Ratio Processes in Diffusion Models
Assumptions and notations 6.9
Local absolute continuity of laws of solutions to SDE, density processes 6.10
Example: Ornstein–Uhlenbeck model 6.11
Quadratic models attached to the solution of an SDE 6.12
A remark on quadratic variation and change of measure 6.12'

6.3 *Time Changes for Brownian Motion with Unknown Drift
Changing time in the model 'scaled Brownian motion with unknown drift' 6.13
Stopping times inducing quadratic experiments which are not mixed normal 6.14
Observation up to some independent time 6.15
Transformation by independent time change 6.16
Example: one-sided stable processes and Mittag–Leffler processes 6.17–6.18

Exercises: 6.1''', 6.14", 6.15', 6.15", 6.18'

6.1 Quadratic and Mixed Normal Experiments

In this chapter we discuss quadratic and in particular mixed normal experiments. We show that the main results of the preceding chapter – the convolution theorem and the minimax theorem – can be generalised to mixed normal experiments, but do not carry over to general quadratic experiments. We present quadratic and mixed normal experiments in an approach which has been sketched by Davies [19], see also

Jeganathan [66, 67] and Le Cam and Yang [84]. Stochastic process examples leading to quadratic or to mixed normal experiments will be given in Sections 6.2 and 6.3.

It seems natural to generalise Definition 5.2 and to allow for random matrices J taking values in the set

$$D^+ := \{ j \in \mathbb{R}^{d \times d} : j \text{ is symmetric and strictly positive definite} \} \in \mathcal{B}(\mathbb{R}^{d \times d}).$$

6.1 Definition. A statistical model $(\Omega, \mathcal{A}, \{P_h : h \in \mathbb{R}^d\})$ is called a *quadratic experiment* $\mathcal{E}(S, J)$ if there exists a pair (S, J) of statistics

$$S : (\Omega, \mathcal{A}) \to (\mathbb{R}^d, \mathcal{B}(\mathbb{R}^d)), \quad J : (\Omega, \mathcal{A}) \to (\mathbb{R}^{d \times d}, \mathcal{B}(\mathbb{R}^{d \times d}))$$

with the following properties (i) and (ii):
  (i) P_0-almost surely, the statistic J takes values in the set D^+;
  (ii) for every $h \in \mathbb{R}^d$,

$$\omega \longrightarrow \exp\left(h^\top S(\omega) - \frac{1}{2} h^\top J(\omega) h \right) =: L^{h/0}(\omega)$$

is a version of the likelihood ratio of P_h with respect to P_0.
The *central statistic* in a quadratic experiment $\mathcal{E}(S, J)$ is defined as

$$(+) \qquad\qquad Z := 1_{\{J \in D^+\}} J^{-1} S.$$

Frequently we will suppress the indicator in (+), and write for short $Z = J^{-1} S$ instead of (+).

6.1' Remark. (a) In a quadratic experiment $\mathcal{E}(S, J)$ we have $P_h \sim P_0$ for all $h \in \mathbb{R}^d$, and the central statistic $Z = J^{-1} S$ is a maximum likelihood estimator for the unknown parameter $h \in \mathbb{R}^d$.
  (b) In contrast to the Gaussian shift experiment $\mathcal{E}(J)$, cf. Proposition 5.3(c) and (d) or Example 5.1(c), general quadratic experiments $\mathcal{E}(S, J)$ are no longer robust against reparameterisation around reference points $0 \ne h_0 \in \mathbb{R}^d$ (examples will be seen in Chapter 8). So there is no analogue to Proposition 5.3(c) and (d) in Definition 6.1. □

6.1" Remark. We have seen in Example 5.1 that a Gaussian shift experiment $\mathcal{E}(J)$ exists for any given deterministic matrix $J \in \mathbb{R}^{d \times d}$. In contrast to this, it is no longer true that arbitrarily prescribed pairs (S, J) induce quadratic experiments once J is random. The following nontrivial condition (∗) is contained in the formulation of part (ii) of Definition 6.1:

$$(*) \qquad\qquad \int e^{h^\top S - \frac{1}{2} h^\top J h} \, dP_0 = 1 \quad \text{for all } h \in \mathbb{R}^d.$$

This is a necessary and sufficient condition in order that a 'candidate' pair (S, J) on a probability space $(\Omega, \mathcal{A}, P_0)$ generates a quadratic experiment $\mathcal{E}(S, J) = (\Omega, \mathcal{A}, \{P_h : h \in \mathbb{R}^d\})$. □

6.1''' Exercise. Consider the statistical model $(\Omega, \mathcal{A}, \{P_\mu : \mu \in \mathbb{R}\})$

$$\Omega := (0, \infty) \times \mathbb{R}, \quad \mathcal{A} := \mathcal{B}(0, \infty) \otimes \mathcal{B}(\mathbb{R}), \quad P_\mu(dt, dx) := \lambda e^{-\lambda t} dt \, K^{(\mu)}(t, dx)$$

where the second factor represents a family of transition probabilities

$$K^{(\mu)}(t, dx) := \mathcal{N}(\mu t, t)(dx), \quad t \in (0, \infty), \, x \in \mathbb{R}$$

from $((0, \infty), \mathcal{B}(0, \infty))$ to $(\mathbb{R}, \mathcal{B}(\mathbb{R}))$, parameterised by $\mu \in \mathbb{R}$. We keep $0 < \lambda < \infty$ fixed.

(a) Write $f_\mu(t, x)$ for the density of P_μ with respect to Lebesgue measure on $(0, \infty) \times \mathbb{R}$ and show from Example 5.1 that

$$f_\mu(t, x) = e^{\mu x - \frac{1}{2}\mu^2 t} f_0(t, x), \quad t \in (0, \infty), \, x \in \mathbb{R}.$$

Thus, with $S(t, x) = x$ and $J(t, x) = t$, we have a quadratic model in the parameter $\mu \in \mathbb{R}$, as defined in Definition 6.1. Moreover, the model $\{P_\mu : \mu \in \mathbb{R}\}$ is in particular mixed normal in the sense of Definition 6.2 since $\mathcal{L}(J|P_\mu)$ does not depend on the parameter $\mu \in \mathbb{R}$.

(b) For a pair (τ, B) where B and τ are *independent*, where $B = (B_t)_{t \geq 0}$ is a one-dimensional standard Brownian motion starting from $B_0 \equiv 0$ and τ an exponential time with parameter λ, the above model arises as

$$\{P_\mu = \mathcal{L}(\tau, B_\tau + \mu \tau) : \mu \in \mathbb{R}\}.$$

(c) From (a), $\{P_\mu : \mu \in \mathbb{R}\}$ is a curved exponential family in $\zeta(\cdot)$ and T where $\zeta : \mathbb{R} \to \mathbb{R}^2$ is the mapping $\mu \to \zeta(\mu) = (-\frac{1}{2}\mu^2, \mu)$ and $T = \mathrm{id}_{|(0,\infty) \times \mathbb{R}}$ the canonical variable on $(0, \infty) \times \mathbb{R}$. □

6.2 Definition. A *mixed normal experiment* $\mathcal{E}(S, J)$ is a quadratic experiment as in Definition 6.1 for which

$$\mathcal{L}(J \mid P_h) = \mathcal{L}(J \mid P_0) \quad \text{does not depend on the parameter } h \in \mathbb{R}^d.$$

In this case we write $P_\bullet^J$ for $\mathcal{L}(J \mid P_h)$, $h \in \mathbb{R}^d$ arbitrary.

Mixed normality can be characterised as follows.

6.3 Proposition. In a quadratic experiment $\mathcal{E}(S, J)$, the following assertions (i)–(iv) are equivalent:

(i) $\mathcal{E}(S, J)$ is mixed normal;

(ii) $P_0^{S|J=j} = \mathcal{N}(0, j)$ for P_0^J-almost all $j \in \mathbb{R}^{d \times d}$;

Section 6.1 Quadratic and Mixed Normal Experiments   151

(iii) for all $h \in \mathbb{R}^d$ we have $P_h^{S|J=j} = \mathcal{N}(jh, j)$ for P_h^J-almost all $j \in \mathbb{R}^{d \times d}$;

(iv) for all $h \in \mathbb{R}^d$ we have $P_h^{Z-h|J=j} = \mathcal{N}(0, j^{-1})$ for P_h^J-almost all $j \in \mathbb{R}^{d \times d}$.

Proof. (1) By Definition 6.2, mixed normality is equivalent to the assertion

(+)
$$E_0(1_B(J)) = E_h(1_B(J)) = E_0\big(1_B(J)\, L^{h/0}\big)$$
$$\text{for all } h \in \mathbb{R}^d \text{ and all } B \in \mathcal{B}(\mathbb{R}^d).$$

Equality between the first and the third term in (+) is equivalent to any one of the following assertions:

- for every $h \in \mathbb{R}^d$, the constant function 1 is a version of
$$E_0(e^{h^\top S - \frac{1}{2} h^\top J h} \mid J = \cdot);$$

- for every $h \in \mathbb{R}^d$ we have $E_0(e^{h^\top S} \mid J = j) = e^{+\frac{1}{2} h^\top j h}$ for P_0^J-almost all $j \in \mathbb{R}^{d \times d}$

- for every $h \in \mathbb{R}^d$ we have $P_0^{S|J=j} = \mathcal{N}(0, j)$ for P_0^J-almost all $j \in \mathbb{R}^{d \times d}$.

Thus assertion (i)$\Longleftrightarrow$(ii) of the proposition is proved; (iii)$\Longleftrightarrow$(iv) is by definition $Z = 1_{\{J \in D^+\}} J^{-1} S$ of the central statistic. We obviously have (iii)$\Longrightarrow$(ii); so the following step (2) will finish the proof.

(2) Proof of (i)+(ii)$\Longrightarrow$(iii): Under (i) and (ii) we have for arbitrary $A \in \mathcal{B}(\mathbb{R}^d)$, $B \in \mathcal{B}(\mathbb{R}^{d \times d})$:

$$E_h(1_B(J)\, 1_A(S))$$
$$= E_0\left(1_B(J)\, 1_A(S)\, L^{h/0}\right)$$
$$= E_0\left(1_B(J)\, 1_A(S)\, e^{h^\top S - \frac{1}{2} h^\top J h}\right)$$
$$= \int P_0^J(dj)\, 1_B(j)\, e^{-\frac{1}{2} h^\top j h} \left[\int P_0^{S|J=j}(ds)\, 1_A(s)\, e^{h^\top s}\right].$$

From (ii) and the definition of the set D^+ preceding Definition 6.1, the last line reads

$$\int P_0^J(dj)\, 1_{B \cap D^+}(j)\, e^{-\frac{1}{2} h^\top j h} \left[\int_{\mathbb{R}^d} ds\, \frac{1}{\sqrt{(2\pi)^d |\det(j)|}}\, e^{-\frac{1}{2} s^\top j^{-1} s}\, 1_A(s)\, e^{h^\top s}\right],$$

and after rearranging terms $-\frac{1}{2} h^\top j h - \frac{1}{2} s^\top j^{-1} s + h^\top s = -\frac{1}{2} (s-jh)^\top j^{-1} (s-jh)$ gives

$$\int P_0^J(dj)\, 1_{B \cap D^+}(j) \int_{\mathbb{R}^d} ds\, \frac{1}{\sqrt{(2\pi)^d |\det(j)|}}\, e^{-\frac{1}{2}(s-jh)^\top j^{-1}(s-jh)}\, 1_A(s).$$

By (i) and Definition 6.2, the last expression equals

$$\int P_0^J(dj)\, 1_{B \cap D^+}(j)\, \mathcal{N}(jh, j)(A) = \int P_h^J(dj)\, 1_{B \cap D^+}(j)\, \mathcal{N}(jh, j)(A).$$

The set D^+ has full measure under P_h^J since $P_h \sim P_0$. So the complete chain of equalities shows

$$E_h (1_B(J) 1_A(S)) = \int P_h^J(dj) 1_B(j) \mathcal{N}(jh, j)(A)$$

where $A \in \mathcal{B}(\mathbb{R}^d)$ and $B \in \mathcal{B}(\mathbb{R}^{d \times d})$ are arbitrary. This gives (iii) and completes the proof. □

6.4 Definition. In a quadratic experiment $\mathcal{E}(S, J)$, for $h \in \mathbb{R}^d$, we call $S - Jh$ the *score in h* and J the *observed information*.

Note that $S - Jh$ is not a score in h in the classical sense of Definition 1.2: if we have

$$M_h := S - Jh = (\nabla \log f)(h, \cdot) \quad \text{on } \Omega,$$

there is so far no information on integrability properties of M_h under P_h. Similarly, the random object J is not Fisher information in h in the classical sense of Definition 1.2.

6.4' Remark. In *mixed normal* experiments $\mathcal{E}(S, J)$, the notion of 'observed information' due to Barndorff-Nielsen (e.g. [3]) gains a new and deeper sense: the observation $\omega \in \Omega$ itself communicates through the observed value $j = J(\omega)$ the amount of 'information' it carries about the unknown parameter. Indeed from Definition 6.2 and Proposition 6.3(iii), the family of conditional probability measures

$$\left\{ P_h^{\bullet | J = j} : h \in \mathbb{R}^d \right\}$$

is a Gaussian shift experiment $\mathcal{E}(j)$ as discussed in Proposition 5.3 for $P_\bullet^J$-almost all $j \in \mathbb{R}^{d \times d}$; this means that we can condition on the observed information. By Proposition 6.3, this interpretation is valid under mixed normality, and does not carry over to general quadratic experiments.

Conditioning on the observed information, the convolution theorem carries over to mixed normal experiments – as noticed by [66] – provided we strengthen the general notion of Definition 5.4 of equivariance and consider estimators which are equivariant conditional on the observed information.

6.5 Definition. In a mixed normal experiment $\mathcal{E}(S, J)$, an estimator κ for the unknown parameter $h \in \mathbb{R}^d$ is termed *strongly equivariant* if the transition kernel

(∗) $\quad (j, A) \longrightarrow P_h^{(\kappa - h) | J = j}(A), \quad j \in D^+, \ A \in \mathcal{B}(\mathbb{R}^d)$

admits a version which does not depend on the parameter $h \in \mathbb{R}^d$.

Section 6.1 Quadratic and Mixed Normal Experiments

In a mixed normal experiment $\mathcal{E}(S, J)$, the central statistic $Z = 1_{\{J \in D^+\}} J^{-1} S$ is a strongly equivariant estimator for the unknown parameter, by Proposition 6.3(iv). In the particular case of a Gaussian shift experiment $\mathcal{E}(J)$, J is deterministic, and the notions of equivariance (Definition 5.4) and strong equivariance (Definition 6.5) coincide: this follows from a reformulation of Definition 6.5.

6.5' Proposition. Consider a mixed normal experiment $\mathcal{E}(S, J)$ and an estimator κ for the unknown parameter $h \in \mathbb{R}^d$. Then κ is strongly equivariant if and only if

(+) $\quad \mathcal{L}((\kappa - h, J) \mid P_h)$ does not depend on the parameter $h \in \mathbb{R}^d$.

Proof. Clearly Definition 6.2 combined with (∗) of Definition 6.5 gives (+). To prove the converse, fix a countable and ∩-stable generator $\mathcal{S}$ of the σ-field $\mathcal{B}(\mathbb{R}^d)$, and assume that Definition 6.2 holds in combination with (+). Then for every $A \in \mathcal{S}$ and every $h \in \mathbb{R}^d$, we have from (+)

$$E_h\left(1_C(J) 1_A(\kappa - h)\right) = E_0\left(1_C(J) 1_A(\kappa)\right)$$

for all $C \in \mathcal{B}(\mathbb{R}^{d \times d})$, and thus

$$\int P_\bullet^J(dj) 1_C(j) P_h^{\kappa-h|J=j}(A) = \int P_\bullet^J(dj) 1_C(j) P_0^{\kappa|J=j}(A)$$

for all $C \in \mathcal{B}(\mathbb{R}^{d \times d})$. This shows that the functions

$$j \longrightarrow P_h^{\kappa-h|J=j}(A) \quad \text{and} \quad j \longrightarrow P_0^{\kappa|J=j}(A)$$

coincide $P_\bullet^J$-almost surely on $\mathbb{R}^{d \times d}$. Thus there is some $P_\bullet^J$-null set $N_{h,A} \in \mathcal{B}(\mathbb{R}^d)$ such that

$$P_h^{\kappa-h|J=j}(A) = P_0^{\kappa|J=j}(A) \quad \text{for all } j \in D^+ \setminus N_{h,A}$$

where D^+ is the set defined before Definition 6.1. Keeping h fixed, the countable union $N_h = \bigcup_{A \in \mathcal{S}} N_{h,A}$ is again a $P_\bullet^J$-null set in $\mathcal{B}(\mathbb{R}^d)$, and we have

$$j \in D^+ \setminus N_h : \quad P_h^{\kappa-h|J=j}(A) = P_0^{\kappa|J=j}(A) \quad \text{for all } A \in \mathcal{S}.$$

$\mathcal{S}$ being a ∩-stable generator of $\mathcal{B}(\mathbb{R}^d)$, we infer:

(×) $\quad$ the probability laws $P_h^{\kappa-h|J=j}(\cdot)$ and $P_0^{\kappa|J=j}(\cdot)$ coincide on $(\mathbb{R}^d, \mathcal{B}(\mathbb{R}^d))$ when $j \in D^+ \setminus N_h$.

Let $K(\cdot, \cdot)$ denote a regular version of the conditional law $(j, A) \to P_0^{\kappa|J=j}(A)$, $j \in D^+$, $A \in \mathcal{B}(\mathbb{R}^d)$. Since $D^+ \setminus N_h$ is a set of full measure under $P_\bullet^J$, assertion (×) shows that $K(\cdot, \cdot)$ is at the same time a regular version of $(j, A) \to P_h^{\kappa-h|J=j}(A)$.

So far, $h \in \mathbb{R}^d$ was fixed but arbitrary: hence $K(\cdot,\cdot)$ is a regular version simultaneously for all conditional laws $(j, A) \to P_h^{\kappa-h|J=j}(A)$, $h \in \mathbb{R}^d$. We thus have (∗) in Definition 6.5 which finishes the proof. □

We state the first main result on estimation in mixed normal experiments, due to Jeganathan [66].

6.6 Convolution Theorem. Consider a strongly equivariant estimator κ in a mixed normal experiment $\mathcal{E}(S, J)$. Then there exist probability measures $\{Q^j : j \in \mathsf{D}^+\}$ on $(\mathbb{R}^d, \mathcal{B}(\mathbb{R}^d))$ such that

for $P_\bullet^J$-almost all $j \in \mathsf{D}^+$:

$$P_h^{(\kappa-h)|J=j} = \mathcal{N}(0, j^{-1}) \star Q^j \quad \text{does not depend on } h \in \mathbb{R}^d,$$

and estimation errors of κ are distributed as

$$\mathcal{L}(\kappa - h \mid P_h) = \int P_\bullet^J(dj) \left[\mathcal{N}(0, j^{-1}) \star Q^j \right]$$

independently of $h \in \mathbb{R}^d$. In addition, $(j, A) \to Q^j(A)$ is obtained as a regular version of the conditional law $P_0^{(\kappa-Z)|J=j}(A)$ where $j \in \mathsf{D}^+$ and $A \in \mathcal{B}(\mathbb{R}^d)$.

Proof. Since κ is a strongly equivariant estimator for the unknown parameter $h \in \mathbb{R}^d$, it is sufficient to determine a family of probability measures $\{Q^j : j \in \mathbb{R}^{d \times d}\}$ on $(\mathbb{R}^d, \mathcal{B}(\mathbb{R}^d))$ such that

for $P_\bullet^J$-almost all $j \in \mathsf{D}^+$: $\quad P_0^{\kappa|J=j}(A) = \mathcal{N}(0, j^{-1}) \star Q^j$.

Prepare a regular version $\widetilde{K}(\cdot,\cdot)$ of the conditional law of the pair (κ, S) given $J = \cdot$ under P_0

$$(j, C) \longrightarrow \widetilde{K}(j, C) := P_0^{(\kappa, S)|J=j}(C), \quad j \in \mathsf{D}^+, \quad C \in \mathcal{B}(\mathbb{R}^d) \otimes \mathcal{B}(\mathbb{R}^d)$$

with second marginals specified according to Proposition 6.3(ii):

$$\widetilde{K}(j, \mathbb{R}^d \times \cdot) = \mathcal{N}(0, j) \quad \text{for all } j \in \mathsf{D}^+.$$

(1) For $t \in \mathbb{R}^d$ fixed, for arbitrary events $B \in \mathcal{B}(\mathbb{R}^{d \times d})$ and arbitrary $h \in \mathbb{R}^d$, we start from Proposition 6.5'

$$E_0\bigl(1_B(J)\, e^{i t^\top \kappa}\bigr) = E_h\bigl(1_B(J)\, e^{i t^\top (\kappa - h)}\bigr)$$

Section 6.1 Quadratic and Mixed Normal Experiments   155

which gives on the right-hand side

$$E_0\left(1_B(J)\,e^{it^\top \kappa}\right)$$
$$= E_0\left(1_B(J)\,e^{it^\top(\kappa-h)}\,L^{h/0}\right) = E_0\left(1_B(J)\,e^{it^\top(\kappa-h)}\,e^{h^\top S - \frac{1}{2}h^\top J h}\right)$$
$$= \int P_0^J(dj)\,1_B(j)\left[\int \widetilde{K}(j,(dk,ds))\,e^{it^\top(k-h)}\,e^{h^\top s - \frac{1}{2}h^\top j h}\right]$$
$$= \int P_\bullet^J(dj)\,1_{B \cap D^+}(j)\left[\int \widetilde{K}(j,(dk,ds))\,e^{it^\top(k-h)}\,e^{h^\top s - \frac{1}{2}h^\top j h}\right]$$

and on the left-hand side

$$E_0\left(1_B(J)\,e^{it^\top \kappa}\right) = \int P_\bullet^J(dj)\,1_{B \cap D^+}(j)\left[\int \widetilde{K}(j,(dk,ds))\,e^{it^\top k}\right]$$

by definition of $\widetilde{K}(\cdot,\cdot)$. We compare the expressions in square brackets in both equalities: since $B \in \mathcal{B}(\mathbb{R}^{d \times d})$ is arbitrary and since $P_\bullet^J$ is concentrated on D^+, the functions

$$j \longrightarrow \int \widetilde{K}(j,(dk,ds))\,e^{it^\top(k-h)}\,e^{h^\top s - \frac{1}{2}h^\top j h}$$

and

$$j \longrightarrow \int \widetilde{K}(j,(dk,ds))\,e^{it^\top k}$$

coincide $P_\bullet^J$-almost surely on D^+, for every fixed value of $h \in \mathbb{R}^d$. Let N_h denote an exceptional $P_\bullet^J$-null set with respect to a fixed value of h in the last formula. The countable union $N := \bigcup_{h \in \mathbb{Q}^d} N_h$ is a $P_\bullet^J$-null set in D^+ with the property

($\diamond$) $\begin{cases} \int_{\mathbb{R}^d \times \mathbb{R}^d} \widetilde{K}(j,(dk,ds))\,e^{it^\top(k-h)}\,e^{h^\top s - \frac{1}{2}h^\top j h} \\ \quad = \int_{\mathbb{R}^d \times \mathbb{R}^d} \widetilde{K}(j,(dk,ds))\,e^{it^\top k} \\ \text{for all } j \in D^+ \setminus N \text{ and for all } h \in \mathbb{Q}^d. \end{cases}$

(2) Now we fix $j \in D^+ \setminus N$. We consider the first integral in ($\diamond$) as a function of h

$$\mathbb{R}^d \ni h \longrightarrow \int \widetilde{K}(j,(dk,ds))\,e^{it^\top(k-h)}\,e^{h^\top s - \frac{1}{2}h^\top j h}$$

which as in the proof of Theorem 5.5 is extended to an analytic function

$$\mathbb{C}^d \ni z \longrightarrow \int \widetilde{K}(j,(dk,ds))\,e^{it^\top(k-z)}\,e^{z^\top s - \frac{1}{2}z^\top j z} =: f_j(z).$$

The second integral in ($\diamond$) equals $f_j(0)$. The function f_j being analytic, assertion ($\diamond$) yields

for all $j \in D^+ \setminus N$: the function $f_j(\cdot)$ is constant on $\mathbb{C}^d$.

In particular, for all $j \in D^+ \setminus N$,
$$f_j(0) = f_j(-i\, j^{-1} t)$$
which gives in analogy to (∗) in the proof of Theorem 5.5

(+) $\quad \int \widetilde{K}(j,(dk,ds))\, e^{it^\top k} = e^{-\frac{1}{2} t^\top j^{-1} t} \int \widetilde{K}(j,(dk,ds))\, e^{it^\top (k-j^{-1}s)}.$

By definition of $\widetilde{K}(\cdot,\cdot)$ we have proved

(++) $\quad$ for all $j \in D^+ \setminus N$:
$$E_0\!\left(e^{it^\top \kappa} \mid J = j\right) = e^{-\frac{1}{2} t^\top j^{-1} t}\, E_0\!\left(e^{it^\top (\kappa - Z)} \mid J = j\right).$$

(3) So far, we have considered some fixed value of $t \in \mathbb{R}^d$. Hence the $P_\bullet^J$-null set N in (++) depends on t. Taking the union of such null sets for all $t \in \mathbb{Q}^d$ we obtain a $P_\bullet^J$-null set $\widetilde{N}$ for which dominated convergence with respect to the argument t in both integrals in (+) establishes

for all $j \in D^+ \setminus \widetilde{N}$:
$$E_0\!\left(e^{it^\top \kappa} \mid J = j\right) = e^{-\frac{1}{2} t^\top j^{-1} t}\, E_0\!\left(e^{it^\top (\kappa - Z)} \mid J = j\right), \quad t \in \mathbb{R}^d.$$

This is an equation between characteristic functions of conditional laws. Introducing a regular version
$$(j, A) \longrightarrow Q^j(A) = P_0^{(\kappa - Z)|J=j}(A)$$
for the conditional law of $\kappa - Z$ given $J = \cdot$ under P_0 we have

for all $j \in D^+ \setminus \widetilde{N}$, all $A \in \mathcal{B}(\mathbb{R}^d)$: $\quad P_0^{\kappa|J=j}(A) = [\mathcal{N}(0, j^{-1}) \star Q^j](A).$

This finishes the proof of the convolution theorem in the mixed normal experiment $\mathcal{E}(S, J)$. $\square$

In combination with Proposition 6.3(iv), the Convolution Theorem 6.6 shows that within the class of all strongly equivariant estimators for the unknown parameter $h \in \mathbb{R}^d$ in a mixed normal experiment $\mathcal{E}(S, J)$, the central statistic $Z = 1_{\{J \in D^+\}} J^{-1} S$ achieves minimal spread of estimation errors, or equivalently, achieves the best possible concentration of an estimator around the true value of the parameter. In analogy to Corollary 5.8 this can be reformulated – again using Anderson's Lemma 5.7 – as follows:

6.6' Corollary. In a mixed normal experiment $\mathcal{E}(S, J)$, with respect to any subconvex loss function, the central statistic Z minimises the risk
$$R(\kappa, h) \geq R(Z, h), \quad h \in \mathbb{R}^d$$
in the class of all strongly equivariant estimators κ for the unknown parameter.

Section 6.1 Quadratic and Mixed Normal Experiments

6.6" Remark. In the Convolution Theorem 6.6 and in Corollary 6.6', the best concentrated distribution for estimation errors

$$\mathcal{L}(Z \mid P_0) = \mathcal{L}\left(\int P_{\bullet}^J(dj) \, \mathcal{N}\left(0, j^{-1}\right)\right)$$

does not necessarily admit finite second moments. As an example, the mixing distribution $P_{\bullet}^J$ might be in dimension $d = 1$ some Gamma law $\Gamma(a, p)$ with shape parameter $a \in (0, 1)$: then

$$E_{P_0}(Z^2) = \int_{(0,\infty)} \Gamma(a, p)(du) \frac{1}{u} = \infty$$

and the central statistic Z does not belong to $L^2(P_0)$. We shall see examples in Chapter 8. When $\mathcal{L}(Z|P_0)$ does not admit finite second moments, comparison of estimators based on squared loss $\ell(x) = x^2$ – or based on polynomial loss functions – does not make sense, thus bounded loss functions are of intrinsic importance in Theorem 6.6 and Corollary 6.6', or in Lemma 6.7 and Theorem 6.8 below. Note that under mixed normality, estimation errors are always compared conditionally on the observed information, never globally.

In mixed normal experiments, one would like to be able to compare the central statistic Z not only to strongly equivariant estimators, but to arbitrary estimators for the unknown parameter $h \in \mathbb{R}^d$. Again we can condition on the observed information to prove in analogy to Lemma 5.9 that arbitrary estimators are 'approximately strongly invariant' under a 'very diffuse' prior, i.e. in the absence of any a priori information except that the unknown parameter should range over large balls centred at the origin. Again B_n denotes the closed ball $\{x \in \mathbb{R}^d : |x| \leq n\}$, and π_n the uniform law on B_n.

6.7 Lemma. In a mixed normal experiment $\mathcal{E}(S, J)$, every estimator η for the unknown parameter $h \in \mathbb{R}^d$ can be associated to probability measures $\{Q_n^j : j \in D^+, n \geq 1\}$ on $(\mathbb{R}^d, \mathcal{B}(\mathbb{R}^d))$ such that

$$d_1\left(\int \pi_n(dh) \, \mathcal{L}(\eta - h|P_h), \int P_{\bullet}^J(dj)\left[\mathcal{N}(0, j^{-1}) \star Q_n^j\right]\right) \longrightarrow 0$$

as $n \to \infty$,

and for every bounded subconvex loss function the risks

$$R(\eta, B_n) = \int \pi_n(dh) \, E_h(\ell(\eta - h))$$

can be represented as $n \to \infty$ in the form

$$R(\eta, B_n) = \int P_0^J(dj) \int \left[\mathcal{N}(0, j^{-1}) \star Q_n^j\right](dv) \, \ell(v) + o(1), \quad n \to \infty.$$

Proof. The second assertion is a consequence of the first since $\ell(\cdot)$ is measurable and bounded, by (+) in Remark 5.8'. We prove the first assertion in analogy to the proof of Lemma 5.9.

(1) The pair (Z, J) is a sufficient statistic in the mixed normal experiment $\mathcal{E}(S, J)$. Thus, for any random variable η taking values in $(\mathbb{R}^d, \mathcal{B}(\mathbb{R}^d))$, there is a regular version of the conditional law of η given $(Z, J) = \cdot$ which does not depend on the parameter $h \in \mathbb{R}^d$. We fix a transition kernel $\widetilde{K}(\cdot, \cdot)$ from $\mathbb{R}^d \times D^+$ to $\mathbb{R}^d$ which provides a common regular version

$$\widetilde{K}((z, j), A) = P_h^{\eta \mid (Z,J)=(z,j)}(A), \quad A \in \mathbb{R}^d, \ (z, j) \in \mathbb{R}^d \times D^+$$

under all values $h \in \mathbb{R}^d$ of the parameter. When we wish to keep $j \in D^+$ fixed, we write for short

$$K^j(z, A) := \widetilde{K}((z, j), A), \quad A \in \mathbb{R}^d, \ z \in \mathbb{R}^d.$$

Sufficiency also allows to consider mixtures

$$Q_n^j(A) := \int \pi_n(dh) \, P_\bullet^{(\eta-Z)\mid(Z,J)=(h,j)}(A), \quad A \in \mathcal{B}(\mathbb{R}^d)$$

for every $j \in D^+$ and every $n \in \mathbb{N}$.

(2) With notations introduced in step (1), we have from Proposition 6.3(iv)

$$\int \pi_n(dh) \, P_h(\eta - h \in A)$$
$$= \int \int \int \pi_n(dh) \, P_h^J(dj) \, P_h^{Z\mid J=j}(dz) \, \widetilde{K}((z, j), A + h)$$
$$= \int P_\bullet^J(dj) \int \pi_n(dh) \int \mathcal{N}(h, j^{-1})(dz) \, K^j(z, A + h)$$
$$= \int P_\bullet^J(dj) \int \mathcal{N}(0, j^{-1})(dx) \left[\int \pi_n(dh) \, K^j(x + h, A + h) \right]$$

in analogy to step (3) of the proof of Lemma 5.9.

(3) In analogy to step (2) in the proof of Lemma 5.9, we have bounds

$$(*) \quad \left| \int \pi_n(dh) \, K^j(x + h, A + h) - Q_n^j(A - x) \right| \leq \frac{\lambda(B_n \triangle (B_n + x))}{\lambda(B_n)}$$

which are uniform in $A \in \mathcal{B}(\mathbb{R}^d)$ and $j \in D^+$. To check this, start from

$$\frac{1}{\lambda(B_n)} \int \lambda(dh) \, 1_{B_n}(h) \, K^j(x + h, A + h)$$
$$= \frac{1}{\lambda(B_n)} \int \lambda(dh') \, 1_{B_n + x}(h') \, K^j(h', A - x + h')$$

Section 6.1 Quadratic and Mixed Normal Experiments

which by definition of $\widetilde{K}(\cdot,\cdot)$ equals

$$\frac{1}{\lambda(B_n)} \int \lambda(dh') \, 1_{B_n+x}(h') \, \widetilde{K}\left((h',j), A - x + h'\right)$$

$$= \frac{1}{\lambda(B_n)} \int \lambda(dh') \, 1_{B_n+x}(h') \, P_\bullet\left(\eta \in (A - x + h') \mid (Z,J) = (h',j)\right)$$

$$= \frac{1}{\lambda(B_n)} \int \lambda(dh') \, 1_{B_n+x}(h') \, P_\bullet\left((\eta - Z) \in (A - x) \mid (Z,J) = (h',j)\right).$$

Now the last expression

$$\frac{1}{\lambda(B_n)} \int \lambda(dh') \, 1_{B_n+x}(h') \, P_\bullet^{(\eta-Z)|(Z,J)=(h',j)}(A - x)$$

can be compared to

$$Q_n^j(A - x) = \frac{1}{\lambda(B_n)} \int \lambda(dh') \, 1_{B_n}(h') \, P_\bullet^{(\eta-Z)|(Z,J)=(h',j)}(A - x)$$

up to the error bound on the right-hand side of (∗), uniformly in $A \in \mathcal{B}(\mathbb{R}^d)$ and $j \in D^+$.

(4) Combining steps (2) and (3), we approximate

$$\int \pi_n(dh) \, P_h(\eta - h \in A) \quad \text{by} \quad \int P_\bullet^J(dj) \int \mathcal{N}(0, j^{-1})(dx) \, Q_n^j(A - x)$$

uniformly in $A \in \mathcal{B}(\mathbb{R}^d)$ as $n \to \infty$, up to error terms

$$\int P_\bullet^J(dj) \int \mathcal{N}(0, j^{-1})(dx) \, \frac{\lambda(B_n \triangle (B_n + x))}{\lambda(B_n)}$$

as in the proof of Lemma 5.9. Using dominated convergence twice, these vanish as $n \to \infty$. □

The following is – together with the Convolution Theorem 6.6 – the second main result on estimation in mixed normal experiments.

6.8 Minimax Theorem. In a mixed normal experiment $\mathcal{E}(S, J)$, the central statistic Z is a minimax estimator for the unknown parameter with respect to any subconvex loss function $\ell(\cdot)$: we have

$$\sup_{h \in \mathbb{R}^d} R(\eta, h) \geq E_0(\ell(Z)) = \int\int P_\bullet^J(dj) \, P_0^{Z|J=j}(dz) \, \ell(z) = \sup_{h \in \mathbb{R}^d} R(Z, h).$$

Proof. Consider any estimator η for the unknown parameter $h \in \mathbb{R}^d$ in $\mathcal{E}(S, J)$, and any subconvex loss function $\ell(\cdot)$. Since $E_0(\ell(Z))$ (finite or not) is the increasing limit of $E_0([\ell \wedge N](Z))$ as $N \to \infty$ where all $\ell \wedge N$ are subconvex loss functions, it is sufficient to prove a lower bound

$$\sup_{h \in \mathbb{R}^d} R(\eta, h) \geq E_0(\ell(Z)) = R(Z, 0)$$

for subconvex loss functions which are bounded. Then we have a chain of inequalities (its first two '$\geq$' are trivial) where we can apply Lemma 6.7:

$$\sup_{h \in \mathbb{R}^d} R(\eta, h) \geq \sup_{h \in B_n} R(\eta, h) \geq \int \pi_n(dh) \, E_h\left(\ell(\eta - h)\right) = R(\eta, B_n)$$

$$= \int P_0^J(dj) \int \left[\mathcal{N}(0, j^{-1}) \star Q_n^j\right](dv) \, \ell(v) + o(1) \quad \text{as} \quad n \to \infty.$$

Now Anderson's Lemma 5.7 allows us to compare integral terms for all n, j fixed, and gives lower bounds

$$\geq \int P_\bullet^J(dj) \int \mathcal{N}(0, j^{-1})(dv) \, \ell(v) + o(1) \quad \text{as} \quad n \to \infty.$$

The last integral does not depend on n, and by Definition 6.2 and Proposition 6.3(iv) equals $E_h(\ell(Z - h)) = R(Z, h)$ for arbitrary $h \in \mathbb{R}^d$. This finishes the proof. $\square$

What happens beyond mixed normality, in the genuinely quadratic case? We know that $Z = J^{-1}S$ is a maximum likelihood estimator for the unknown parameter, by Remark 6.1', we know that for $J(\omega) \in \mathbb{D}^+$ the log-likelihood surface

$$h \longrightarrow \Lambda^{h/0}(\omega) = h^\top S(\omega) - \frac{1}{2} h^\top J(\omega) h$$

$$= -\frac{1}{2}(h - Z(\omega))^\top J(\omega) (h - Z(\omega)) + \text{expressions not depending on } h$$

has the shape of a parabola which opens towards $-\infty$ and which admits a unique maximum at $Z(\omega)$, but this is no optimality criterion. For quadratic experiments which are not mixed normal, satisfactory optimality results seem unknown. For squared loss, Gushchin [38, Thm. 1, assertion 3] proves admissibility of the ML estimator Z under random norming by J (which makes the randomly normed estimation errors at h coincide with the score $S - Jh$ at h) in dimension $d = 1$. He also has Cramér–Rao type results for restricted families of estimators. Beyond the setting of mixed normality, results which allow to distinguish an optimal estimator from its competitors simultaneously under a sufficiently large class of loss functions – such as those in the convolution theorem or in the minimax theorem – seem unknown.

6.2 *Likelihood Ratio Processes in Diffusion Models

Statistical models for diffusion processes provide natural examples for quadratic experiments. For laws of solutions of stochastic differential equations, we consider first the problem of local absolute continuity or local equivalence of probability measures and the structure of the density processes (this relies on Liptser and Shiryayev [88] and Jacod and Shiryaev [64]), then we specialise to settings where quadratic statistical models arise. Since this section is preceded by an asterisk *, a reader not interested in stochastic processes may skip this and go directly to Chapter 7.

As in Notation 5.13, $(C, \mathcal{C}, \mathbb{G})$ is the canonical path space for $\mathbb{R}^d$-valued stochastic processes with continuous paths; $\mathbb{G}$ is the (right-continuous) filtration generated by the canonical process $\eta = (\eta_t)_{t \geq 0}$. Local absolute continuity and local equivalence of probability measures were introduced in Definition 5.11; density processes were defined in Theorem 5.12.

6.9 Assumptions and Notations for Section 6.2. (a) We consider d-dimensional stochastic differential equations (SDE)

(I) $\qquad dX_t = b(t, X_t)\,dt + \sigma(t, X_t)\,dW_t, \quad t \geq 0, \quad X_0 \equiv x_0$

(II) $\qquad dX'_t = b'(t, X'_t)\,dt + \sigma(t, X'_t)\,dW_t, \quad t \geq 0, \quad X_0 \equiv x_0$

driven by m-dimensional Brownian motion W. SDE (I) and (II) share the same diffusion coefficient

$$\sigma : [0,\infty) \times \mathbb{R}^d \longrightarrow \mathbb{R}^{d \times m}$$

but have different drift coefficients

$$b, b' : [0,\infty) \times \mathbb{R}^d \longrightarrow \mathbb{R}^d.$$

Both equations (I) and (II) have the same starting value $x_0 \in \mathbb{R}^d$. All coefficients are measurable in (t, x); we assume Lipschitz continuity in the second variable

$$|b(t, x) - b(t, y)| + |b'(t, x) - b'(t, y)| + \|\sigma(t, x) - \sigma(t, y)\| \leq K |x - y|$$

(for $t \geq 0$ and $x, y \in \mathbb{R}^d$) together with linear growth

$$|b(t, x)|^2 + |b'(t, x)|^2 + \|\sigma(t, x)\|^2 \leq K(1 + |x|^2)$$

where K is some global constant. The $d \times d$-matrices

$$c(t, x) := \sigma(t, x)\sigma^\top(t, x)$$

will be important in the sequel.

(c) Under the assumptions of (b), equations (I) and (II) have unique strong solutions (see e.g. [69, Chap. 5.2.B]) on the probability space $(\Omega, \mathcal{A}, \mathbb{F}, P)$ carrying the driving Brownian motion W, with $\mathbb{F}$ the filtration generated by W. We will be interested in the laws of the solutions

$$Q := \mathcal{L}(X \mid P) \quad \text{and} \quad Q' := \mathcal{L}(X' \mid P)$$

on the canonical path space $(C, \mathcal{C}, \mathbb{G})$.

(d) We write m^X and $m^{X'}$ for the $(P, \mathbb{F})$-local martingale parts of X and X'

$$m_t^X = X_t - X_0 - \int_0^t b(s, X_s)\, ds, \quad m_t^{X'} = X_t' - X_0' - \int_0^t b(s, X_s')\, ds.$$

Their angle brackets are the predictable processes

$$\langle m^X \rangle_t = \int_0^t c(s, X_s)\, ds, \quad \langle m^{X'} \rangle_t = \int_0^t c(s, X_s')\, ds$$

taking values in the space of symmetric and non-negative definite $d \times d$-matrices.

We shall explain in Remark 6.12' below why equations (I) and (II) have been assumed to share the same diffusion coefficient. The next result is proved with the arguments which [88, Sect. 7] use on the way to their Theorems 7.7 and 7.18. See also [64, pp. 159–160, 179–181, and 187]).

6.10 Theorem. Let the drift coefficients b in (I) and b' in (II) be such that a measurable function

$$\gamma : [0, \infty) \times \mathbb{R}^d \longrightarrow \mathbb{R}^d$$

exists which satisfies the following conditions (+) and (++):

(+) $\quad b'(t, x) = b(t, x) + c(t, x)\gamma(t, x), \quad t \geq 0, \, x \in \mathbb{R}^d,$

(++) $\quad \int_0^t \left(\gamma^\top c\, \gamma \right)(s, \eta_s)\, ds < \infty \quad$ for all $t < \infty$, Q- and Q'-almost surely.

(a) Then the probability laws Q' and Q on $(C, \mathcal{C})$ are locally equivalent relative to $\mathbb{G}$.
(b) The density process of Q' with respect to Q relative to $\mathbb{G}$ is the $(Q, \mathbb{G})$-martingale

$$L = (L_t)_{t \geq 0}, \quad L_t = \exp\left\{ \int_0^t \gamma^\top (s, \eta_s)\, dm_s^Q - \frac{1}{2} \int_0^t \left(\gamma^\top c\, \gamma \right)(s, \eta_s)\, ds \right\}$$

where m^Q denotes the local martingale part of the canonical process η under Q:

$$m^Q = (m_t^Q)_{t \geq 0}, \quad m_t^Q = \eta_t - \eta_0 - \int_0^t b(s, \eta_s)\, ds$$

and where the local martingale

$$\int \gamma^\top (s, \eta_s)\, dm_s^Q =: M$$

under Q has angle bracket

$$\langle M \rangle_t = \int_0^t \left(\gamma^\top c\, \gamma \right)(s, \eta_s)\, ds, \quad t \geq 0.$$

Section 6.2 *Likelihood Ratio Processes in Diffusion Models   163

(c) The process L in (b) is the exponential

$$L_t = \exp\left\{M_t - \frac{1}{2}\langle M\rangle_t\right\}, \quad t \geq 0$$

of M under Q, in the sense of a solution $(L_t)_{t\geq 0}$ under Q to the SDE

$$L_t = 1 + \int_0^t L_t\, dM_t, \quad t \geq 0.$$

A usual notation for the exponential of M under Q is $\mathcal{E}_Q(M)$.

Proof. (1) On $(C, \mathcal{C}, \mathbb{G})$ define stopping times

$$\rho_n := \inf\left\{t > 0 : \int_0^t \left(\gamma^\top c\, \gamma\right)(s, \eta_s)\, ds > n\right\}, \quad n \geq 1$$

(which may take the value $+\infty$ with positive probability under Q or under Q') and write

$$\gamma^{(n)}(s, \eta_s) := \gamma(s, \eta_s)\, 1_{[[0, \rho_n[[}(s), \quad n \geq 1.$$

By assumption (++) which is symmetric in Q and Q', we have

$$\rho_n \uparrow \infty \qquad Q\text{- and } Q'\text{-almost surely.}$$

(2) For every $n \in \mathbb{N}$, on $(\Omega, \mathcal{A}, \mathbb{F}, P)$, we also have unique strong solutions for equations $(\mathrm{II}^{(n)})$:

$$(\mathrm{II}^{(n)}) \quad dX_t^{(n)} = [b + c\gamma^{(n)}](t, X_t^{(n)})\, dt + \sigma(t, X_t^{(n)})\, dW_t, \quad t \geq 0, \quad X_0 \equiv x_0.$$

By unicity, $X^{(n)}$ coincides on $[[0, \sigma_n]]$ with the solution X' to SDE (II) where σ_n is the $\mathbb{F}$-stopping time $\sigma_n := \rho_n \circ X'$. If we write $Q^{(n)}$ for the law $\mathcal{L}(X^{(n)}|P)$ on $(C, \mathcal{C}, \mathbb{G})$, then the laws $Q^{(n)}$ and Q' coincide in restriction to the σ-field $\mathcal{G}_{\rho_n}$ of events up to time ρ_n.

(3) On $(C, \mathcal{C}, \mathbb{G})$, the Novikov criterion [61, p. 152] applies – as a consequence of the stopping in step (1) – and grants that for fixed $n \in \mathbb{N}$

$$\check{L}_t^{(n)} = \exp\left\{\int_0^t [\gamma^{(n)}]^\top(s, \eta_s)\, dm_s^Q - \frac{1}{2}\int_0^t \left([\gamma^{(n)}]^\top c\, [\gamma^{(n)}]\right)(s, \eta_s)\, ds\right\}, \quad t \geq 0$$

is a Q-martingale. Thus $\check{L}^{(n)}$ defines a probability measure $\check{Q}^{(n)}$ on $(C, \mathcal{C}, \mathbb{G})$ via $\check{Q}^{(n)}(A) = \int_A \check{L}_t^{(n)}\, dQ$ if $A \in \mathcal{G}_t$. Writing β for arbitrary directions in $\mathbb{R}^d$, Girsanov's theorem [64, Thm. III.3.11] then states that

$$\beta^\top m^Q - \left\langle \beta^\top m^Q, \int_0^\cdot [\gamma^{(n)}]^\top(s, \eta_s)\, dm_s^Q\right\rangle, \quad 1 \leq i \leq d$$

is a $\check{Q}^{(n)}$-local martingale whose angle bracket under $\check{Q}^{(n)}$ coincides with the angle bracket of $\beta^\top m^Q$ under Q. For the d-dimensional process m^Q this shows that

$$m^Q - \int_0^\cdot [c\,\gamma^{(n)}](s,\eta_s)\,ds$$

is a $\check{Q}^{(n)}$-local martingale whose angle bracket under $\check{Q}^{(n)}$ coincides with the angle bracket of m^Q under Q. Since $\mathcal{C}$ is generated by the coordinate projections, this signifies that $\check{Q}^{(n)}$ is the unique law on $(C,\mathcal{C},\mathbb{G})$ such that the canonical process η under $\check{Q}^{(n)}$ is a solution to SDE ($\mathrm{II}^{(n)}$). As a consequence, the two laws coincide: $Q^{(n)}$ from step (2) is the same law as $\check{Q}^{(n)}$, on $(C,\mathcal{C},\mathbb{G})$.

(4) For every n fixed, step (3) shows the following: the laws $Q^{(n)}$ and Q are locally equivalent relative to $\mathbb{G}$ since the density process of $Q^{(n)}$ with respect to Q relative to $\mathbb{G}$

$$L_{\bullet \wedge \rho_n} = \check{L}^{(n)}$$

is strictly positive Q-almost surely. Recall also that the laws $Q^{(n)}$ and Q' coincide in restriction to $\mathcal{G}_{\rho_n}$.

(5) The following argument proves $Q' \overset{\mathrm{loc}}{\ll} Q$ relative to $\mathbb{G}$. For every $t < \infty$,

$$\begin{aligned} A \in \mathcal{G}_t \quad &\Longrightarrow \quad Q'(A) = P\left(X'_{\bullet \wedge t} \in A\right) \\ &\leq P\left(X'_{\bullet \wedge t} \in A,\ \sigma_n > t\right) + P(\sigma_n \leq t) \\ &\leq P\left(X^{(n)}_{\bullet \wedge t} \in A\right) + P(\sigma_n \leq t) \\ &= Q^{(n)}(A) + Q'(\rho_n \leq t) \\ &= Q^{(n)}(A) + o(1) \quad \text{as } n \to \infty \end{aligned}$$

since X' and $X^{(n)}$ coincide up to time $\sigma_n = \rho_n \circ X'$ as above, and since $\rho_n \uparrow \infty$ Q'-almost surely. Now, $Q^{(n)}$ and Q being locally equivalent relative to $\mathbb{G}$ for all n, by step (4), the above gives

$$A \in \mathcal{G}_t,\ Q(A) = 0 \quad \Longrightarrow \quad Q^{(n)}(A) = 0\ \forall n \geq 1 \quad \Longrightarrow \quad Q'(A) = 0$$

which proves $Q' \overset{\mathrm{loc}}{\ll} Q$ relative to $\mathbb{G}$.

(6) We prove that L is the density process of Q' with respect to Q relative to $\mathbb{G}$. For $A \in \mathcal{G}_t$, the event $A \cap \{t < \rho_n\}$ belongs to $\mathcal{G}_{\rho_n^-}$, the σ-field of events strictly before time ρ_n (for a $\mathbb{G}$-stopping time S, $\mathcal{G}_{S-}$ is defined as the smallest σ-field containing the system

$$\mathcal{G}_0 \cup \{G \cap \{s < S\} : G \in \mathcal{G}_s,\ 0 \leq s < \infty\},$$

cf. [98, p. 17], and is contained in $\mathcal{G}_S$). Thus $A \cap \{t < \rho_n\} \in \mathcal{G}_{\rho_n}$: as a consequence,

$$\begin{aligned} Q'(A) &= \lim_{n \to \infty} Q'(A \cap \{t < \rho_n\}) = \lim_{n \to \infty} Q^{(n)}(A \cap \{t < \rho_n\}) \\ &= \lim_{n \to \infty} \check{Q}^{(n)}(A \cap \{t < \rho_n\}) \end{aligned}$$

which by the above – and since $\rho_n \uparrow \infty$ Q-almost surely – equals

$$\lim_{n \to \infty} \int_{A \cap \{t < \rho_n\}} \check{L}_t^{(n)} dQ = \lim_{n \to \infty} \int_{A \cap \{t < \rho_n\}} L_{t \wedge \rho_n} dQ$$

$$= \lim_{n \to \infty} \int_{A \cap \{t < \rho_n\}} L_t \, dQ = \int_A L_t \, dQ.$$

This identifies L as the density process of Q' with respect to Q relative to $\mathbb{G}$.

(7) By step (5), we know $Q' \overset{\text{loc}}{\ll} Q$ relative to $\mathbb{G}$. By step (6), $L = (L_t)_{t \geq 0}$ is a strictly positive Q-martingale, so this strengthens to $Q' \overset{\text{loc}}{\sim} Q$ relative to $\mathbb{G}$. Now parts (a) and (b) of the theorem are proved. Part (c) is an application of the Ito formula. □

The Ornstein–Uhlenbeck model as a special case of Theorem 6.10 (take $b(t, x) \equiv 0$, $\sigma(t, x) \equiv 1$, and $\gamma(t, x) = hx$ for $h \in \mathbb{R}$) is a well-known prototype of quadratic models, in dimension $d = 1$.

6.11 Example. On the canonical path space $(C, \mathcal{C}, \mathbb{G})$ of Notation 5.13 with canonical process η, write Q_0 for Wiener measure. Let Q_h denote the law of the solution to the Ornstein–Uhlenbeck SDE

$$dX_t = h X_t \, dt + dB_t, \quad X_0 \equiv 0$$

for every $h \in \mathbb{R}$. Then according to Theorem 6.10, all laws Q_h are locally equivalent relative to $\mathbb{G}$, and the density process of Q_h with respect to Q_0 relative to $\mathbb{G}$ is

$$L^{h/0} = \left(L_t^{h/0}\right)_{t \geq 0}, \quad L_t^{h/0} := \exp\left\{h \int_0^t \eta_s \, dm_s^{(0)} - \frac{1}{2} h^2 \int_0^t \eta_s^2 \, ds\right\}$$

with $m^{(0)}$ the martingale part of η under Q_0. For $0 < t < \infty$ fixed, observation of η up to time t

$$\mathcal{E}(S, J) = \left(C, \mathcal{G}_t, \{Q_h^{(t)} : h \in \mathbb{R}\}\right), \quad Q_h^{(t)} \text{ the restriction of } Q_h \text{ to } \mathcal{G}_t$$

yields a quadratic experiment in the sense of Definition 6.1, with score in 0 and observed information given by

$$S := \int_0^t \eta_s \, dm_s^{(0)}, \quad J := \int_0^t \eta_s^2 \, ds.$$

This is not a mixed normal model: first, $\mathcal{L}(J|Q_h)$ depends on $h \in \mathbb{R}$, second, from Ito formula and $\mathcal{L}((\eta, m^{(0)})|Q_0) = \mathcal{L}((B, B))$ where B denotes Brownian motion, the law

$$\mathcal{L}(S|Q_0) = \mathcal{L}\left(\int_0^t B_s \, dB_s\right) = \mathcal{L}\left(\frac{1}{2}(B_t^2 - t)\right)$$

is concentrated on the half-line $[-\frac{t}{2}, \infty)$. Thus, according to Definition 6.2 and Proposition 6.3, this quadratic model $\mathcal{E}(S, J)$ is not a mixed normal model. □

Scores of similar structure such as $\int B_s dB_s$ motivated Jeganathan to view quadratic models as 'Brownian functional models' (see [67], and the references there). The following example shows that on $(C, \mathcal{C}, \mathbb{G})$, we can attach quadratic statistical models to solutions of stochastic differential equations in a natural way, under the sole restriction that angle brackets of local martingale parts of the observed η should be invertible.

6.12 Example. Assume that $c(t, x) \in \mathbb{R}^{d \times d}$ is invertible for all t and all x. Fix a drift coefficient b for SDE (I) and write Q for the law of the solution of (I). For SDE (II), introduce a parameter $h \in \mathbb{R}^d$, define drift functions

$$b_h(t, x) := b(t, x) + c(t, x) h, \quad h \in \mathbb{R}^d$$

which depend on the parameter, and write Q_h for the law of the solution of (II) with $b' = b_h$. Clearly $b_h \neq b_{h'}$ for $h \neq h'$ by assumption on $c(\cdot, \cdot)$. We thus have a statistical model

$$\{ Q_h : h \in \mathbb{R}^d \} \quad \text{on} \quad (C, \mathcal{C}, \mathbb{G}) \quad \text{with} \quad Q_0 := Q.$$

Note that for every $h \in \mathbb{R}^d$, with $\gamma(\cdot, \cdot) \equiv h$ constant, the assumptions (+) and (++) of Theorem 6.10 are satisfied, and Theorem 6.10 gives the following.

(a) We have

$$Q_h \overset{\text{loc}}{\sim} Q_0 \quad \text{relative to} \quad \mathbb{G},$$

and the density process of Q_h with respect to Q_0 relative to $\mathbb{G}$ is

$$L_t^{h/0} = \exp\left\{ h^\top m_t^{(0)} - \frac{1}{2} h^\top \langle m^{(0)} \rangle_t h \right\} = \mathcal{E}_t\left(h^\top m^{(0)} \right)$$

where $m^{(0)}$ is the $(\mathbb{G}, Q_0)$-martingale part of the canonical process η. By assumption on $c(\cdot, \cdot)$,

$$h^\top \langle m^{(0)} \rangle_t h = \int_0^t h^\top c(s, \eta_s) h \, ds$$

is strictly positive for every $h \in \mathbb{R}^d$. Hence the random matrix $\langle m^{(0)} \rangle_t$ defined on $(C, \mathcal{C})$ is invertible for every $t \geq 0$. As a consequence, for every $0 < t < \infty$, the model

$$\{ Q_h^t : h \in \mathbb{R}^d \} = \mathcal{E}(S, J) : \quad S := m_t^{(0)}, \quad J := \langle m^{(0)} \rangle_t$$

(with Q_h^t the restriction of Q_h to $\mathcal{G}_t$) is a quadratic model satisfying all assumptions of Definition 6.1.

(b) Fix any $\mathbb{G}$-stopping time τ on $(C, \mathcal{C})$ which has the property

$$0 < \tau < \infty \quad Q_h\text{-almost surely for every } h \in \mathbb{R}^d .$$

Then according to (∗) in Theorem 5.12(c), we can replace the deterministic time t in (a) by τ. This yields a statistical model

$$\{Q_h^\tau : h \in \mathbb{R}^d\} = \mathcal{E}(S, J) : \quad S := m_\tau^0, \quad J := \langle m^0 \rangle_\tau$$

(with Q_h^τ the restriction of Q_h to $\mathcal{G}_\tau$) which again is quadratic in the sense of Definition 6.1. □

It remains to explain why equations (I) and (II) in Assumption 6.9 have been assumed to share the same diffusion coefficient. The reason is that for measures $Q \overset{\text{loc}}{\sim} Q'$ on $(C, \mathcal{C}, \mathbb{G})$, under time-continuous observation of the canonical process η, the local martingale part of η can not be modified:

6.12' Remark. Consider Assumptions 6.9(a) in the special case $d = m = 1$, and replace $\sigma(\cdot, \cdot)$ in equation (II) by $\sigma'(\cdot, \cdot) : [0, \infty) \times \mathbb{R} \to \mathbb{R}$ such that Lipschitz and linear growth conditions as in Assumptions 6.9 hold for both pairs (b, σ) and (b', σ'). For X solution to (I) and X' solution to (II), fix a common starting point $x \in \mathbb{R}$, and consider the laws $Q = \mathcal{L}(X|P)$ and $Q' = \mathcal{L}(X'|P')$ on $(C, \mathcal{C}, \mathbb{G})$ as in Assumptions 6.9. Assume that Q and Q' are locally equivalent with respect to $\mathbb{G}$. Consider a sequence of partitions

$$\Pi_n := \{t_{n,i} : i \in \mathbb{N}_0\}, \quad 0 = t_{n,0} < t_{n,1} < \cdots < t_{n,\ell_n} = n,$$

$$\sup_{1 \leq i \leq \ell_n} |t_{n,i} - t_{n,i-1}| \longrightarrow 0$$

as $n \to \infty$, and empirical quadratic variations

$$V_n(t) := \sum_{i : t_{n,i} < t} \left(\eta_{t_{n,i+1} \wedge t} - \eta_{t_{n,i}} \right)^2, \quad t \geq 0.$$

We have the following convergences in probability as $n \to \infty$, cf. [98, p. 122–125], and [64, p. 55]):

(×)
$$V_n(t) \longrightarrow \int_0^t [\sigma(s, \eta_s)]^2 \, ds \quad \text{under } Q,$$

$$V_n(t) \longrightarrow \int_0^t [\sigma'(s, \eta_s)]^2 \, ds \quad \text{under } Q'.$$

Next, select some subsequence $(n_k)_k$ such that (×) holds almost surely for every time t which is rational, and then consider events in $\mathcal{G}_t$

$$A_t := \left\{ \lim_{k \to \infty} V_{n_k}(t) = \int_0^t [\sigma(s, \eta_s)]^2 \, ds \right\},$$

$$A'_t := \left\{ \lim_{k \to \infty} V_{n_k}(t) = \int_0^t [\sigma'(s, \eta_s)]^2 \, ds \right\}.$$

Here A_t is a set of full Q-measure, and A'_t a set of full Q'-measure. From $Q \stackrel{\text{loc}}{\sim} Q'$ relative to $\mathbb{G}$ we deduce that the set $A_t \cap A'_t$ has full measure under both Q and Q'. This holds for all rational t. Hence under both laws Q and Q', the processes

$$(\times\times) \qquad \langle m^Q \rangle = \left(\int_0^t [\sigma(s, \eta_s)]^2 \, ds \right)_{t \geq 0}, \quad \langle m^{Q'} \rangle = \left(\int_0^t [\sigma'(s, \eta_s)]^2 \, ds \right)_{t \geq 0}$$

are indistinguishable. Thus, under time-continuous observation of the canonical process η and locally absolutely continuous change of measure $Q \stackrel{\text{loc}}{\sim} Q'$ on $(C, \mathcal{C}, \mathbb{G})$, the local martingale part of η necessarily remains unchanged. In the sense of $(\times\times)$, the diffusion coefficient plays the role of an 'observable' quantity under time-continuous observation. □

6.3 *Time Changes for Brownian Motion with Unknown Drift

The present section is on time changes in the model 'scaled Brownian motion with unknown drift' of Section 5.2. We shall see that time changes which are independent of Brownian motion lead to mixed normal models, other time changes to quadratic models which are not mixed normal.

6.13 Example. We continue Example 5.15, with all notations as there: $(C, \mathcal{C}, \mathbb{G})$ is the canonical path space for $\mathbb{R}^d$-valued processes with continuous paths, B a d-dimensional standard Brownian motion starting in 0, the matrix $\Lambda \in \mathbb{R}^{d \times d}$ is symmetric and strictly positive definite, with square root $\Lambda^{1/2}$; we consider the model

$$(C, \mathcal{C}, \mathbb{G}, \{Q_h : h \in \mathbb{R}^d\}), \quad Q_h = \mathcal{L}((\Lambda^{1/2} B_t + (\Lambda h)t)_{t \geq 0}), \quad h \in \mathbb{R}^d.$$

With η the canonical process on $(C, \mathcal{C})$ and $m^{(h)}$ the Q_h-martingale part of η, Example 5.15 states that all Q_h are locally equivalent relative to $\mathbb{G}$, and that

$$L^{h'/h} = \left(L_t^{h'/h} \right)_{t \geq 0}, \quad L_t^{h'/h} := \exp\left\{ (h' - h)^\top m_t^{(h)} - \frac{1}{2} (h' - h)^\top \Lambda t \, (h' - h) \right\}$$

is the density process of $Q_{h'}$ with respect to Q_h relative to $\mathbb{G}$, where $\mathcal{L}(m^{(h)} | Q_h) = \mathcal{L}(\Lambda^{1/2} B)$ does not depend on the parameter $h \in \mathbb{R}^d$.

Consider a $\mathbb{G}$-stopping time τ with the property

$$(*) \qquad 0 < \tau < \infty \quad Q_h\text{-almost surely, for every } h \in \mathbb{R}^d.$$

By Theorem 5.12(c) and (d) – compare to Remark 6.1" – condition $(*)$ guarantees that the 'candidate pair'

$$(S, J) = (m_\tau^{(0)}, \Lambda \tau) = (\eta_\tau, \Lambda \tau)$$

Section 6.3 *Time Changes for Brownian Motion with Unknown Drift   169

generates a quadratic experiment as defined in Definition 6.1

$$\mathcal{E}(S, J) = \left(C , \mathcal{G}_\tau , \{Q_h^\tau : h \in \mathbb{R}^d\} \right)$$

where $Q_h^\tau := (Q_h)_{|\mathcal{G}_\tau}$ are equivalent probability measures, $h \in \mathbb{R}^d$, and where

$$L_\tau^{h'/h} = \exp\left((h' - h)^\top m_\tau^{(h)} - \frac{1}{2}(h' - h)^\top \Lambda\tau \, (h' - h) \right)$$

provides a version of the density of $Q_{h'}^\tau$ with respect to Q_h^τ. The model $\mathcal{E}(S, J)$ represents scaled Brownian motion with unknown drift observed up to time τ.

So far, we have a rather restricted number of time transformations given by $\mathbb{G}$-stopping times.

6.14 Example. Consider Example 6.13 in the one-dimensional case $d = 1$, with scaling factor $\Lambda := 1$. For $a > 0$ and $b > 0$ fixed, we have $\mathbb{G}$-stopping times

$$\tau_{a,b} := T_a \wedge S_b , \quad T_a := \inf\{r > 0 : \eta_r > a\}, \quad S_b := \inf\{r > 0 : \eta_r < -b\}.$$

Writing for short $\tau := \tau_{a,b}$ and considering first the Wiener measure $P := Q_0$, we have

$$(+) \quad E_P(\tau) = ab , \quad P(\eta_\tau = a) = \frac{b}{b+a} , \quad P(\eta_\tau = -b) = \frac{a}{b+a}$$

from [112, Prop. (3.8) and Exer. 3° of (3.11), Chap. II].

(a) We prove that $(*)$ in Example 6.13 holds for $\tau = T_a \wedge S_b$. Consider Q_0 first; clearly $Q_0 (0 < \tau < \infty) = 1$. The canonical process η starting almost surely in 0 under Q_0, the sequence $(\eta_{\tau \wedge N})_{N \in \mathbb{N}}$ takes values almost surely in the interval $[-b, a]$: hence, for $h \in \mathbb{R}$ arbitrary,

$$L_{\tau \wedge N}^{h/0} = e^{h \eta_{\tau \wedge N} - \frac{1}{2}h^2(\tau \wedge N)}, \quad N \geq 1$$

is a uniformly integrable martingale under Q_0. Applying Theorem 5.12(d), the condition $(*)$ in Example 6.13 is verified.

(b) In the quadratic experiment $\mathcal{E}(S, J)$ induced by τ, with $S = \eta_\tau$ and $J = \tau$, conditional laws

$$(\circ) \quad Q_0^{S|J=j} = Q_0^{\eta_\tau | \tau = j}, \quad j \in (0, \infty)$$

are supported by the two-point-set $\{a, -b\}$, by definition of $\tau = T_a \wedge S_b$. This is incompatible with Proposition 6.3(ii). Hence $\mathcal{E}(S, J)$ is not a mixed normal experiment. □

6.14' Exercise. This exercise complements Example 6.14. With all notations as in Example 6.14, we consider the one-dimensional case $d = 1$ with scaling factor $\Lambda := 1$. We focus on the $\mathbb{G}$-stopping times T_a, for $a > 0$ fixed.

(a) Since $\eta_0 = 0$ Q_0-almost surely, the law of the iterated logarithm grants $Q_0(\{0 < T_a < \infty\}) = 1$.

(b) Write for short $P := Q_0$ and $M_t := \max_{0 \le s \le t} \eta_s$. Write Φ for the distribution function of $\mathcal{N}(0, 1)$. For a, t in $(0, \infty)$, use the reflection principle and rescaling

$$P(T_a < t) = P(M_t > a) = 2P(\eta_t > a) = 2P(\eta_1 > a/\sqrt{t}) = 2(1 - \Phi(a/\sqrt{t}))$$

to determine the density of $\mathcal{L}(T_a|P)$:

$$f_a(t) = \frac{a}{\sqrt{2\pi}} t^{-\frac{3}{2}} e^{-\frac{1}{2}\frac{a^2}{t}}, \quad t, a \in (0, \infty).$$

Determine the Laplace transform of $\mathcal{L}(T_a|P)$

$$E_P\left(e^{-\lambda T_a}\right) = e^{-a\sqrt{2\lambda}}, \quad \lambda \ge 0$$

from the statistical argument in Theorem 5.12(d): for positive drift parameter $h > 0$ we do have the equivalent assertions

$$Q_h(\{0 < T_a < \infty\}) = 1 \iff E_{Q_0}\left(L_{T_a}^{h/0}\right) = 1, \quad h > 0$$

where $L^{h/0}$ is the density process of Q_h with respect to Q_0 as in Examples 5.14 or 6.13; on the right-hand side, exploit $\eta_{T_a} = a$ Q_0-almost surely; finally, change variables $\lambda := \frac{1}{2}h^2$.

(c) For positive drift $h > 0$, determine the Laplace transform of $\mathcal{L}(T_a \mid Q_h)$

$$E_{Q_h}\left(e^{-\lambda T_a}\right) = e^{-a(\sqrt{h^2 + 2\lambda} - h)}, \quad \lambda \ge 0$$

using (b) and change of measure $E_{Q_h}(e^{-\lambda T_a}) = E_{Q_0}(e^{-\lambda T_a} L_{T_a}^{h/0})$.

(d) For negative drift $h < 0$, prove that

$$Q_h(\{0 < T_a < \infty\}) = e^{2ah} < 1, \quad h < 0$$

(hint: combine an argument as in step (3) of the proof of Theorem 5.12 with application of (b) above to $\widetilde{h} := |h| > 0$).

(e) Deduce the following from (d). For $a > 0$ fixed, the candidate pair of statistics $S = \eta_{T_a}$, $J = T_a$ under Q_0 does not generate a statistical experiment which satisfies all assumptions of Definition 6.1, cf. Remark 6.1". As a consequence, 'observing Brownian motion with unknown drift up to time T_a' does not lead to a quadratic experiment in the sense of Definition 6.1. □

More interesting time transformations for the model 'scaled Brownian motion with unknown drift' are at hand if we extend the probability space and consider stopping times which are independent from Brownian motion. This will lead to mixed normal experiments. We will formulate mixed normality in two variants: the first collects all independent variables, processes, stopping times in one initial σ-field, the second variant keeps track of the temporal dynamics of the time transformation.

6.15 Observing Scaled Brownian Motion with Unknown Drift up to some Independent Random Time. We extend the experiment

$$\left(C, \mathcal{C}, \mathbb{G}, \{Q_h : h \in \mathbb{R}^d\}\right), \quad Q_h := \mathcal{L}\left(\left(\Lambda^{1/2} B_t + (\Lambda h) t\right)_{t \ge 0}\right), \quad h \in \mathbb{R}^d$$

Section 6.3 *Time Changes for Brownian Motion with Unknown Drift

considered in Example 5.15 in order to obtain random variables, processes or stopping times τ which are independent from Brownian motion B. Prepare some space $(\Omega'', \mathcal{A}'', P'')$ carrying τ such that

(∗) $\tau : \Omega'' \to [0, \infty]$ is $\mathcal{A}''$-measurable, and $0 < \tau < \infty$ P''-almost surely.

The object τ in (∗) might be defined in terms of other processes or random variables living on $(\Omega'', \mathcal{A}'', P'')$. With the notations of Example 5.15, introduce a product space

(×) $\left(H, \mathcal{H}, \mathbb{H}, \{\widehat{Q}_h : h \in \mathbb{R}^d\}\right)$, $\widehat{Q}_h := Q_h \otimes P''$

where

$$H := C \times \Omega'', \quad \mathcal{H} := \mathcal{C} \otimes \mathcal{A}'', \quad \mathcal{H}_t := \mathcal{G}_t \otimes \mathcal{A}'', \quad \mathbb{H} = (\mathcal{H}_t)_{t \geq 0}.$$

This constructions lifts the canonical process η from $(C, \mathcal{C})$ to $(H, \mathcal{H})$, and lifts random variables from $(\Omega'', \mathcal{A}'')$ to $(H, \mathcal{H})$. In particular, any $[0, \infty]$-valued random variable on $(\Omega'', \mathcal{A}'')$ lifted to the extension will be $\mathcal{H}_0$-measurable, and thus can be used as $\mathbb{H}$-stopping time. On $(H, \mathcal{H})$, objects lifted from $(C, \mathcal{C})$ and objects lifted from $(\Omega'', \mathcal{A}'')$ will be independent under all $\widehat{Q}_h := Q_h \otimes P''$, $h \in \mathbb{R}^d$, and the law of objects stemming from $(\Omega'', \mathcal{A}'')$ will not depend on $h \in \mathbb{R}^d$. Due to this independence structure, density processes of $\widehat{Q}_{h'}$ with respect to $\widehat{Q}_h$ relative to $\mathbb{H}$ are obtained by lifting the density processes given in Example 5.15 from $(C, \mathcal{C}, \mathbb{G})$. Thus (∗) and (∗∗) of Example 5.15 remain unchanged provided we redefine on $(H, \mathcal{H}, \mathbb{H})$

$$m^{(h)} := (\eta_t - \Lambda h t)_{t \geq 0} \quad \text{martingale part of } \eta \text{ under } \widehat{Q}_h.$$

In this sense, the model (×) is a simple extension of the statistical model in Example 5.15. On $(H, \mathcal{H})$, we write again τ for the $\mathbb{H}$-stopping time obtained by lifting (∗) above to $(H, \mathcal{H}, \mathbb{H})$. Then condition (∗) of Theorem 5.12(c) is satisfied

$$0 < \tau < \infty \quad \widehat{Q}_h\text{-almost surely, for every } h \in \mathbb{R}^d$$

since the laws $\mathcal{L}(\tau \mid \widehat{Q}_h)$ do not depend on $h \in \mathbb{R}^d$, and conditional laws

$$(\widehat{Q}_h)^{\eta_\tau \mid \tau = t} = \mathcal{N}([\Lambda t] h, \Lambda t)$$

have the structure required in Proposition 6.3. Write $\widehat{Q}_h^\tau$ for the restriction of $\widehat{Q}_h$ to the σ-field $\mathcal{H}_\tau$ of all events up to time τ. By Theorem 5.12(c), these are equivalent probability laws, and the likelihood of $\widehat{Q}_{h'}^\tau$ with respect to $\widehat{Q}_h^\tau$ on $\mathcal{H}_\tau$ is given by

$$L_\tau^{h'/h} = \exp\left\{(h' - h)^\top m_\tau^{(h)} - \frac{1}{2} (h' - h)^\top \Lambda\tau \, (h' - h)\right\},$$

for all $h', h \in \mathbb{R}^d$. Thus the quadratic experiment

$$\mathcal{E}(S, J) = \left(H, \mathcal{H}_\tau, \{\widehat{Q}_h^\tau : h \in \mathbb{R}^d\}\right), \quad S = \eta_\tau = m_\tau^{(0)}, \quad J = \Lambda\tau$$

is a mixed normal experiment in the sense of Definition 6.2. It describes time-continuous observation of scaled Brownian motion with unknown drift up to the independent time τ. The central statistic is

$$Z := \Lambda^{-1} \begin{pmatrix} \eta_\tau^{(1)} / \tau \\ \vdots \\ \eta_\tau^{(d)} / \tau \end{pmatrix}$$

where $\eta^{(1)}, \ldots, \eta^{(d)}$ are the components of the process η. □

6.15' Exercise. In the setting of 6.15, let $(\Omega'', \mathcal{A}'', P'')$ carry a Poisson process $(N_t)_{t\geq 0}$ with parameter $\lambda > 0$. Let τ denote the time of the k-th jump of $(N_t)_{t\geq 0}$. In the particular case $k = 1$ we recover the example of Exercise 6.1''', with observation up to an independent exponential time. □

6.15'' Exercise. In the setting of 6.15, fix $\lambda > 0$ and let $(\Omega'', \mathcal{A}'', P'')$ carry a Gamma process, i.e. a PIIS $(\zeta_t)_{t\geq 0}$ (process with stationary and independent increments) where $\mathcal{L}(\zeta_{r_2} - \zeta_{r_1} | P'') = \Gamma(r_2 - r_1, \lambda)$ for $0 \leq r_1 < r_2 < \infty$. Here $\Gamma(a, \lambda)$ denotes the Gamma law with density $f_{a,\lambda}(x) = 1_{(0,\infty)}(x) \frac{\lambda^a}{\Gamma(a)} x^{a-1} e^{-\lambda x}$, $x \in \mathbb{R}$. The state of the process ζ at time $t = 1$ is thus exponentially distributed with parameter λ; defining $\tau := \zeta_k$ we have an alternative to Exercise 6.15', for $k \in \mathbb{N}$. Beyond this, we can use the process $t \to \zeta_t$ as a time change for Brownian motion. Note that Gamma processes (cf. Bertoin [10, p. 73]) are obtained as integrals

$$\int_0^t \int_{(0,\infty)} x\, \mu(ds, dx), \quad t \geq 0$$

where $\mu(ds, dx)$ denotes Poisson random measure with intensity $ds\, x^{-1} e^{-\lambda x} dx$ on $(0, \infty) \times (0, \infty)$. Hence they are increasing càdlàg processes, increasing only by jumps, and admitting an infinite number of small jumps (positive and summable) over every finite time interval. Accordingly, time-changed Brownian motion $(B(\zeta_t))_{t\geq 0}$ has càdlàg paths. □

The construction in 6.15 does not reflect adequately the temporal dynamics of a process of time change in cases where such a process may play a key role; the following construction remedies to this.

6.16 Independent Time Transformation for 'Scaled Brownian Motion with Unknown Drift'. Consider a probability space $(\Omega', \mathcal{A}', P')$ carrying processes $B = (B_t)_{t\geq 0}$ and $A = (A_t)_{t\geq 0}$ as follows:

(i) B is a d-dimensional standard Brownian motion with $B_0 \equiv 0$,

(ii) for all $\omega \in \Omega'$, paths $t \to A_t(\omega)$ are càdlàg non-decreasing, with $A_0(\omega) = 0$ and $\lim_{t\to\infty} A_t(\omega) = +\infty$,

(iii) B and A are independent under P'.

Section 6.3 *Time Changes for Brownian Motion with Unknown Drift

Write $\mathbb{F} = (\mathcal{F}_t)_{t\geq 0}$ for the right-continuous filtration generated by the pair (B, A). Then

$$\phi_t := \inf\{v : A_v > t\}, \quad 0 < t < \infty$$

are $\mathbb{F}$-stopping times which by (ii) have the property

$$0 < \phi_t(\omega) < \infty \text{ for all } 0 < t < \infty \text{ and all } \omega \in \Omega'.$$

We put $\phi_0 = 0$. Then all paths $t \to \phi_t$ are càdlàg non-decreasing, thus $B \circ \phi := (B_{\phi_t})_{t\geq 0}$ is a càdlàg process. For $\Lambda \in \mathbb{R}^{d\times d}$ deterministic and strictly positive definite, the statistical experiment

$$(\diamond) \qquad \mathcal{Q}_h := \mathcal{L}\left(\left(\Lambda^{1/2} B_{\phi_t} + (\Lambda h) \phi_t, \phi_t \right)_{t\geq 0} \mid P' \right), \quad h \in \mathbb{R}^d$$

is mixed normal in the sense that for all $0 < t < \infty$ fixed, the restrictions $\mathcal{Q}_h^t$ of $\mathcal{Q}_h$ to the σ-field of events up to time t form an experiment $\{\mathcal{Q}_h^t : h \in \mathbb{R}^d\}$ which is mixed normal in the sense of Definition 6.2; it represents *scaled Brownian motion with unknown drift time-transformed by the level crossing times of an independent increasing process A.*

Let us state the necessary details more carefully before giving a proof. Write D for the space of d-dimensional càdlàg functions $[0, \infty) \to \mathbb{R}^d$ equipped with Skorohod topology and Borel σ-field $\mathcal{D}$ (see [64, p. 292]). $(D, \mathcal{D})$ is a Polish space, and $\mathcal{D}$ coincides with the σ-field generated by the coordinate projections. With notation $\eta = (\eta_t)_{t\geq 0}$ for the canonical process on $(D, \mathcal{D})$, write $\mathcal{G}_t := \bigcap_{r>t} \sigma(\eta_s : 0 \leq s \leq r)$. Next, in dimension 1, we define $(D_+, \mathcal{D}_+)$ as the restriction of the Skorohod space of one-dimensional càdlàg functions to the closed subspace of non-decreasing functions h starting at $h(0) = 0$ (cf. [64, p. 306]): then $(D_+, \mathcal{D}_+)$ is again Polish. With notation $\zeta = (\zeta_t)_{t\geq 0}$ for the canonical process on $(D_+, \mathcal{D}_+)$, write $\mathcal{G}_t^+ := \bigcap_{r>t} \sigma(\zeta_s : 0 \leq s \leq r)$. Then the laws $\mathcal{Q}_h$ defined by $(\diamond)$ live on the product space $(H, \mathcal{H}, \mathbb{H})$

$$(\diamond\diamond) \quad H := D \times D_+, \quad \mathcal{H} := \mathcal{D} \otimes \mathcal{D}_+, \quad \mathcal{H}_t := \mathcal{G}_t \otimes \mathcal{G}_t^+, \quad \mathbb{H} := (\mathcal{H}_t)_{t\geq 0}.$$

We write (η, ζ) for the canonical process on $(H, \mathcal{H}, \mathbb{H})$ and have the following:

(α) All laws $\mathcal{Q}_h$, $h \in \mathbb{R}^d$, are locally equivalent relative to $\mathbb{H}$.

(β) The density process of $\mathcal{Q}_h$ with respect to $\mathcal{Q}_0$ relative to $\mathbb{H}$ is

$$L^{h/0} = (L_t^{h/0})_{t\geq 0}, \quad L_t^{h/0} := \exp\left\{ h^\top \eta_t - \frac{1}{2} h^\top \Lambda \zeta_t\, h \right\}.$$

(γ) The density process of $\mathcal{Q}_{h'}$ with respect to $\mathcal{Q}_h$ relative to $\mathbb{H}$ is

$$L^{h'/h} = (L_t^{h'/h})_{t\geq 0}, \quad L_t^{h'/h} := \exp\left\{ (h'-h)^\top m_t^{(1,h)} - \frac{1}{2}(h'-h)^\top \Lambda \zeta_t\, (h'-h) \right\}$$

for $h', h \in \mathbb{R}^d$, with notation

$$m^{(1,h)} := \eta - \eta_0 - [\Lambda h]\zeta$$

for the Q_h-local martingale part of the first component η of the canonical process (η, ζ) on $(H, \mathcal{H}, \mathbb{H})$. Here $\mathcal{L}(\zeta \mid Q_h) = \mathcal{L}(\phi \mid P')$ does not depend on the parameter $h \in \mathbb{R}^d$, and

$$\mathcal{L}\left(m^{(1,h)} \mid Q_h\right) = \mathcal{L}\left(\Lambda^{1/2} B \circ \phi \mid P'\right) \quad \text{for all } h \in \mathbb{R}^d$$

where Brownian motion $(B_t)_t$ and time change $t \to \phi_t$ are independent, by assumption.

(δ) For $0 < t < \infty$ fixed, the model $(H, \mathcal{H}_t, \{(Q_h)|_{\mathcal{H}_t} : h \in \mathbb{R}^d\})$ is a mixed normal model $\mathcal{E}(S, J)$ with $S = \eta_t$ and $J = \Lambda \zeta_t$ in application of Proposition 6.3. The central statistic in $\mathcal{E}(S, J)$ is

$$Z := \Lambda^{-1} \begin{pmatrix} \eta_t^{(1)}/\zeta_t \\ \vdots \\ \eta_t^{(d)}/\zeta_t \end{pmatrix}$$

with $\eta^{(1)}, \ldots, \eta^{(d)}$ the components of η.

Below, we shall prove the assertions (α)–(δ) in three steps.

(1) Introduce one more factor $(D_+, \mathcal{D}_+)$ and consider first the path space

$$(\widetilde{H}, \widetilde{\mathcal{H}}, \widetilde{\mathbb{H}}): \quad \widetilde{H} := D \times D_+ \times D_+, \quad \widetilde{\mathcal{H}} := \mathcal{D} \otimes \mathcal{D}_+ \otimes \mathcal{D}_+,$$
$$\widetilde{\mathbb{H}} = (\widetilde{\mathcal{H}}_t)_{t \geq 0}, \quad \widetilde{\mathcal{H}}_t := \mathcal{G}_t \otimes \mathcal{G}_t^+ \otimes \mathcal{G}_t^+$$

with canonical process $(\widetilde{\eta}, \widetilde{\zeta}, \widetilde{\xi})$, equipped with the family of laws

$$\widetilde{Q}_h := \mathcal{L}\left(\left(\Lambda^{1/2} B + (\Lambda h)\,\mathrm{id},\ \mathrm{id},\ A\right) \mid P'\right), \quad h \in \mathbb{R}^d$$

where id is the deterministic process taking value t at time t. We define $\widetilde{\mathbb{H}}$-stopping times

$$\widetilde{\phi}_t := \inf\{v : \widetilde{\xi}_v > t\}, \quad 0 < t < \infty$$

on $(\widetilde{H}, \widetilde{\mathcal{H}})$. By the properties of the increasing process A under P', we have

(∗) $\qquad 0 < \widetilde{\phi}_t < \infty \quad \widetilde{Q}_h\text{-almost surely}$

for all $h \in \mathbb{R}^d$ and all $0 < t < \infty$, and $\mathcal{L}(\widetilde{\phi}_t \mid \widetilde{Q}_h)$ does not depend on $h \in \mathbb{R}^d$. Put $\widetilde{\phi}_0 \equiv 0$ and note that paths $t \to \widetilde{\phi}_t$ and $t \to \widetilde{\eta}_{\widetilde{\phi}_t}$ are càdlàg on $[[0, \Delta[[$ where $\Delta := \inf\{t : \widetilde{\phi}_t = \infty\}$ equals $+\infty$ $\widetilde{Q}_h$-almost surely for all $h \in \mathbb{R}^d$.

(2) By definition of the laws $\widetilde{Q}_h$ on $(\widetilde{H}, \widetilde{\mathcal{H}}, \widetilde{\mathbb{H}})$, Brownian motion B being independent from the increasing process A under P' (and trivially independent from the second

component which is deterministic), Example 5.15 immediately extends to $(\widetilde{H}, \widetilde{\mathcal{H}}, \widetilde{\mathbb{H}})$: thus

$$\widetilde{Q}_h \stackrel{\text{loc}}{\sim} \widetilde{Q} \quad \text{for all } h \in \mathbb{R}^d,$$

and the density process of $\widetilde{Q}_h$ with respect to $\widetilde{Q}_0$ relative to $\widetilde{\mathbb{H}}$ is

$$\widetilde{L}^{h/0} = (\widetilde{L}_t^{h/0})_{t \geq 0}, \quad \widetilde{L}_t^{h/0} := \exp\left\{h^\top \widetilde{\eta}_t - \frac{1}{2} h^\top \Lambda t \, h\right\}.$$

(3) In the statistical model of step (2), we can change time according to Theorem 5.12(c) – this hinges on the property (∗) for all stopping times $\widetilde{\phi}_t$, $0 < t < \infty$, in step (1) – and consider mappings

$$\Psi : \quad (\widetilde{\eta}, \widetilde{\zeta}, \widetilde{\xi}) \longrightarrow \Psi((\widetilde{\eta}, \widetilde{\zeta}, \widetilde{\xi})) := (\widetilde{\eta} \circ \widetilde{\phi}, \widetilde{\phi})$$

from $(\widetilde{H}, \widetilde{\mathcal{H}})$ to $(H, \mathcal{H})$ which have the properties

$$\Psi^{-1}(\mathcal{H}_t) = \widetilde{\mathcal{H}}_{\widetilde{\phi}_t}, \quad 0 < t < \infty, \quad \mathcal{L}(\Psi \mid \widetilde{Q}_h) = Q_h, \quad h \in \mathbb{R}^d.$$

Thus properties (α), (β), (γ) and (δ) for the statistical model ($\diamond$) and ($\diamond\diamond$) hold as consequences of step (2). □

The following is a special case of Construction 6.16: we specify the increasing process $A = (A_t)_{t \geq 0}$ at the start of Construction 6.16 as a one-sided stable process (stable subordinator) with index $\alpha \in (0, 1)$. This is of particular importance for null recurrent Markov processes, under a regular variation condition, see Chapter 8. We give the necessary definitions first, and then return to the statistical model.

6.17 Remark. (1) For $0 < \alpha < 1$, the one-sided stable process $S^{(\alpha)}$ with index α is defined from independent and stationary increments having Laplace transforms

$$E\left(e^{-\lambda (S_{r_2}^\alpha - S_{r_1}^\alpha)}\right) = e^{-(r_2 - r_1)\lambda^\alpha}, \quad \lambda \geq 0, \quad 0 \leq r_1 < r_2 < \infty.$$

This process is a functional of Poisson random measure $\mu(ds, dx)$ on $(0, \infty) \times (0, \infty)$ with intensity $\nu(ds, dx) = ds \, \frac{\alpha}{\Gamma(1-\alpha)} x^{-\alpha - 1} \, dx$ on $(0, \infty) \times (0, \infty)$

$$S_t^{(\alpha)} = \int_0^t \int_{(0,\infty)} x \, \mu(ds, dx), \quad t \geq 0,$$

(see [62], or [10, p. 73)]). Paths of $S^{(\alpha)}$ are càdlàg and strictly increasing – they have positive and summable jumps, increase only by jumps, and have an infinite number of infinitesimally small jumps over every finite time interval – such that $S_0^{(\alpha)} = 0$ and $\lim_{t \to \infty} S_t^{(\alpha)} = \infty$.

(2) Let $V^{(\alpha)}$ denote the process inverse to $S^{(\alpha)}$, i.e. the process of level crossing times

$$V_t^{(\alpha)} := \inf\{v > 0 : S_v^{(\alpha)} > t\}, \quad 0 \leq t < \infty.$$

The process $V^{(\alpha)}$ is called the Mittag–Leffler process of index $0 < \alpha < 1$. Paths of $V^{(\alpha)}$ are continuous and non-decreasing such that $V_0^{(\alpha)} = 0$ and $\lim_{t \to \infty} V_t^{(\alpha)} = \infty$. Laplace transforms of $V_t^{(\alpha)}$ are given by

$$E\left(e^{-\lambda V_t^{(\alpha)}}\right) = \sum_{n=0}^{\infty} \frac{(-\lambda)^n}{\Gamma(1+n\alpha)} t^{n\alpha}, \quad \lambda \geq 0, \, t \geq 0$$

and $\mathcal{L}(V_t^{(\alpha)})$ admits finite moments of arbitrary order. See [24, p. 453], [12], or [131]. □

6.18 Example. At the start of Construction 6.16, for $0 < \alpha < 1$, let us take the increasing process A as the one-sided stable process $S^{(\alpha)}$ with index α, thus $\phi = V^{(\alpha)}$ in Construction 6.16 is the Mittag–Leffler process of index α. We put $\Lambda := I$ for simplicity. In ($\diamond$) in Construction 6.16 we consider the mixed normal statistical model

$$Q_h := \mathcal{L}\left(\left(B(V_t^{(\alpha)}) + h\, V_t^{(\alpha)}, \, V_t^{(\alpha)}\right)_{t \geq 0}\right), \quad h \in \mathbb{R}^d.$$

Brownian motion B and the Mittag–Leffler process $V^{(\alpha)}$ are independent. Corresponding to observation over the time interval $[0, t]$, likelihoods in (γ) of Construction 6.16 are of type

$$L_t^{h'/h} = \exp\left\{(h' - h)^\top B(V_t^{(\alpha)}) - \frac{1}{2}(h' - h)^\top V_t^{(\alpha)} I\,(h' - h)\right\}$$

and the central statistics in (δ) of Construction 6.16 is of type

$$Z := \frac{1}{V_t^{(\alpha)}} B(V_t^{(\alpha)}).$$

Thus the best concentrated distribution for estimation errors, in the sense of the Convolution Theorem 6.6, its Corollary 6.6' and of the Minimax Theorem 6.8, is

(*) $$\mathcal{L}(Z|Q_0) = \int_{(0,\infty)} \mathcal{L}(V_t^{(\alpha)})(du)\, \mathcal{N}\left(0, \frac{1}{u} I\right).$$

We remark that the law (*) does not admit finite second moments (see Exercise 6.18') for $0 < \alpha < 1$. □

Under mixed normality, recall that comparison of estimation errors works conditionally on the observed information. As strengthened in Remark 6.6", the best concentrated distribution according to the Convolution Theorem 6.6, its Corollary 6.6' and the Minimax Theorem 6.8 is not required to have finite second or higher moments; Example 6.18 again illustrates this fact.

Section 6.3 *Time Changes for Brownian Motion with Unknown Drift

6.18' Exercise. We prove that the law (∗) in Example 6.18 does not admit finite second moments:

With notations of Remark 6.17, for $0 < \alpha < 1$, deduce from the definition of $V^{(\alpha)}$ as process inverse of $S^{(\alpha)}$ and from scaling properties of $S^{(\alpha)}$ that

$$\mathcal{L}(V_1^{(\alpha)}) = \mathcal{L}\left(\left[\frac{1}{S_1^{(\alpha)}}\right]^\alpha\right);$$

this representation is sometimes used as a definition of a Mittag–Leffler law. Write now

$$(+) \qquad \int_{(0,\infty)} \mathcal{L}(V_1^{(\alpha)})(du)\,\frac{1}{u} = E\left(\left[\frac{1}{V_1^{(\alpha)}}\right]\right) = E\left([S_1^{(\alpha)}]^\alpha\right)$$

and check that the integral in (+) equals $+\infty$:

For $0 < \alpha < 1$, let G denote the distribution function of $S_1^{(\alpha)}$, then

$$1 - G(t) \sim \frac{1}{\Gamma(1-\alpha)}\,t^{-\alpha} \quad \text{as } t \to \infty$$

by [24, p. 448] or [12, p. 361)]. For $0 < p \le \alpha$, we obtain

$$E\left([S_1^{(\alpha)}]^p\right) < \infty \quad \text{for } p < \alpha, \quad E\left([S_1^{(\alpha)}]^p\right) = \infty \quad \text{for } p = \alpha$$

from a representation

$$E\left([S_1^{(\alpha)}]^p\right) = -\lim_{N \to \infty} \int_0^N x^p\,d(1 - G(x))$$

via partial integration. □

Chapter 7

Local Asymptotics of Type LAN, LAMN, LAQ

Topics for Chapter 7:

7.1 Local Asymptotics of Type LAN, LAMN, LAQ
LAQ, LAMN, LAN in ϑ 7.1
Notations relative to local asymptotics   7.1'
Contiguity of local alternatives   7.2–7.3
Markov extensions of statistical models   7.4
Rescaled estimation errors in the local experiment at ϑ / estimators in the limit experiment   7.5
Uniform convergence of risks of estimators over shrinking neighbourhoods of ϑ 7.6
Linking rescaled estimation errors at ϑ to the central sequence at ϑ 7.7

7.2 Asymptotic Optimality of Estimators in the LAN or LAMN Setting
Regularity of estimator sequences as asymptotic (strong) equivariance   7.8
Example: linking rescaled estimation errors to the central sequence   7.9
LAN: Hájek's convolution theorem   7.10(a)
LAMN: Jeganathan's convolution theorem   7.10(b)
Characterising efficiency at ϑ 7.11
LAMN or LAN: local asymptotic minimax theorem   7.12
A remark on efficient sequences, and an example   7.13–7.13'

7.3 Le Cam's One-step Modification of Estimators
Conditions for one-step modification   7.14
Local scale with estimated parameter   7.15
Discretisation of estimators   7.16
Information with estimated parameter   7.17
Score with estimated parameter   7.18
The one-step modification   7.19

7.4 The Case of i.i.d. Observations
LAN via Le Cam's second lemma for i.i.d. observations   7.20
Examples for efficient estimator sequences   7.21–7.22

Exercises: 7.9', 7.9", 7.10", 7.22', 7.22"

Given a sequence of statistical experiments $(\mathcal{E}_n)_n$ parameterised by the same d-dimensional parameter $\xi \in \Theta$, *local asymptotics* means that we fix a parameter value $\vartheta \in \Theta$ and consider shrinking neighbourhoods of ϑ in the following way:

(i) for some choice of *local scale* $(\delta_n(\vartheta))_n$ tending to 0 in a specific way, we reparameterise with respect to ϑ in $\mathcal{E}_n$ by writing $\xi = \vartheta + \delta_n(\vartheta)\, h$, and then focus on h as a *local parameter at ϑ* at stage n of the asymptotics;

(ii) as $n \to \infty$, *local experiments* parameterised by h tend in a suitable sense to a *limit experiment* whose parameter h ranges over the full space $\mathbb{R}^d$;

(iii) as $n \to \infty$, statistical properties which characterise the limit experiment carry over to the local experiments at ϑ, in the following sense: in an approximating local experiment at ϑ, at stage n of the asymptotics, they hold true 'approximately' when n is large.

Quite different types of limit experiment – exhibiting essentially different statistical properties – can be considered in such a framework. Also, sets of conditions which at stage n of the asymptotics link the approximating local experiment at ϑ to the limit experiment can take various forms. For convergence of experiments in general, see Strasser [121]. Shiryaev and Spokoiny [122] explore a spectrum of different routes linking approximating local experiments at ϑ as $n \to \infty$ to a limit experiment.

The present chapter is on local asymptotics where the limit experiment is a Gaussian shift, a mixed normal or a quadratic experiment as considered in Chapters 5 and 6, and where the local experiments at ϑ will inherit the statistical properties of a Gaussian shift as $n \to \infty$, of a mixed normal or a quadratic limit experiment. Our main references are Le Cam [81], Hàjek [40], Davies [19], Jeganathan [66,67], Le Cam and Yang [84]; we follow in particular the approach outlined by [19].

The main results of this chapter, for a mixed normal limit experiment, are the local asymptotic versions of the convolution theorem (in 7.10 and 7.11 below) and of the minimax theorem (in 7.12 below). We shall look at the special case of i.i.d. observations in Section 7.4. Some stochastic process examples will then be considered in a separate Chapter 8.

7.1 Local Asymptotics of Type LAN, LAMN, LAQ

The following notations will be used throughout the chapter. The parameter space Θ is an open subset of $\mathbb{R}^d$, and we have a sequence $(\mathcal{E}_n)_n$ of statistical experiments which are parameterised by Θ

$$\mathcal{E}_n := (\Omega_n, \mathcal{A}_n, \{P_{n,\vartheta} : \vartheta \in \Theta\}), \quad n \geq 1.$$

We fix a reference point $\vartheta \in \Theta$. Relative to this point, we consider a norming sequence $(\delta_n(\vartheta))_n$ decreasing to 0. Our $\delta_n = \delta_n(\vartheta)$ are strictly positive real numbers; everything would work similarly with matrices in $\mathbb{R}^{d\times d}$, strictly positive definite and decreasing in the sense of half-ordering of positive definite matrices to $0 \in \mathbb{R}^{d \times d}$. We write again $\mathsf{D}^+ \in \mathcal{B}(\mathbb{R}^{d\times d})$ as in Definition 6.1 for the set of symmetric and strictly positive definite matrices.

A main example for this setting is as follows: for sequences of i.i.d. random variables whose law depends on a parameter ϑ, consider experiments $\mathcal{E}_n$ corresponding to observation of the first n variables under unknown $\vartheta \in \Theta$ as in Assumption 4.10(b). In this case, assuming smoothness of the parameterisation at ϑ in the sense of Assumption 4.10(a), we take local scale at ϑ as $\delta_n(\vartheta) = n^{-1/2}$. Then the following should be compared to Le Cam's Second Lemma 4.11 for i.i.d. observations:

7.1 Definition. (a) A sequence of experiments $(\mathcal{E}_n)_n$ as above is called *locally asymptotically quadratic at ϑ (LAQ)* if relative to ϑ there are pairs of statistics

$$(S_n, J_n) = (S_n(\vartheta), J_n(\vartheta)) : \Omega_n \longrightarrow \mathbb{R}^d \times \mathbb{R}^{d\times d} \quad \mathcal{A}_n\text{-measurable}, \quad n \geq 1$$

for every $n \geq 1$, $P_{n,\vartheta}$-almost surely, $J_n(\vartheta)$ takes values in D^+

and norming constants

$$\delta_n = \delta_n(\vartheta) \text{ decreasing to } 0 \text{ as } n \to \infty$$

such that the following properties (i) and (ii) are satisfied:
(i) at ϑ, quadratic expansions

$$\log \frac{d\, P_{n,\vartheta+\delta_n h_n}}{d\, P_{n,\vartheta}} = h_n^\top S_n - \frac{1}{2} h_n^\top J_n h_n + o_{(P_{n,\vartheta})}(1), \quad n \to \infty$$

hold true, for arbitrary bounded sequences $(h_n)_n$ in $\mathbb{R}^d$ (then, Θ being open, we have $\vartheta + \delta_n(\vartheta)h_n \in \Theta$ when n is large enough);
(ii) depending on ϑ, a quadratic experiment exists

$$\mathcal{E}_\infty = \mathcal{E}(S, J) = (\Omega, \mathcal{A}, \{P_h : h \in \mathbb{R}^d\}),$$
$$\mathcal{E}_\infty = \mathcal{E}_\infty(\vartheta), \ S = S(\vartheta), \ J = J(\vartheta), \ P_h = P_h(\vartheta)$$

as defined in Definition 6.1 such that weak convergence of pairs

$$\mathcal{L}(S_n, J_n \mid P_{n,\vartheta}) \longrightarrow \mathcal{L}(S, J \mid P_0) = \mathcal{L}(S(\vartheta), J(\vartheta) \mid P_0(\vartheta))$$

(weakly in $\mathbb{R}^d \times \mathbb{R}^{d\times d}$) holds as $n \to \infty$.
(b) In particular, $(\mathcal{E}_n)_n$ is called *locally asymptotically mixed normal at ϑ (LAMN)* if the limit experiment $\mathcal{E}_\infty = \mathcal{E}(S, J)$ in (a.ii) is a mixed normal experiment as defined in Definition 6.2 and Proposition 6.3.

Section 7.1 Local Asymptotics of Type LAN, LAMN, LAQ

(c) In particular, $(\mathcal{E}_n)_n$ is called *locally asymptotically normal at ϑ (LAN)* if the limit experiment $\mathcal{E}_\infty = \mathcal{E}(S, J)$ in (a.ii) is a Gaussian shift experiment $\mathcal{E}(J)$ as defined in Definition 5.2.

In the LAN case, $J = J(\vartheta)$ is deterministic; recall that a Gaussian shift experiment $\mathcal{E}(J)$ exists for every $J \in \mathbb{D}^+$. Since convergence in law of $J_n = J_n(\vartheta)$ under $P_{n,\vartheta}$ to a deterministic limit $J = J(\vartheta)$ is equivalent to convergence in probability, we can write under LAN

$$J_n = J + o_{P_{n,\vartheta}}(1) \quad \text{as } n \to \infty$$

and can replace in this case the statistics $J_n = J_n(\vartheta)$ in part (a) of Definition 7.1 by the deterministic quantity $J = J(\vartheta)$ simultaneously for all $n \geq 1$.

For local asymptotics at some reference point $\vartheta \in \Theta$ we shall always use the following notations:

7.1' Definition. (a) For all $n \geq 1$, we write for short

$$L_{n,\vartheta}^{h/0} := \frac{d\,P_{n,\vartheta + \delta_n(\vartheta)h}}{d\,P_{n,\vartheta}}, \quad \Lambda_{n,\vartheta}^{h/0} = \log L_{n,\vartheta}^{h/0}$$

in the experiments $\mathcal{E}_n$, with all conventions on likelihood ratios as explained in Notations 3.1; Θ being open, we have *local parameter spaces*

$$\Theta_{\vartheta,n} := \{h \in \mathbb{R}^d : h \text{ such that } \vartheta + \delta(\vartheta)h \text{ belongs to } \Theta\}$$

which as $n \to \infty$ form a sequence of open sets increasing to $\mathbb{R}^d$. At stage $n \geq 1$ of the asymptotics, we call

$$\mathcal{E}_{\vartheta,n} := \left(\Omega_n, \mathcal{A}_n, \{P_{n,\vartheta+\delta_n(\vartheta)h} : h \in \Theta_{\vartheta,n}\}\right)$$

a *local model at ϑ with local scale $\delta_n(\vartheta)$ and local parameter h*.

(b) When LAQ holds at ϑ, we call $\mathcal{E}_\infty = \mathcal{E}_\infty(\vartheta)$ of Definition 7.1(a.ii) the *limit experiment at ϑ*, and write

$$L^{h/0} := \frac{dP_h}{dP_0}, \quad \Lambda^{h/0} = \log L^{h/0}, \quad h \in \mathbb{R}^d$$

in $\mathcal{E}_\infty$. According to Definitions 6.4 and 6.2, the limit experiment has score $S - Jh = S(\vartheta) - J(\vartheta)h$ in $h \in \mathbb{R}^d$, the observed information in the sense of Definition 6.4 is $J = J(\vartheta)$. Note that in the limit experiment $\mathcal{E}_\infty(\vartheta)$, the law of the observed information $\mathcal{L}(J|P_h) = \mathcal{L}(J(\vartheta) \mid P_h(\vartheta))$ may depend on $h \in \mathbb{R}^d$.

(c) The central statistic in the limit experiment $\mathcal{E}_\infty$ is $Z = Z(\vartheta)$, $Z = 1_{\{J \in \mathbb{D}^+\}} J^{-1} S$, by Definition 6.1. When LAQ holds at ϑ, we call the sequence $(Z_n)_n$ defined by

$$Z_n = Z_n(\vartheta), \quad Z_n := 1_{\{J_n \in \mathbb{D}^+\}} J_n^{-1} S_n, \quad n \geq 1$$

the *central sequence at ϑ*.

The name 'central sequence' suggests a benchmark sequence: in fact, we will see below that it allows us to judge estimation errors at the reference point ϑ under LAMN simultaneously with respect to a broad variety of loss functions. In a setting of i.i.d. observations, we will give examples for LAN in Section 7.4; stochastic process examples for LAN, LAMN or LAQ will be discussed in Chapter 8. For local scale at ϑ, besides the well-known $\delta_n(\vartheta) = n^{-1/2}$ encountered under certain conditions, various other rates – slower or faster – will occur in the stochastic process examples of Chapter 8. In particular, we underline the following: in a given sequence of models $(\mathcal{E}_n)_n$, we may have at different points $\vartheta \in \Theta$ different rates $\delta_n(\vartheta) \downarrow 0$ and different limit experiments $\mathcal{E}_\infty(\vartheta)$.

We begin to discuss the statistical implications of the LAQ setting. Note that Definition 7.1 did not require equivalence or absolute continuity for probability measures $P_{n,\xi'}, P_{n,\xi}, \xi' \neq \xi$, in the experiments $\mathcal{E}_n$ at the pre-limiting stage $n < \infty$. In particular, log-likelihoods $\Lambda_{n,\vartheta}^{h/0}$ in $\mathcal{E}_n$ may take the values $\pm\infty$ with positive $P_{n,\vartheta}$-probability: recall the definition of $\overline{\mathbb{R}}$-tightness from Notations 3.1 and 3.1'.

7.2 Proposition. When LAQ holds at ϑ, for bounded sequences $(h_n)_n$ in $\mathbb{R}^d$, log-likelihoods $(\Lambda_{n,\vartheta}^{h_n/0})_n$ and likelihoods $(L_{n,\vartheta}^{h_n/0})_n$ under $P_{n,\vartheta}$ are $\overline{\mathbb{R}}$-tight as $n \to \infty$, and convergence of $(h_n)_n$ to a limit $h \in \mathbb{R}^d$ implies

$$\Lambda_{n,\vartheta}^{h_n/0} - \Lambda_{n,\vartheta}^{h/0} = o_{(P_{n,\vartheta})}(1), \quad L_{n,\vartheta}^{h_n/0} - L_{n,\vartheta}^{h/0} = o_{(P_{n,\vartheta})}(1), \quad n \to \infty.$$

Proof. (1) For convergent sequences $h_n \to h$, tightness in $\mathbb{R}^d \times \mathbb{R}^{d \times d}$ under $(P_{n,\vartheta})_n$ of the pairs $(S_n, J_n)_n$ in virtue of Definition 7.1(a.ii) implies that the quantities $h_n^\top J_n (h_n - h)$ and $(h_n - h)^\top J_n h$ vanish as $n \to \infty$ in $P_{n,\vartheta}$-probability. For bounded sequences $(h_n)_n$, both $h_n^\top J_n h_n$ and $h_n^\top S_n$ remain tight under $P_{n,\vartheta}$ as $n \to \infty$. Thus we obtain for convergent sequences $h_n \to h$

$$(+) \qquad \left(h_n^\top S_n - \frac{1}{2} h_n^\top J_n h_n\right) - \left(h^\top S_n - \frac{1}{2} h^\top J_n h\right) = o_{(P_{n,\vartheta})}(1)$$

as $n \to \infty$. For bounded sequences $(h_n)_n$ in $\mathbb{R}^d$, we have

$$(++) \qquad \mathcal{L}\left(h_n^\top S_n - \frac{1}{2} h_n^\top J_n h_n \mid P_{n,\vartheta}\right), \quad n \geq 1, \quad \text{is tight in } \mathbb{R}.$$

(2) Combining (+) with the quadratic expansions in Definition 7.1(a.i) of the log-likelihoods gives

$$\Lambda_{n,\vartheta}^{h_n/0} - \Lambda_{n,\vartheta}^{h/0} = o_{(P_{n,\vartheta})}(1) \quad \text{for convergent sequences } h_n \to h.$$

(3) The log-likelihoods $(\Lambda_{n,\vartheta}^{h_n/0})_n$ form a sequence of $\overline{\mathbb{R}}$-valued random variables, cf. Notations 3.1 and 3.1'. For bounded sequences $(h_n)_n$, (++) combined with the

quadratic expansions in Definition 7.1(a.i) yields

(*) $\qquad \left(\Lambda_{n,\vartheta}^{h_n/0}\right)_n$ is $\mathbb{R}$-tight under $(P_{n,\vartheta})_n$.

The likelihoods $(L_{n,\vartheta}^{h_n/0})_n$ are $[0,\infty]$-valued, and Proposition 3.3 shows

(**) $\qquad \left(L_{n,\vartheta}^{h_n/0}\right)_n$ is $\mathbb{R}$-tight under $(P_{n,\vartheta})_n$.

(4) In order to prove the last assertion of the proposition, we consider functions $\frac{1}{N} \wedge \exp(\cdot) \vee N$ for large N which are bounded and uniformly continuous, and deduce

$$L_{n,\vartheta}^{h_n/0} - L_{n,\vartheta}^{h/0} = o_{(P_{n,\vartheta})}(1)$$

for convergent sequences $h_n \to h$ from (**) and (*) and step (2). $\qquad \square$

Contiguity of probability measures in local models at ϑ will play a key role in the following. For bounded sequences $(h_n)_n$, we consider $P_{n,\vartheta+\delta_n(\vartheta)h_n}$ as *local alternatives* to $P_{n,\vartheta}$ when $n \to \infty$, and make use of the results in Chapter 3.

7.3 Proposition. Under LAQ at ϑ we have mutual contiguity

($\diamond$) $\qquad \left(P_{n,\vartheta+\delta_n(\vartheta)h_n}\right)_n \triangleleft \triangleright (P_{n,\vartheta})_n$

for arbitrary bounded sequences $(h_n)_n$. For convergent sequences $h_n \to h$ we have weak convergence

$$\mathcal{L}(\Lambda_{n,\vartheta}^{h_n/0}, S_n, J_n \mid P_{n,\vartheta+\delta_n(\vartheta)h_n}) \longrightarrow \mathcal{L}(\Lambda^{h/0}, S, J \mid P_h)$$
$$(\text{in } \mathbb{R} \times \mathbb{R}^d \times \mathbb{R}^{d \times d}, \text{ as } n \to \infty).$$

Proof. (1) Consider first convergent sequences $h_n \to h$. Combining Definition 7.1 of LAQ at ϑ with Proposition 7.2 and Notation 3.1' we have

$$\mathcal{L}(\Lambda_{n,\vartheta}^{h_n/0} \mid P_{n,\vartheta}) \longrightarrow \mathcal{L}\left(h^\top S - \frac{1}{2} h^\top J h \mid P_0\right) = \mathcal{L}(\Lambda^{h/0} \mid P_0)$$

(weakly in $\mathbb{R}$, for $n \to \infty$)

where $\Lambda^{h/0}$ is the log-likelihood ratio of P_h with respect to P_0 in the quadratic limit experiment $\mathcal{E}_\infty(\vartheta)$. By Definition 6.1 and Remark 6.1', probability measures in the limit experiment are equivalent. Hence Le Cam's first lemma applies and establishes ($\diamond$) in this case, cf. Lemma 3.5 and Remark 3.5'.

(2) Directly from Definition 3.2, mutual contiguity ($\diamond$) holds if and only if any subsequence $(n_k)_k$ of $\mathbb{N}$ contains a further subsequence $(n_{k_\ell})_\ell$ along which

$$\left(P_{n_{k_\ell},\vartheta+\delta_{n_{k_\ell}}(\vartheta)h_{n_{k_\ell}}}\right)_\ell \triangleleft \triangleright \left(P_{n_{k_\ell},\vartheta}\right)_\ell$$

holds. When $(h_n)_n$ is a bounded sequence, selecting further subsequences if necessary, we can always assume that $(h_{n_{k_\ell}})_\ell$ converges. Hence it is sufficient to establish $(\diamond)$ for convergent sequences $(h_m)_m$ which was done in step (1). Thus $(\diamond)$ is proved in general.

(3) We recall some consequences of mutual continuity $(\diamond)$: for bounded sequences $(h_n)_n$ and for $\mathcal{A}_n$-measurable variables $(Y_n)_n$ taking values in $\overline{\mathbb{R}}^q$, we can consider events $A_n := \{|Y_n| > \varepsilon\}$ and have

$$Y_n = o_{P_{n,\vartheta}}(1) \iff Y_n = o_{P_{n,\vartheta+\delta_n h_n}}(1) \, ;$$

if we consider events $A_n := \{|Y_n| > K\}$, the ε-δ-characterisation of mutual contiguity in Proposition 3.7 shows that

$$(Y_n)_n \text{ under } (P_{n,\vartheta})_n \text{ is } \mathbb{R}^q\text{-tight} \iff (Y_n)_n \text{ under } (P_{n,\vartheta+\delta_n h_n})_n \text{ is } \mathbb{R}^q\text{-tight}.$$

In particular, the quadratic expansions under $P_{n,\vartheta}$ which define LAQ in Definition 7.1 remain valid under $P_{n,\vartheta+\delta_n h_n}$ whenever $(h_n)_n$ is bounded: when LAQ holds at ϑ we have

$$(\diamond) \qquad \Lambda^{h_n/0}_{n,\vartheta} = h_n^\top S_n - \frac{1}{2} h_n^\top J_n h_n + o_{(P_{n,\vartheta+\delta_n h_n})}(1) \quad \text{as } n \to \infty$$

for bounded sequences $(h_n)_n$.

(4) Now we can prove the second part of the proposition. Under LAQ at ϑ, weak convergence of pairs

$$\mathcal{L}(S_n, J_n \mid P_{n,\vartheta}) \longrightarrow \mathcal{L}(S, J \mid P_0) \quad (\text{in } \mathbb{R}^d \times \mathbb{R}^{d \times d}, \text{ as } n \to \infty)$$

implies by the continuous mapping theorem and by representation of log-likelihood ratios in Definition 7.1

$$\mathcal{L}(\Lambda^{h/0}_{n,\vartheta}, S_n, J_n \mid P_{n,\vartheta}) \longrightarrow \mathcal{L}(\Lambda^{h/0}, S, J \mid P_0)$$
$$(\text{weakly in } \mathbb{R} \times \mathbb{R}^d \times \mathbb{R}^{d \times d}, \text{ as } n \to \infty)$$

for fixed $h \in \mathbb{R}^d$. Proposition 7.2 allows to extend this to convergent sequences $h_n \to h$:

$$\mathcal{L}(\Lambda^{h_n/0}_{n,\vartheta}, S_n, J_n \mid P_{n,\vartheta}) \longrightarrow \mathcal{L}(\Lambda^{h/0}, S, J \mid P_0)$$
$$(\text{weakly in } \mathbb{R} \times \mathbb{R}^d \times \mathbb{R}^{d \times d}, \text{ for } n \to \infty).$$

Here $\Lambda^{h_n/0}_{n,\vartheta}$ is the log-likelihood ratio of $P_{n,\vartheta+\delta_n h_n}$ with respect to $P_{n,\vartheta}$ in $\mathcal{E}_n(\vartheta)$, and $\Lambda^{h/0}$ the log-likelihood ratio of P_h with respect to P_0 in $\mathcal{E}_\infty(\vartheta)$. Now we apply Le Cam's third lemma, cf. Lemma 3.6 and Remark 3.6', and obtain weak convergence under contiguous alternatives

$$\mathcal{L}(\Lambda^{h_n/0}_{n,\vartheta}, S_n, J_n \mid P_{n,\vartheta+\delta_n h_n}) \longrightarrow \mathcal{L}(\Lambda^{h/0}, S, J \mid P_h)$$
$$(\text{weakly in } \mathbb{R} \times \mathbb{R}^d \times \mathbb{R}^{d \times d}, \text{ for } n \to \infty).$$

This finishes the proof of the proposition. $\square$

Section 7.1 Local Asymptotics of Type LAN, LAMN, LAQ

Among competing sequences of estimators for the unknown parameter in the sequence of experiments $(\mathcal{E}_n)_n$, one would like to identify – if possible – sequences which are 'asymptotically optimal'. Fix $\vartheta \in \Theta$. In a setting of local asymptotics at ϑ with local scale $(\delta_n(\vartheta))_n$, the basic idea is as follows. To any estimator T_n for the unknown parameter in $\mathcal{E}_n$ associate

$$U_n = U_n(\vartheta) := \delta_n^{-1}(\vartheta)(T_n - \vartheta)$$

which is the rescaled estimation error or T_n at ϑ; clearly

$$U_n - h = \delta_n^{-1}(\vartheta)(T_n - [\vartheta + \delta_n(\vartheta)h])$$

for $h \in \Theta_{n,\vartheta}$. Thus, ϑ being fixed, we may consider $U_n = U_n(\vartheta)$ as an estimator for the local parameter h in the local experiment $\mathcal{E}_n(\vartheta) = \{P_{n,\vartheta+\delta_n(\vartheta)h} : h \in \Theta_{n,\vartheta}\}$ at ϑ. The aim will be to associate to such sequences $(U_n)_n$ *limit objects* $U = U(\vartheta)$ *in the limit experiment* $\mathcal{E}_\infty(\vartheta)$, in a suitably strong sense. Then one may first discuss properties of the estimator U in the limit experiment $\mathcal{E}_\infty(\vartheta)$ and then convert these into properties of U_n as estimator for the local parameter $h \in \Theta_{n,\vartheta}$ in the local experiment $\mathcal{E}_n(\vartheta)$. This means that we will study properties of T_n in shrinking neighbourhoods of ϑ defined from local scale $\delta_n(\vartheta) \downarrow 0$. Lemma 7.5 below is the technical key to this approach. It requires the notion of a Markov extension of a statistical experiment.

7.4 Definition. Consider a statistical experiment $(E, \mathcal{E}, \mathcal{P})$, another measurable space $(E', \mathcal{E}')$ and a transition probability $K(y, dy')$ from $(E, \mathcal{E})$ to $(E', \mathcal{E}')$. Equip the product space

$$(\widehat{E}, \widehat{\mathcal{E}}), \quad \widehat{E} = E \times E', \quad \widehat{\mathcal{E}} = \mathcal{E} \otimes \mathcal{E}'$$

with probability laws

$$P_K(dy, dy') := P(dy)\, K(y, dy'), \quad P \in \mathcal{P}.$$

This yields a statistical model

$$\widehat{\mathcal{P}} := \{P_K : P \in \mathcal{P}\} \quad \text{on} \quad (\widehat{E}, \widehat{\mathcal{E}})$$

with the following properties:
 (i) for every pair Q, P and every version L of the likelihood ratio of P with respect to Q in the experiment $(E, \mathcal{E}, \mathcal{P})$, $\widehat{L}$ defined by

$$\widehat{L}(y, y') := L(y) \text{ on } (E \times E', \mathcal{E} \otimes \mathcal{E}')$$

is a version of the likelihood ratio of P_K with respect to Q_K in the experiment $(\widehat{E}, \widehat{\mathcal{E}}, \widehat{\mathcal{P}})$;

(ii) in the model $(\widehat{E}, \widehat{\mathcal{E}}, \widehat{\mathcal{P}})$, statistics U taking values in $(E', \mathcal{E}')$ and realising the prescribed laws

$$PK(A') := \int_E P(dy)\, K(y, A')\,, \quad A' \in \mathcal{E}',$$

under the probability measure $P_K \in \widehat{\mathcal{P}}$

exist: on $(\widehat{E}, \widehat{\mathcal{E}})$, we simply define the random variable U as the projection $\widehat{E} \ni (y, y') \to y' \in E'$ on the second component which gives $P_K(U \in A') = PK(A')$ for $A' \in \mathcal{E}'$.

The experiment $(\widehat{E}, \widehat{\mathcal{E}}, \widehat{\mathcal{P}})$ is called a Markov extension of $(E, \mathcal{E}, \mathcal{P})$; clearly every statistic Y already available in the original experiment is also available in $(\widehat{E}, \widehat{\mathcal{E}}, \widehat{\mathcal{P}})$ via lifting $Y(y, y') := Y(y)$.

We comment on this definition. By Definition 7.4(i), $(\widehat{E}, \widehat{\mathcal{E}}, \widehat{\mathcal{P}})$ is *statistically the same experiment as* $(E, \mathcal{E}, \mathcal{P})$ since 'likelihoods remain unchanged': for arbitrary $P^{(0)}$ and $P^{(1)}, \ldots, P^{(\ell)}$ in $\mathcal{P}$, writing $L^{(i)}$ for the likelihood ratio of $P^{(i)}$ with respect to $P^{(0)}$ in $(E, \mathcal{E})$, and $\widehat{L}^{(i)}$ for the likelihood ratio of $P_K^{(i)}$ with respect to $P_K^{(0)}$ in $(\widehat{E}, \widehat{\mathcal{E}}, \widehat{\mathcal{P}})$, we have (cf. Strasser [121, 53.10 and 25.6]) by construction

$$\mathcal{L}\big((L^{(1)}, \ldots, L^{(\ell)}) \mid P^{(0)}\big) = \mathcal{L}\big((\widehat{L}^{(1)}, \ldots, \widehat{L}^{(\ell)}) \mid P_K^{(0)}\big).$$

Whereas $(E', \mathcal{E}')$-valued random variables having law PK under $P \in \mathcal{P}$ might not exist in the original experiment, their existence is granted on the Markov extension $(\widehat{E}, \widehat{\mathcal{E}}, \widehat{\mathcal{P}})$ of $(E, \mathcal{E}, \mathcal{P})$ by Definition 7.4(ii).

7.5 Lemma. Fix $\vartheta \in \Theta$ such that LAQ holds at ϑ. Relative to ϑ, consider $\mathcal{A}_n$-measurable mappings $U_n = U_n(\vartheta) : \Omega_n \to \mathbb{R}^k$ with the property

$$\mathcal{L}(U_n \mid P_{n,\vartheta})\,, \quad n \geq 1\,, \quad \text{is tight in } \mathbb{R}^k\,.$$

Then there are subsequences $(n_l)_l$ and $\mathcal{A}$-measurable mappings $U : \Omega \to \mathbb{R}^k$ in (if necessary, Markov extensions of) the limit experiment $\mathcal{E}_\infty = \mathcal{E}(S, J)$, with the following property:

$$\begin{cases} \text{weak convergence in } \mathbb{R}^d \times \mathbb{R}^{d \times d} \times \mathbb{R}^k \\ \mathcal{L}\big(S_{n_l}, J_{n_l}, U_{n_l} \mid P_{n_l, \vartheta + \delta_{n_l} h_l}\big) \to \mathcal{L}(S, J, U \mid P_h) \quad \text{as} \quad l \to \infty \\ \text{holds for arbitrary } h \text{ and convergent sequences } h_l \to h\,. \end{cases}$$

Proof. (1) By LAQ at ϑ we have tightness of $\mathcal{L}(S_n, J_n \mid P_{n,\vartheta}), n \geq 1$; by assumption, we have tightness of $\mathcal{L}(U_n \mid P_{n,\vartheta}), n \geq 1$. Thus

$$\mathcal{L}(S_n, J_n, U_n \mid P_{n,\vartheta})\,, \quad n \geq 1\,, \quad \text{is tight in } \mathbb{R}^d \times \mathbb{R}^{d \times d} \times \mathbb{R}^k\,.$$

Section 7.1 Local Asymptotics of Type LAN, LAMN, LAQ

Extracting weakly convergent subsequences, let $(n_l)_l$ denote a subsequence of $\mathbb{N}$ and μ_0 a probability law on $\mathbb{R}^d \times \mathbb{R}^{d \times d} \times \mathbb{R}^k$ such that

(∗) $\qquad \mathcal{L}\left(S_{n_l}, J_{n_l}, U_{n_l} \mid P_{n,\vartheta}\right) \longrightarrow \mu_0, \quad l \to \infty.$

For convergent sequences $h_l \to h$ we deduce from the quadratic expansion of log-likelihoods in Definition 7.1 together with Proposition 7.2

(+) $\qquad \mathcal{L}\left(\Lambda_{n_l,\vartheta}^{h_l/0}, S_{n_l}, J_{n_l}, U_{n_l} \mid P_{n,\vartheta}\right) \longrightarrow \widetilde{\mu}_0, \quad l \to \infty$

where the law $\widetilde{\mu}_0$ on $\mathbb{R} \times \mathbb{R}^d \times \mathbb{R}^{d \times d} \times \mathbb{R}^k$ is the image measure of μ_0 under the mapping

$$(s, j, u) \longrightarrow \left(h^\top s - \frac{1}{2} h^\top j h, s, j, u\right).$$

Mutual contiguity (◇) in Proposition 7.3 and Le Cam's Third Lemma 3.6 transform (+) into weak convergence

$$\mathcal{L}\left(\Lambda_{n_l,\vartheta}^{h_l/0}, S_{n_l}, J_{n_l}, U_{n_l} \mid P_{n_l,\vartheta+\delta_{n_l} h_l}\right) \longrightarrow \widetilde{\mu}_h, \quad l \to \infty$$

where $\widetilde{\mu}_h$ is a probability law on $\mathbb{R} \times \mathbb{R}^d \times \mathbb{R}^{d \times d} \times \mathbb{R}^k$ defined from $\widetilde{\mu}_0$ by

$$\widetilde{\mu}_h(d\lambda, ds, dj, du) = e^\lambda \widetilde{\mu}_0(d\lambda, ds, dj, du).$$

Projecting on the components (s, j, u), we thus have proved for any limit point $h \in \mathbb{R}^d$ and any sequence $(h_l)_l$ converging to h

(++) $\qquad \mathcal{L}\left(S_{n_l}, J_{n_l}, U_{n_l} \mid P_{n_l,\vartheta+\delta_{n_l} h_l}\right) \longrightarrow \mu_h, \quad l \to \infty$

where – with μ_0 as in (∗) above – the probability measures μ_h are defined from μ_0 by

(+++) $\quad \mu_h(ds, dj, du) := e^{h^\top s - \frac{1}{2} h^\top j h} \mu_0(ds, dj, du) \quad$ on $\mathbb{R}^d \times \mathbb{R}^{d \times d} \times \mathbb{R}^k.$

Note that the statistical model $\{\mu_h : h \in \mathbb{R}^d\}$ arising in (+++) is attached to the particular accumulation point μ_0 for $\{\mathcal{L}(S_n, J_n, U_n \mid P_{n,\vartheta}) : n \geq 1\}$ which was selected in (∗) above, and that different accumulation points for $\{\mathcal{L}(S_n, J_n, U_n \mid P_{n,\vartheta}) : n \geq 1\}$ lead to different models (+++).

(2) We construct a Markov extension of the limit experiment $\mathcal{E}(S, J)$ carrying a random variable U which allows to identify μ_h in (++) as $\mathcal{L}(S, J, U \mid P_h)$, for all $h \in \mathbb{R}^d$.

Let $\check{s}$, $\check{j}$, $\check{u}$ denote the projections which map $(s, j, u) \in \mathbb{R}^d \times \mathbb{R}^{d \times d} \times \mathbb{R}^k$ to either one of its coordinates $s \in \mathbb{R}^d$ or $j \in \mathbb{R}^{d \times d}$ or $u \in \mathbb{R}^k$. In the statistical model $\{\mu_h : h \in \mathbb{R}^d\}$ fixed by (+++), the pair $(\check{s}, \check{j})$ is a sufficient statistic. By sufficiency, conditional distributions given $(\check{s}, \check{j})$ of $\mathbb{R}^k$-valued random variables admit regular versions which do not depend on the parameter $h \in \mathbb{R}^d$. Hence there is a transition probability $K(\cdot,\cdot)$ from $\mathbb{R}^d \times \mathbb{R}^{d \times d}$ to $\mathbb{R}^k$ such that

(∗∗) $$K((s,j),du) =: \mu_\bullet^{\check{u}|(\check{s},\check{j})=(s,j)}(du)$$

provides a common determination of all conditional laws $\mu_h^{\check{u}|(\check{s},\check{j})=(s,j)}(du)$, $h \in \mathbb{R}^d$. Defining

(×) $$\widehat{\Omega} := \Omega \times \mathbb{R}^k, \quad \widehat{\mathcal{A}} := \mathcal{A} \otimes \mathcal{B}(\mathbb{R}^k),$$
$$\widehat{P}_h(d\omega, du) := P_h(d\omega) K((S(\omega), J(\omega)), du), \quad h \in \mathbb{R}^d$$

we have a Markov extension

$$\left(\widehat{\Omega}, \widehat{\mathcal{A}}, \{\widehat{P}_h : h \in \mathbb{R}^d\}\right)$$

of the original limit experiment $\mathcal{E}_\infty(\vartheta) = \mathcal{E}(S,J) = (\Omega, \mathcal{A}, \{P_h : h \in \mathbb{R}^d\})$ of Definition 7.1. Exploiting again sufficiency, we can put the laws μ_h of (++) and (+++) in the form

$$\mu_h(ds,dj,du) = \mu_h^{(\check{s},\check{j})}(ds,dj) \, \mu_h^{\check{u}|(\check{s},\check{j})=(s,j)}(du)$$
$$= \mu_h^{(\check{s},\check{j})}(ds,dj) \, \mu_\bullet^{\check{u}|(\check{s},\check{j})=(s,j)}(du)$$
$$= \mu_h^{(\check{s},\check{j})}(ds,dj) \, K((s,j),du)$$

for all $h \in \mathbb{R}^d$. Combining (∗) and (++) with Proposition 7.3, we can identify the last expression with

$$P_h^{(S,J)}(ds,dj) \, K((s,j),du)$$

for all $h \in \mathbb{R}^d$. The Markov extension (×) of the original limit experiment allows us to write this as

$$\widehat{P}_h^{(S,J,U)}(ds,dj,du)$$

where U denotes the projection $U(\omega, u) := u$ from $\widehat{\Omega}$ to $\mathbb{R}^k$. Hence we have proved

$$\mu_h(ds,dj,du) = \widehat{P}_h^{(S,J,U)}(ds,dj,du) \quad \text{for all } h \in \mathbb{R}^d$$

which is the assertion of the lemma. □

From now on, we need no longer distinguish carefully between the original limit experiment and its Markov extension. From (∗) and (∗∗) in the last proof, we see that any accumulation point of $\{\mathcal{L}(S_n, J_n, U_n | P_{n,\vartheta}) : n \geq 1\}$ can be written in the form $\mathcal{L}(S,J|P_0)(ds,dj)K((s,j),du)$ for some transition probability from $\mathbb{R}^d \times \mathbb{R}^{d \times d}$ to $\mathbb{R}^k$. In this sense, statistical models $\{\mu_h : h \in \mathbb{R}^d\}$ which can arise in (++) and (+++) correspond to transition probabilities $K(\cdot,\cdot)$ from $\mathbb{R}^d \times \mathbb{R}^{d \times d}$ to $\mathbb{R}^k$.

Section 7.1 Local Asymptotics of Type LAN, LAMN, LAQ

7.6 Theorem. Assume LAQ at ϑ. For any estimator sequence $(T_n)_n$ for the unknown parameter in $(\mathcal{E}_n)_n$, let $U_n = U_n(\vartheta) = \delta_n^{-1}(T_n - \vartheta)$ denote the rescaled estimation errors of T_n at ϑ, $n \geq 1$. Assume joint weak convergence in $\mathbb{R}^d \times \mathbb{R}^{d \times d} \times \mathbb{R}^d$

$$\mathcal{L}(S_n, J_n, U_n \mid P_{n,\vartheta}) \longrightarrow \mathcal{L}(S, J, U \mid P_0) \quad \text{as} \quad n \to \infty$$

where U is a statistic in the (possibly Markov extended) limit experiment $\mathcal{E}_\infty(\vartheta) = \mathcal{E}(S, J)$. Then we have for arbitrary convergent sequences $h_n \to h$

(+) $$\mathcal{L}(S_n, J_n, U_n - h_n \mid P_{n,\vartheta+\delta_n h_n}) \longrightarrow \mathcal{L}(S, J, U - h \mid P_h)$$
$$\text{as} \quad n \to \infty$$

(weak convergence in $\mathbb{R}^d \times \mathbb{R}^{d \times d} \times \mathbb{R}^d$), and from this

(++) $$\sup_{|h| \leq C} \left| E_{n,\vartheta+\delta_n h} \left(\ell \left(\delta_n^{-1}(T_n - (\vartheta + \delta_n h)) \right) \right) - E_h(\ell(U - h)) \right|$$
$$\longrightarrow 0 \quad \text{as} \quad n \to \infty$$

for bounded and continuous loss functions $\ell(\cdot)$ on $\mathbb{R}^d$ and for arbitrarily large constants $C < \infty$.

Proof. (1) Assertion (+) of the theorem corresponds to the assertion of Lemma 7.5, under the stronger assumption of joint weak convergence of (S_n, J_n, U_n) under $P_{n,\vartheta}$ as $n \to \infty$, without selecting subsequences. For loss functions $\ell(\cdot)$ in $\mathcal{C}_b(\mathbb{R}^d)$, (+) contains the following assertion: for arbitrary convergent sequences $h_n \to h$, as $n \to \infty$,

(×) $$E_{n,\vartheta+\delta_n h_n}\left(\ell\left(\delta_n^{-1}(T_n - (\vartheta + \delta_n h_n))\right)\right) = E_{n,\vartheta+\delta_n h_n}(\ell(U_n - h_n))$$
$$\longrightarrow E_h(\ell(U - h)) .$$

(2) We prove that in the limit experiment $\mathcal{E}_\infty(\vartheta) = \mathcal{E}(S, J)$, for $\ell(\cdot)$ continuous and bounded,

$$h \longrightarrow E_h(\ell(U - h)) \quad \text{is continuous on } \mathbb{R}^d.$$

Consider convergent sequences $h_n \to h$. The structure in Definition 6.1 of likelihoods in the limit experiment implies pointwise convergence of $L^{h_n/0}$ as $n \to \infty$ to $L^{h/0}$; these are non-negative, and $E_0(L^{h/0}) = 1 = E_0(L^{h_n/0})$ holds for all n. This gives (cf. [20, Nr. 21 in Chap. II])

the sequence $L^{h_n/0}$ under P_0, $n \geq 1$, is uniformly integrable.

For $\ell(\cdot)$ continuous and bounded, we deduce

the sequence $L^{h_n/0} \ell(U - h_n)$ under P_0, $n \geq 1$, is uniformly integrable

which contains the assertion of step (2): it is sufficient to write as $n \to \infty$

$$E_{h_n}(\ell(U - h_n)) = E_0\left(L^{h_n/0} \ell(U - h_n)\right)$$
$$\longrightarrow E_0\left(L^{h/0} \ell(U - h)\right) = E_h\left(\ell(U - h)\right).$$

(3) Now it is easy to prove (++): in the limit experiment, thanks to step (2), we can rewrite assertion (×) of step (1) in the form

(××) $\quad E_{n,\vartheta+\delta_n h_n}(\ell(U_n - h_n)) - E_{h_n}(\ell(U - h_n)) = o(1), \quad n \to \infty$

for arbitrary convergent sequences $h_n \to h$. For large constants $C < \infty$ define

$$\alpha_n(C) := \sup_{|h| \leq C} \left| E_{n,\vartheta+\delta_n h}(\ell(U_n - h)) - E_h(\ell(U - h)) \right|, \quad n \geq 1.$$

Assume that for some C the sequence $(\alpha_n(C))_n$ does not tend to 0. Then there is a sequence $(h_n)_n$ in the closed ball $\{|h| \leq C\}$ and a subsequence $(n_k)_k$ of $\mathbb{N}$ such that for all k

$$\left| E_{n_k,\vartheta+\delta_{n_k} h_{n_k}}(\ell(U_{n_k} - h_{n_k})) - E_{h_{n_k}}(\ell(U - h_{n_k})) \right| > \varepsilon$$

for some $\varepsilon > 0$. The corresponding $(h_{n_k})_k$ taking values in a compact, we can find some further subsequence $(n_{k_\ell})_\ell$ and a limit point $\check{h}$ such that convergence $h_{n_{k_\ell}} \to \check{h}$ holds as $\ell \to \infty$, whereas

$$\left| E_{n_{k_\ell},\vartheta+\delta_{n_{k_\ell}} h_{n_{k_\ell}}}(\ell(U_{n_{k_\ell}} - h_{n_{k_\ell}})) - E_{h_{n_{k_\ell}}}(\ell(U - h_{n_{k_\ell}})) \right| > \varepsilon$$

still holds for all ℓ. This is in contradiction to (××). We thus have $\alpha_n(C) \to 0$ as $n \to \infty$. $\square$

We can rephrase Theorem 7.6 as follows: when LAQ holds at ϑ, any estimator sequence $(T_n)_n$ for the unknown parameter in $(\mathcal{E}_n)_n$ satisfying a joint convergence condition

(∗) $\quad \mathcal{L}(S_n, J_n, U_n \mid P_{n,\vartheta}) \longrightarrow \mathcal{L}(S, J, U \mid P_0), \quad U_n = \delta_n^{-1}(T_n - \vartheta)$

works over shrinking neighbourhoods of ϑ, defined through local scale $\delta_n = \delta_n(\vartheta) \downarrow 0$, as well as the limit object U viewed as estimator for the unknown parameter $h \in \mathbb{R}^d$ in the limit experiment $\mathcal{E}_\infty(\vartheta) = \mathcal{E}(S, J)$ of Definition 7.1, irrespective of choice of loss functions in the class $\mathcal{C}_b(\mathbb{R}^d)$, and thus irrespective of any particular way of penalising estimation errors.

Of particular interest are sequences $(T_n)_n$ coupled to the central sequence at ϑ by

(◇) $\quad \delta_n^{-1}(\vartheta)(T_n - \vartheta) = Z_n(\vartheta) + o_{(P_{n,\vartheta})}(1), \quad n \to \infty$

where $Z_n = 1_{\{J_n \in D^+\}} J_n^{-1} S_n$ is as in Definition 7.1'. In the limit experiment, P_0-almost surely, J takes values in D^+ (Definitions 7.1 and 6.1). By the continuous mapping theorem, the joint convergence condition (∗) then holds with $U := J^{-1}S = Z$ on the right-hand side, for sequences satisfying (◇), and Theorem 7.6 reads as follows:

7.7 Corollary. Under LAQ in ϑ, any sequence $(T_n)_n$ with the coupling property $(\diamond)$ satisfies

$$\sup_{|h|\leq C} \left| E_{n,\vartheta+\delta_n h} \left(\ell \left(\delta_n^{-1} (T_n - (\vartheta + \delta_n h)) \right) \right) - E_h \left(\ell(Z - h) \right) \right| \longrightarrow 0, \quad n \to \infty$$

for continuous and bounded loss functions $\ell(\cdot)$ and for arbitrary constants $C < \infty$. This signifies that $(T_n)_n$ works over shrinking neighbourhoods of ϑ, defined through local scale $\delta_n = \delta_n(\vartheta) \downarrow 0$, as well as the maximum likelihood estimator Z in the limit model $\mathcal{E}_\infty(\vartheta) = \mathcal{E}(S, J)$.

Recall that in Corollary 7.7, the laws $\mathcal{L}(Z - h | P_h)$ may depend on the parameter $h \in \mathbb{R}^d$, and may not admit finite higher moments (cf. Remark 6.6"; examples will be seen in Chapter 8). Note also that the statement of Corollary 7.7 – the 'best' result as long as we do not assume more than LAQ – should not be mistaken as an optimality criterion: Corollary 7.7 under $(\diamond)$ is simply a result on risks of estimators at ϑ – in analogy to Theorem 7.6 under the joint convergence condition $(*)$ – which does not depend on a particular choice of a loss function, and which is uniform over shrinking neighbourhoods of ϑ defined through local scale $\delta_n = \delta_n(\vartheta) \downarrow 0$. We do not know about the optimality of maximum likelihood estimators in general quadratic limit models (recall the remark following Theorem 6.8).

7.2 Asymptotic optimality of estimators in the LAN or LAMN setting

In local asymptotics at ϑ of type LAMN or LAN, we can do much better than Corollary 7.7. We first consider estimator sequences which are *regular at* ϑ, following a terminology well established since Hájek [40]: in a local asymptotic sense, this corresponds to the definition of strong equivariance in Definition 6.5 and Proposition 6.5' for a mixed normal limit experiment, and to the definition of equivariance in Definition 5.4 for a Gaussian shift. The aim is to pass from criteria for optimality which we have in the limit experiment (Theorems 6.6 and 6.8 in the mixed normal case, and Theorems 5.5 and 5.10 for the Gaussian shift) to criteria for *local asymptotic optimality* for estimator sequences $(T_n)_n$ in $(\mathcal{E}_n)_n$ at ϑ. The main results are Theorems 7.10, 7.11 and 7.12.

7.8 Definition. For $n \geq 1$, consider estimators T_n for the unknown parameter $\vartheta \in \Theta$ in $\mathcal{E}_n$.

(a) (Hájek [40]) If LAN holds at ϑ, the sequence $(T_n)_n$ is termed *regular at* ϑ if there is a probability measure $F = F(\vartheta)$ on $\mathbb{R}^d$ such that for every $h \in \mathbb{R}^d$

$$\mathcal{L}\left(\delta_n^{-1} (T_n - (\vartheta + \delta_n h)) \mid P_{n,\vartheta+\delta_n h} \right) \longrightarrow F$$

(weak convergence in $\mathbb{R}^d$ as $n \to \infty$)

where the limiting law F does not depend on the value of the local parameter $h \in \mathbb{R}^d$.

(b) (Jeganathan [66]) If LAMN holds at ϑ, $(T_n)_n$ is termed *regular at* ϑ if there is some probability measure $\widetilde{F} = \widetilde{F}(\vartheta)$ on $\mathbb{R}^{d\times d} \times \mathbb{R}^d$ such that for every $h \in \mathbb{R}^d$

$$\mathcal{L}\left(J_n, \delta_n^{-1}(T_n - (\vartheta + \delta_n h)) \mid P_{n,\vartheta+\delta_n h}\right) \longrightarrow \widetilde{F}$$

(weakly in $\mathbb{R}^{d\times d} \times \mathbb{R}^d$ as $n \to \infty$)

where the limiting law $\widetilde{F}$ does not depend on $h \in \mathbb{R}^d$.

Thus 'regular' is a short expression for 'locally asymptotically equivariant' in the LAN case, and for 'locally asymptotically strongly equivariant' in the LAMN case.

7.9 Example. Under LAMN or LAN at ϑ, estimator sequences $(T_n)_n$ in $(\mathcal{E}_n)_n$ linked to the central sequence $(Z_n)_n$ at ϑ by the coupling condition of Corollary 7.7

$$(\diamond) \qquad \delta_n^{-1}(\vartheta)(T_n - \vartheta) = Z_n(\vartheta) + o_{(P_{n,\vartheta})_n}(1), \quad n \to \infty$$

are regular at ϑ. This is seen as follows. As in the remarks preceding Corollary 7.7, rescaled estimation errors $U_n := \delta_n^{-1}(\vartheta)(T_n - \vartheta)$ satisfy a joint convergence condition with $U = Z$ on the right-hand side:

$$\mathcal{L}(S_n, J_n, U_n \mid P_{n,\vartheta}) \longrightarrow \mathcal{L}(S, J, Z \mid P_0), \quad n \to \infty,$$

from which by Theorem 7.6, for every $h \in \mathbb{R}^d$,

$$\mathcal{L}\left(S_n, J_n, (U_n - h) \mid P_{n,\vartheta+\delta_n h}\right) \longrightarrow \mathcal{L}(S, J, (Z - h) \mid P_h), \quad n \to \infty.$$

When LAN holds at ϑ, $F := \mathcal{L}(Z - h \mid P_h)$ does not depend on $h \in \mathbb{R}^d$, see Proposition 5.3(b). When LAMN holds at ϑ, $\widetilde{F} := \mathcal{L}(J, Z - h \mid P_h)$ does not depend on h, see Definition 6.2 together with Proposition 6.3(iv). This establishes regularity at ϑ of sequences $(T_n)_n$ which satisfy condition $(\diamond)$, under LAN or LAMN. □

7.9' Exercise. Let $\mathcal{E}$ denote the location model $\{F_\xi(\cdot) = F_0(\cdot - \xi) : \xi \in \mathbb{R}\}$ generated by the doubly exponential distribution $F_0(x) = \frac{1}{2} e^{-|x|} dx$ on $(\mathbb{R}, \mathcal{B}(\mathbb{R}))$. Write $\mathcal{E}_n$ for the n-fold product experiment. Prove the following, for every reference point $\vartheta \in \Theta$:

(a) Recall from Exercise 4.1''' that $\mathcal{E}$ is L^2-differentiable at $\xi = \vartheta$, and use Le Cam's Second Lemma 4.11 to establish LAN at ϑ with local scale $\delta_n(\vartheta) = 1/\sqrt{n}$.

(b) The median of the first n observations (which is the maximum likelihood estimator in this model) yields a regular estimator sequence at ϑ. The same holds for the empirical mean, or for arithmetic means between upper and lower empirical α-quantiles, $0 < \alpha < \frac{1}{2}$ fixed. Also Bayesians with 'uniform over $\mathbb{R}$ prior' as in Exercise 5.4'

$$T_n^* := \frac{\int_{-\infty}^\infty \xi\, L_n^{\xi/0}\, d\xi}{\int_{-\infty}^\infty L_n^{\xi/0}\, d\xi}, \quad n \geq 1$$

are regular in the sense of Definition 7.8. Check this using only properties of a location model. □

7.9" Exercise. We continue Exercise 7.9', with notations and assumptions as there. Focus on the empirical mean $T_n = \frac{1}{n}\sum_{i=1}^{n} X_i$ as estimator for the unknown parameter. Check that the sequence $(T_n)_n$, regular by Exercise 7.9'(b), induces the limit law $F = \mathcal{N}(0,2)$ in Definition 7.8(a).

Then give an alternative proof for regularity of $(T_n)_n$ based on the LAN property: from joint convergence of

$$\left(\frac{1}{\sqrt{n}} \sum_{i=1}^{n} \text{sgn}(X_i - \vartheta), \; \frac{1}{\sqrt{n}} \sum_{i=1}^{n} (X_i - \vartheta) \right) \quad \text{under } P_{n,\vartheta}$$

specify the two-dimensional normal law which arises as limit distribution for

$$\mathcal{L}\left(\Lambda_{n,\vartheta}^{h/0}, \sqrt{n}(T_n - \vartheta) \mid P_{n,\vartheta} \right) \quad \text{as } n \to \infty,$$

finally determine the limit law for

$$\mathcal{L}\left(\Lambda_{n,\vartheta}^{h/0}, \sqrt{n} \left(T_n - (\vartheta + h/\sqrt{n}) \right) \mid P_{n,\vartheta+h/\sqrt{n}} \right) \quad \text{as } n \to \infty$$

using Le Cam's Third Lemma 3.6 in the particular form of Proposition 3.6". □

In the LAN case, the following is known as Hájek's convolution theorem.

7.10 Convolution Theorem. Assume LAMN or LAN at ϑ, and consider a sequence $(T_n)_n$ of estimators for the unknown parameter in $(\mathcal{E}_n)_n$ which is regular at ϑ.

(a) (Hájek [40]) When LAN holds at ϑ, any limit distribution F arising in Definition 7.8(a) can be written as

$$F = \mathcal{N}\left(0, J^{-1}\right) \star Q$$

for some probability law Q on $\mathbb{R}^d$ as in Theorem 5.5.

(b) (Jeganathan [66]) When LAMN holds at ϑ, any limit law $\widetilde{F}$ in Definition 7.8(b) admits a representation

$$\widetilde{F}(A) = \iint P_\bullet^J(dj) \left[\mathcal{N}(0, j^{-1}) \star Q^j \right](du) \, 1_A(j, u), \quad A \in \mathcal{B}(\mathbb{R}^{d \times d} \times \mathbb{R}^d)$$

for some family of probability laws $\{Q^j : j \in D^+\}$ as in Theorem 6.6.

Proof. We prove (b) first. With notation $U_n := \delta_n^{-1}(T_n - \vartheta)$, regularity means that for some law $\widetilde{F}$,

$$\mathcal{L}\left(J_n, U_n - h \mid P_{n,\vartheta + \delta_n h}\right) \longrightarrow \widetilde{F}, \quad n \to \infty$$

(weakly in $\mathbb{R}^d \times \mathbb{R}^{d \times d}$) where $\widetilde{F}$ does not depend on $h \in \mathbb{R}^d$. Selecting subsequences according to Lemma 7.5, we see that $\widetilde{F}$ has a representation

(+) $\quad \widetilde{F} = \mathcal{L}(J, U - h \mid P_h) \quad$ not depending on $h \in \mathbb{R}^d$

where U is a statistic in the (possibly Markov extended) limit experiment $\mathcal{E}(S, J)$. Then U in (+) is a strongly equivariant estimator for the parameter $h \in \mathbb{R}^d$ in the mixed normal limit experiment $\mathcal{E}(S, J)$, and the Convolution Theorem 6.6 applies to U and gives the assertion. To prove (a), which corresponds to deterministic J, we use a simplified version of the above, and apply Boll's Convolution Theorem 5.5. □

Recall from Definition 5.6 that loss functions on $\mathbb{R}^d$ are subconvex if all levels sets are convex and symmetric with respect to the origin in $\mathbb{R}^d$. Recall from Anderson's Lemma 5.7 that for $\ell(\cdot)$ subconvex,

$$\int \ell(u) [\mathcal{N}(0, j^{-1}) * Q'](du) \geq \int \ell(u) \mathcal{N}(0, j^{-1})(du)$$

for every $j \in \mathbb{D}^+$ and any law Q' on $\mathbb{R}^d$. Anderson's lemma shows that best concentrated limit distributions in the Convolution Theorem 7.10 are characterised by

(7.10′) $\quad Q = \epsilon_0$ under LAN at ϑ,
$\quad\quad\quad Q^j = \epsilon_0$ for P_0-almost all $j \in \mathbb{R}^{d \times d}$ under LAMN at ϑ.

Estimator sequences $(T_n)_n$ in $(\mathcal{E}_n)_n$ which are regular at ϑ and attain in the convolution theorem the limit distribution (7.10') are called *efficient at* ϑ.

7.10″ Exercise. In the location model generated from the two-sided exponential distribution, continuing Exercises 7.9' and 7.9″, check from Exercise 7.9″ that the sequence of empirical means is not efficient. □

In some problems we might find efficient estimators directly, in others not. Under some additional conditions – this will be the topic of Section 7.3 – we can apply a method which allows us to construct efficient estimator sequences. We have the following characterisation.

7.11 Theorem. Consider estimators $(T_n)_n$ in $(\mathcal{E}_n)_n$ for the unknown parameter. Under LAMN or LAN at ϑ, the following assertions (i) and (ii) are equivalent:

(i) the sequence $(T_n)_n$ is regular and efficient at ϑ ;

(ii) the sequence $(T_n)_n$ has the coupling property (◊) of example 7.9 (or of Corollary 7.7):

$$\delta_n^{-1}(\vartheta)(T_n - \vartheta) = Z_n(\vartheta) + o_{(P_{n,\vartheta})}(1), \quad n \to \infty.$$

Proof. We consider the LAMN case (the proof under LAN is then a simplified version). Consider a sequence $(T_n)_n$ which is regular at ϑ, and write $U_n = \delta_n^{-1}(\vartheta)(T_n - \vartheta)$.

The implication (ii)$\Longrightarrow$(i) follows as in Example 7.9 where we have in particular under (ii)

$$\mathcal{L}\left(J_n, (U_n - h) \mid P_{n,\vartheta + \delta_n h}\right) \quad \longrightarrow \quad \mathcal{L}(J, (Z - h) \mid P_h)$$

Section 7.2 Asymptotic optimality of estimators in the LAN or LAMN setting

for every h. But LAMN at ϑ implies according to Definition 6.2 and Proposition 6.3

$$\mathcal{L}(J,(Z-h) \mid P_h)(A) = \mathcal{L}(J,Z \mid P_0)(A) = \int P_\bullet^J(dj)\, \mathcal{N}(0, j^{-1})(du)\, 1_A(j,u)$$

for Borel sets A in $\mathbb{R}^{d \times d} \times \mathbb{R}^d$. According to (7.10') above, we have (i).

To prove (i)$\Longrightarrow$(ii), we start from the regularity assumption

$$\mathcal{L}(J_n, (U_n - h) \mid P_{\vartheta + \delta_n h}) \longrightarrow \widetilde{F}, \quad n \to \infty$$

for arbitrary h where the limit law $\widetilde{F}$ does not depend on $h \in \mathbb{R}^d$. Fix any subsequence of the natural numbers. Applying Lemma 7.5 along this subsequence, we find a further subsequence $(n_l)_l$ and a statistic U in the limit experiment $\mathcal{E}(S,J)$ (if necessary, after Markov extension) such that

$$\mathcal{L}(S_{n_l}, J_{n_l}, U_{n_l} \mid P_{\vartheta+\delta_{n_l} h}) \longrightarrow \mathcal{L}(S, J, U \mid P_h), \quad l \to \infty$$

for all h, or equivalently by definition of $(Z_n)_n$

(o) $\quad \mathcal{L}((Z_{n_l}-h), J_{n_l}, (U_{n_l}-h) \mid P_{\vartheta+\delta_{n_l} h}) \longrightarrow \mathcal{L}((Z-h), J, (U-h) \mid P_h),$
$$l \to \infty$$

for every $h \in \mathbb{R}^d$. Again, regularity yields

$$\mathcal{L}(J, (U-h) \mid P_h) = \widetilde{F} = \mathcal{L}(J, U \mid P_0) \quad \text{does not depend on } h \in \mathbb{R}^d$$

and allows to view U as a strongly equivariant estimator in the limit experiment $\mathcal{E}_\infty(\vartheta) = \mathcal{E}(S,J)$ which is mixed normal. The Convolution Theorem 6.6 yields a representation

$$\widetilde{F}(A) = \int P_\bullet^J(dj) \int [\mathcal{N}(0, j^{-1}) \star Q^j](du)\, 1_A(j,u), \quad A \in \mathcal{B}(\mathbb{R}^{d \times d} \times \mathbb{R}^d).$$

Now we exploit the efficiency assumption for $(T_n)_n$ at ϑ: according to (7.10') above we have

$$Q^j = \epsilon_0 \quad \text{for } P_\bullet^J\text{-almost all } j \in \mathbb{R}^{d \times d}.$$

Since the Convolution Theorem 6.6 identifies $(j, B) \longrightarrow Q^j(B)$ as a regular version of the conditional distribution $P_0^{U-Z \mid J=j}(B)$, the last line establishes

$$U = Z \quad P_0\text{-almost surely.}$$

Using (o) and the continuous mapping theorem, this gives

$$\mathcal{L}(U_{n_l} - Z_{n_l} \mid P_{n_l, \vartheta}) \longrightarrow \mathcal{L}(U - Z \mid P_0) = \epsilon_0, \quad l \to \infty.$$

But convergence in law to a constant limit is equivalent to stochastic convergence, thus

(oo) $\quad U_{n_l} = Z_{n_l} + o_{(P_{n_l, \vartheta})}(1), \quad l \to \infty.$

We have proved that every subsequence of the natural numbers contains some further subsequence $(n_l)_l$ which has the property (∞∞): this gives

$$U_n = Z_n + o_{(P_{n,\vartheta})}(1), \quad n \to \infty$$

which is (ii) and finishes the proof. □

However, there might be interesting estimator sequences $(T_n)_n$ for the unknown parameter in $(\mathcal{E}_n)_n$ which are not regular as required in the Convolution Theorem 7.10, or we might be unable to prove regularity: when LAMN or LAN holds at a point ϑ, we wish to include these in comparison results.

7.12 Local Asymptotic Minimax Theorem. Assume that LAMN or LAN holds at ϑ, consider arbitrary sequences of estimators $(T_n)_n$ for the unknown parameter in $(\mathcal{E}_n)_n$, and arbitrary loss functions $\ell(\cdot)$ which are continuous, bounded and subconvex.

(a) A *local asymptotic minimax bound*

$$\liminf_{c \to \infty} \liminf_{n \to \infty} \sup_{|h| \le c} E_{n,\vartheta+\delta_n h}\left(\ell\left(\delta_n^{-1}\left(T_n - (\vartheta + \delta_n h)\right)\right)\right) \ge E_0\left(\ell(Z)\right)$$

holds whenever $(T_n)_n$ has estimation errors at ϑ which are tight at rate $(\delta_n(\vartheta))_n$:

$$\mathcal{L}\left(\delta_n^{-1}(\vartheta)(T_n - \vartheta) \mid P_{n,\vartheta}\right), \quad n \ge 1, \quad \text{is tight in } \mathbb{R}^d.$$

(b) Sequences $(T_n)_n$ satisfying the coupling property (◇) of Example 7.9

$$\delta_n^{-1}(\vartheta)(T_n - \vartheta) = Z_n(\vartheta) + o_{(P_{n,\vartheta})}(1), \quad n \to \infty$$

attain the local asymptotic minimax bound at ϑ. One has under this condition

$$\lim_{n \to \infty} \sup_{|h| \le c} E_{n,\vartheta+\delta_n h}\left(\ell\left(\delta_n^{-1}\left(T_n - (\vartheta + \delta_n h)\right)\right)\right) = E_0\left(\ell(Z)\right)$$

for arbitrary choice of a constant $0 < c < \infty$.

Proof. We give the proof for the LAMN case (again, the proof under LAN is a simplified version), and write $U_n = \delta^{-1}(T_n - \vartheta)$ for the rescaled estimation errors of T_n at ϑ.

(1) Fix $c \in \mathbb{N}$. The loss function $\ell(\cdot)$ being non-negative and bounded,

$$\liminf_{n \to \infty} \sup_{|h| \le c} E_{n,\vartheta+\delta_n h}\left(\ell(U_n - h)\right)$$

is necessarily finite. Select a subsequence of the natural numbers along which 'liminf' in the last line can be replaced by 'lim', then – using Lemma 7.5 – pass to some further subsequence $(n_l)_l$ and some statistic U in the limit experiment $\mathcal{E}_\infty(\vartheta) = \mathcal{E}(S, J)$ (if

Section 7.2 Asymptotic optimality of estimators in the LAN or LAMN setting

necessary, after Markov extension) such that the following holds for arbitrary limit points h and convergent sequences $h_l \to h$:

$$\mathcal{L}(S_{n_l}, J_{n_l}, U_{n_l} - h_l \mid P_{n_l, \vartheta + \delta_{n_l} h_l}) \longrightarrow \mathcal{L}(S, J, U - h \mid P_h), \quad l \to \infty.$$

From this we deduce as in Theorem 7.6 as $l \to \infty$

(+) $\quad \sup_{|h| \leq c} \left| E_{n_l, \vartheta + \delta_{n_l} h} \left(\ell(U_{n_l} - h) \right) - E_h \left(\ell(U - h) \right) \right| \longrightarrow 0$

since $\ell \in \mathcal{C}_b$. Recall that $h \to E_h(\ell(U - h))$ is continuous, see step (2) in the proof of Theorem 7.6. Write π_c for the uniform law on the closed ball B_c in $\mathbb{R}^d$ centred at 0 with radius c, as in Lemma 6.7 and in the proof of Theorem 6.8. Then by uniform convergence according to (+)

$$\sup_{|h| \leq c} E_{n_l, \vartheta + \delta_{n_l} h} \left(\ell(U_{n_l} - h) \right) \longrightarrow$$

$$\sup_{|h| \leq c} E_h(\ell(U - h)) \geq \int \pi_c(dh) \, E_h(\ell(U - h))$$

as $l \to \infty$. The last '$\geq$' is a trivial bound for an integral with respect to π_c.

(2) Now we exploit Lemma 6.7. In the mixed normal limit experiment $\mathcal{E}_\infty(\vartheta) = \mathcal{E}(S, J)$, arbitrary estimators U for the parameter $h \in \mathbb{R}^d$ can be viewed as being approximately strongly equivariant under a very diffuse prior, i.e. in absence of any a priori information on h except that h should range over large balls centred at the origin: there are probability laws $\{Q_c^j : j \in D^+, c \in \mathbb{N}\}$ such that

$$d_1 \left(\int \pi_c(dh) \, \mathcal{L}(U - h \mid P_h), \int P_\bullet^J(dj) \left[\mathcal{N}(0, j^{-1}) \star Q_c^j \right] \right) \longrightarrow 0$$

as c increases to ∞, with $d_1(\cdot, \cdot)$ the total variation distance. As a consequence, $\ell(\cdot)$ being bounded,

$$R(U, B_c) = \int \pi_c(dh) \, E_h(\ell(U - h))$$

takes the form

$$R(U, B_c) = \int P_\bullet^J(dj) \int \left[\mathcal{N}(0, j^{-1}) \star Q_c^j \right] (du) \, l(u) + \rho(c)$$

where remainder terms $\rho(c)$ involve an upper bound for $\ell(\cdot)$ and vanish as c increases to ∞. In the last line, at every stage $c \in \mathbb{N}$ of the asymptotics, Anderson's Lemma 5.7 allows for a lower bound

$$R(U, B_c) \geq \int P_\bullet^J(dj) \int \mathcal{N}(0, j^{-1})(du) \, l(u) + \rho(c)$$

since $\ell(\cdot)$ is subconvex. According to Definition 6.2 and Proposition 6.3, the law appearing on the right-hand side is $\mathcal{L}(Z \mid P_0)$, and we arrive at

$$R(U, B_c) \geq E_0(\ell(Z)) + \rho(c) \quad \text{where} \quad \lim_{c \to \infty} \rho(c) = 0.$$

(3) Combining steps (1) and (2) we have for $c \in \mathbb{N}$ fixed

$$\liminf_{n\to\infty} \sup_{|h|\leq c} E_{n,\vartheta+\delta_n h}\left(\ell\left(U_n - h\right)\right) = \lim_{l\to\infty} \sup_{|h|\leq c} E_{n_l,\vartheta+\delta_{n_l} h}\left(\ell\left(U_{n_l} - h\right)\right)$$
$$= \sup_{|h|\leq c} E_h\left(\ell\left(U - h\right)\right) \geq R(U, B_c)$$

(this U depends on the choice of the subsequence at the start) where for c tending to ∞

$$\liminf_{c\to\infty} R(U, B_c) \geq E_0(\ell(Z)).$$

Both assertions together yield the local asymptotic minimax bound in part (a) of the theorem.

(4) For estimator sequences $(T_n)_n$ satisfying condition ($\diamond$) in 7.9, (+) above can be strengthened to

$$\sup_{|h|\leq c} \left| E_{n,\vartheta+\delta_n h}\left(\ell\left(U_n - h\right)\right) - E_h\left(\ell(Z - h)\right) \right| \quad \longrightarrow \quad 0, \quad n \to \infty$$

for fixed c, without any need to select subsequences (Corollary 7.7). In the mixed normal limit experiment, $E_h\left(\ell(Z - h)\right) = E_0\left(\ell(Z)\right)$ does not depend on h. Exploiting this we can replace the conclusion of step (1) above by the stronger assertion

$$\sup_{|h|\leq c} E_{n,\vartheta+\delta_n h}\left(\ell\left(U_n - h\right)\right) \quad \longrightarrow \quad E_0(\ell(Z))$$

as $n \to \infty$, for arbitrary fixed value of c. This is part (b) of the theorem. □

7.13 Remark. Theorem 7.12 shows in particular that whenever we try to find estimator sequences $(T_n)_n$ which attain a local asymptotic minimax bound at ϑ, we may restrict our attention to sequences which have the coupling property ($\diamond$) of Example 7.9 (or of Corollary 7.7). We rephrase this statement according to Theorem 7.11: under LAMN or LAN at ϑ, in order to attain the local asymptotic minimax bound of Theorem 7.12, we may focus – within the class of all possible estimator sequences – on those which are regular and efficient at ϑ in the sense of the convolution theorem. □

Under LAMN or LAN at ϑ plus some additional conditions, we can construct efficient estimator sequences explicitely by 'one-step-modification' starting from any preliminary estimator sequence which converges at rate $\delta_n(\vartheta)$ at ϑ. This will be the topic of Section 7.3.

7.13' Example. In the set of all probability measures on $(\mathbb{R}, \mathcal{B}(\mathbb{R}))$, let us consider a one-parametric path $\mathcal{E} = \{P_\xi : |\xi| < 1\}$ in direction $\text{sgn}(\cdot)$ through the law $P_0 := \mathcal{R}(-1, +1)$, the uniform distribution with support $(-1, +1)$, defined as in Examples 1.3 and 4.3 by

$$g(x) := \text{sgn}(x), \quad P_\xi(dx) := (1 + \xi g(x)) P_0(dx), \quad x \in (-1, +1).$$

(1) Write $\Theta = (-1, +1)$. According to Example 4.3, the model $\mathcal{E}$ is L^2-differentiable at $\xi = \vartheta$ with derivative

$$V_\vartheta(x) = \frac{g}{1 + \vartheta g}(x) = \frac{-1}{1-\vartheta} 1_{\{x<0\}} + \frac{1}{1+\vartheta} 1_{\{x>0\}}, \quad x \in (-1, +1).$$

at every reference point $\vartheta \in \Theta$ (it is irrelevant how the derivative is defined on Lebesgue null sets). As in Theorem 4.11, Le Cam's second lemma yields LAN at ϑ with local scale $\delta_n(\vartheta) = 1/\sqrt{n}$, with score at ϑ

$$S_n(\vartheta)(X_1, \ldots, X_n) = \frac{1}{\sqrt{n}} \sum_{i=1}^n V_\vartheta(X_i)$$

at stage n of the asymptotics, and with Fisher information given by

$$J_\vartheta = E_\vartheta(V_\vartheta^2) = \frac{1}{1-\vartheta^2}, \quad \vartheta \in \Theta.$$

(2) With $\widehat{F}_n(\cdot)$ the empirical distribution function based on the first n observations, the score in $\mathcal{E}_n$

$$S_n(\vartheta) = \frac{-1}{1-\vartheta} \sqrt{n} \widehat{F}_n(0) + \frac{1}{1+\vartheta} \sqrt{n}(1 - \widehat{F}_n(0))$$

can be written in the form

(+) $\qquad S_n(\vartheta) = J_\vartheta \cdot \sqrt{n}\,(T_n - \vartheta) \quad$ for all $\vartheta \in \Theta$

when we estimate the unknown parameter $\vartheta \in \Theta$ by

$$T_n := 1 - 2\widehat{F}_n(0).$$

From representation (+), for every $\vartheta \in \Theta$, rescaled estimation errors of $(T_n)_n$ coincide with the central sequence $(Z_n(\vartheta))_n$. From Theorems 7.11, 7.10 and 7.12, for every $\vartheta \in \Theta$, the sequence $(T_n)_n$ thus attains the best concentrated limit distribution (7.10') in the convolution theorem, and attains the local asymptotic minimax bound of Theorem 7.12(b). □

An unsatisfactory point with the last example is that n-fold product models $\mathcal{E}_n$ in Example 7.13' are in fact classical exponential families

$$dP_{n,\xi} = (1+\xi)^{n[1-\widehat{F}_n(0)]} (1-\xi)^{n\widehat{F}_n(0)} dP_{n,0} = (1+\xi)^n \exp\{\zeta(\xi)\check{T}_n\} dP_{n,0},$$

$$\xi \in \Theta$$

in $\zeta(\xi) := \log(\frac{1-\xi}{1+\xi})$ and $\check{T}_n := n\widehat{F}_n(0)$. Hence we shall generalise it (considering different one-parametric paths through a uniform law which are not exponential families) in Example 7.21 below.

7.3 Le Cam's One-step Modification of Estimators

In this section, we go back to the LAQ setting of Section 7.1 and construct estimator sequences $(\widetilde{T}_n)_n$ with the coupling property ($\diamond$) of Corollary 7.7

($\diamond$) $\qquad \delta^{-1}(\vartheta)(\widetilde{T}_n - \vartheta) = Z_n(\vartheta) + o_{(P_{n,\vartheta})}(1) , \quad n \to \infty$

starting from any preliminary estimator sequence $(T_n)_n$ which converges at rate $\delta_n(\vartheta)$ at ϑ. This 'one-step modification' is explicit and requires only few further conditions in addition to LAQ at ϑ; the main result is Theorem 7.19 below.

In particular, when LAMN or LAN holds at ϑ, one-step modification yields optimal estimator sequences locally asymptotically at ϑ, via Example 7.9 and Theorem 7.11: the modified sequence will be regular and efficient in the sense of the Convolution Theorem 7.10, and will attain the local asymptotic minimax bound of Theorem 7.12. In the general LAQ case, we only have the following: the modified sequence will work over shrinking neighbourhoods of ϑ as well as the maximum likelihood estimator Z in the limit experiment $\mathcal{E}_\infty(\vartheta)$, according to Corollary 7.7 where $\mathcal{L}(Z - h|P_h)$ may depend on h.

We follow Davies [19] for this construction. We formulate the conditions which we need simultaneously for all points $\vartheta \in \Theta$, such that the one-step modifications $(\widetilde{T}_n)_n$ of $(T_n)_n$ will have the desired properties simultaneously at all points $\vartheta \in \Theta$. This requires some compatibility between quantities defining LAQ at ϑ and LAQ at ϑ' whenever ϑ' is close to ϑ.

7.14 Assumptions for Section 7.3. We consider a sequence of experiments

$$\mathcal{E}_n = (\Omega_n, \mathcal{A}_n, \{P_{n,\vartheta} : \vartheta \in \Theta\}) , \quad n \geq 1, \quad \Theta \subset \mathbb{R}^d \text{ open}$$

enjoying the following properties (A)–(D):

(A) At every point $\vartheta \in \Theta$ we have LAQ as in Definition 7.1(a), with sequences

$$(\delta_n(\vartheta))_n , \quad (S_n(\vartheta))_n , \quad (J_n(\vartheta))_n$$

depending on ϑ, and a quadratic limit experiment

$$\mathcal{E}_\infty(\vartheta) = \mathcal{E}(S(\vartheta), J(\vartheta))$$

depending on ϑ.

(B) Local scale: (i) For every $n \geq 1$ fixed, $\delta_n(\cdot) : \Theta \to (0, \infty)$ is a measurable mapping which is bounded by 1 (this is no loss of generality: we may always replace $\delta_n(\cdot)$ by $\delta_n(\cdot) \wedge 1$).

(ii) For every $\vartheta \in \Theta$ fixed, we have for all $0 < c < \infty$

$$\sup_{|h| \leq c} \left| \frac{\delta_n(\vartheta + \delta_n(\vartheta) h)}{\delta_n(\vartheta)} - 1 \right| \longrightarrow 0 \quad \text{as} \quad n \to \infty .$$

Section 7.3 Le Cam's One-step Modification of Estimators

(C) <u>Score and observed information:</u> For every $\vartheta \in \Theta$ fixed, in restriction to the set of dyadic numbers $\mathcal{S} := \{\alpha 2^{-k} : k \in \mathbb{N}_0, \alpha \in \mathbb{Z}^d\}$ in $\mathbb{R}^d$, we have

(i) $\displaystyle\sup_{\xi \in \mathcal{S} \cap \Theta, |\xi - \vartheta| \le c \delta_n(\vartheta)} \left| S_n(\xi) - \left\{ S_n(\vartheta) - J_n(\vartheta) \left[\delta_n^{-1}(\vartheta)(\xi - \vartheta) \right] \right\} \right|$
$= o_{P_{n,\vartheta}}(1), \quad n \to \infty$

(ii) $\displaystyle\sup_{\xi \in \mathcal{S} \cap \Theta, |\xi - \vartheta| \le c \delta_n(\vartheta)} |J_n(\xi) - J_n(\vartheta)| = o_{P_{n,\vartheta}}(1), \quad n \to \infty$

for arbitrary choice of a constant $0 < c < \infty$.

(D) <u>Preliminary estimator sequence:</u> We have some preliminary estimator sequence $(T_n)_n$ for the unknown parameter in $(\mathcal{E}_n)_n$ which at all points $\vartheta \in \Theta$ is tight at the rate specified by (A):

for every $\vartheta \in \Theta$: $\quad \mathcal{L}\left(\delta_n^{-1}(\vartheta) (T_n - \vartheta) \mid P_{n,\vartheta} \right), \quad n \ge 1, \quad$ is tight in $\mathbb{R}^d$.

Continuity $\vartheta \to \delta_n(\vartheta)$ of local scale in the parameter should not be imposed (a stochastic process example is given in Chapter 8). Similarly, we avoid to impose continuity of the score or of the observed information in the parameter, not even measurability. This is why we consider, on the left-hand sides in (C), only parameter values belonging to a countably dense subset. Under the set of Assumptions 7.14, the first steps are to define a local scale with an estimated parameter, observed information with an estimated parameter, and a score with an estimated parameter.

7.15 Proposition. (a) Define a *local scale with estimated parameter* by

$$D_n := \delta_n(T_n) \in (0, 1], \quad n \ge 1.$$

Then we have for every $\vartheta \in \Theta$

$$\frac{D_n}{\delta_n(\vartheta)} = 1 + o_{P_{n,\vartheta}}(1), \quad n \to \infty.$$

(b) For every $n \ge 1$, define a $\mathbb{N}_0$-valued random variable $\kappa(n)$ by

$$\kappa(n) = k \text{ if and only if } D_n \in (2^{-(k+1)}, 2^{-k}], \ k \in \mathbb{N}_0.$$

Then for every $\vartheta \in \vartheta$, the two sequences

$$\frac{2^{-\kappa(n)}}{\delta_n(\vartheta)} \quad \text{and} \quad \frac{\delta_n(\vartheta)}{2^{-\kappa(n)}}$$

are tight in $\mathbb{R}$ under $P_{n,\vartheta}$ as $n \to \infty$.

Proof. According to Assumption 7.14(B.i), D_n is a random variable on $(\Omega_n, \mathcal{A}_n)$ taking values in $(0, 1]$. Consider the preliminary estimator T_n of Assumption 7.14(D).

Fix $\vartheta \in \Theta$ and write $U_n(\vartheta) := \delta_n^{-1}(\vartheta)(T_n - \vartheta)$ for the rescaled estimation errors at ϑ. Combining tightness of $\mathcal{L}(U_n(\vartheta)|P_{n,\vartheta})$ as $n \to \infty$ according to Assumption 7.14(D) with a representation $D_n = \delta_n(T_n) = \delta_n(\vartheta + \delta_n(\vartheta)U_n(\vartheta))$ and with Assumption 7.14(B.ii) we obtain (a). From $2^{-(\kappa(n)+1)} \le D_n \le 2^{-\kappa(n)}$ we get $\frac{1}{2}D_n \le 2^{-(\kappa(n)+1)} \le D_n$, thus (b) follows from (a). □

For every $k \in \mathbb{N}_0$, cover $\mathbb{R}^d$ with half-open cubes

$$C(k,\alpha) := \underset{i=1}{\overset{d}{\times}} \left[\alpha_i 2^{-k}, (\alpha_i + 1)2^{-k}\right[, \quad \alpha = (\alpha_1,\ldots,\alpha_d) \in \mathbb{Z}^d.$$

Write $\mathcal{Z}(k) := \{\alpha \in \mathbb{Z}^d : C(k,\alpha) \subset \Theta\}$ for the collection of those which are contained in Θ. Fix any default value ϑ_0 in $\mathcal{S} \cap \Theta$. From the preliminary estimator sequence $(T_n)_n$ and from local scale $(D_n)_n$ with estimated parameter, we define a *discretisation* $(G_n)_n$ of $(T_n)_n$ as follows: for $n \ge 1$,

$$G_n := \begin{cases} \alpha 2^{-\kappa(n)} & \text{if } \alpha \in \mathcal{Z}(\kappa(n)) \text{ and } T_n \in C(\kappa(n),\alpha) \\ \vartheta_0 & \text{else} \end{cases}$$

$$= \sum_{k=0}^{\infty} \left[\vartheta_0 + \sum_{\alpha \in \mathcal{Z}(k)} (\alpha 2^{-k} - \vartheta_0) 1_{C(k,\alpha)}(T_n) \right] 1_{]2^{-(k+1)}, 2^{-k}]}(D_n).$$

Clearly G_n is an estimator for the unknown parameter $\vartheta \in \Theta$ in $\mathcal{E}_n$, taking only countably many values. We have to check that passing from $(T_n)_n$ to $(G_n)_n$ does not modify the tightness rates.

7.16 Proposition. $(G_n)_n$ is a sequence of $\mathcal{S} \cap \Theta$-valued estimators satisfying

$$\{\mathcal{L}(\delta_n^{-1}(\vartheta)(G_n - \vartheta) \mid P_{n,\vartheta}) : n \ge 1\} \text{ is tight in } \mathbb{R}^d$$

for every $\vartheta \in \Theta$.

Proof. Fix $\vartheta \in \Theta$. The above construction specifies 'good' events

$$B_n := \left\{ T_n \in \bigcup_{\alpha \in \mathcal{Z}(\kappa(n))} C(\kappa(n), \alpha) \right\} \in \mathcal{A}_n$$

together with a default value $G_n = \vartheta_0$ on B_n^c. Since Θ is open and rescaled estimation errors of $(T_n)_n$ at ϑ are tight at rate $(\delta_n(\vartheta))_n$, since $(D_n)_n$ or $(e^{-\kappa(n)})_n$ defined in Proposition 7.15(b) are random tightness rates which are equivalent to $(\delta_n(\vartheta))_n$ under $(P_{n,\vartheta})_n$, we have by construction

$$\lim_{n \to \infty} P_{n,\vartheta}(B_n) = 1$$

Section 7.3 Le Cam's One-step Modification of Estimators

together with

$$|G_n - T_n| < \sqrt{d} \cdot 2^{-\kappa(n)} \quad \text{on } B_n, \text{ for every } n \in \mathbb{N}$$

(for arbitrary k and for $\alpha \in \mathbb{Z}^d$, the half-open cube $C(k, \alpha)$ has lower left endpoint $\alpha 2^{-k}$ and diameter $\sqrt{d} \cdot 2^{-k}$). Combining the two last lines we have

$$G_n = T_n + o_{P_{n,\vartheta}}(\delta_n(\vartheta)) \quad \text{as } n \to \infty$$

which is the assertion. □

7.17 Proposition. *The discretised sequence $(G_n)_n$ in Proposition 7.16 allows us to define for every $n \in \mathbb{N}$ information with an estimated parameter*

$$\widehat{J}_n := J_n(G_n) = \sum_{\xi \in \mathcal{S} \cap \Theta} \left[J_n(\xi) \cdot 1_{\{G_n = \xi\}} \right]$$

together with its 'inverse'

$$\widehat{K}_n(\omega) := \begin{cases} [\widehat{J}_n(\omega)]^{-1} & \text{if } \widehat{J}_n(\omega) \text{ belongs to } D^+ \\ I_d & \text{else} \end{cases}$$

such that the following (i) and (ii) hold for every $\vartheta \in \Theta$:

(i) $\quad \widehat{J}_n = J_n(\vartheta) + o_{P_{n,\vartheta}}(1) \quad \text{as } n \to \infty,$

(ii) $\quad \widehat{K}_n = J_n^{-1}(\vartheta) + o_{P_{n,\vartheta}}(1) \quad \text{and} \quad \widehat{K}_n J_n(\vartheta) = I_d + o_{P_{n,\vartheta}}(1), \quad n \to \infty.$

Here I_d denotes the d-dimensional identity matrix.

Proof. Note first that $J_n(\vartheta)$ under $P_{n,\vartheta}$ takes values in D^+ almost surely, by definition of LAQ in Definition 7.1, but the same statement is not clear for $J_n(\xi)$ under $P_{n,\vartheta}$ (we did not require equivalence of laws $P_{n,\vartheta}$, $P_{n,\xi}$ for $\xi \neq \vartheta$). Since G_n takes values in the countable set $\mathcal{S} \cap \Theta$, the random variable $\widehat{J}_n$ on $(\Omega_n, \mathcal{A}_n)$ is well defined and $\mathbb{R}^{d \times d}$-valued; then also $\widehat{K}_n$ is well defined since D^+ is a Borel set in $\mathbb{R}^{d \times d}$.

(1) For every $\vartheta \in \Theta$, we deduce (i) from part (C.ii) of Assumption 7.14, via a representation

$$\widehat{J}_n = J_n(G_n) = J_n(\vartheta + \delta_n(\vartheta)\check{U}_n(\vartheta)), \quad \check{U}_n(\vartheta) := \delta_n^{-1}(\vartheta)(G_n - \vartheta)$$

where G_n is $\mathcal{S} \cap \Theta$-valued, and where $\mathcal{L}(\check{U}_n(\vartheta)|P_{n,\vartheta})$ is tight as $n \to \infty$ by Proposition 7.16.

(2) Fix $\vartheta \in \Theta$. From LAQ at ϑ combined with (i), we have weak convergence

$$\mathcal{L}(\widehat{J}_n, J_n(\vartheta) \mid P_{n,\vartheta}) \longrightarrow \mathcal{L}(J(\vartheta), J(\vartheta) \mid P_0)$$

in $\mathbb{R}^{d\times d}\times \mathbb{R}^{d\times d}$ as $n \to \infty$. The mapping on $\mathbb{R}^{d\times d}$

$$\psi : j \longrightarrow \begin{cases} j^{-1} & \text{if } \det(j) \neq 0 \\ I_d & \text{else} \end{cases}$$

is continuous on the open set $\{j \in \mathbb{R}^{d\times d} : \det(j) \neq 0\}$ which contains D^+. Thus, by assumption of Definition 6.1 on the limit experiment $\mathcal{E}_\infty(\vartheta) = \mathcal{E}(S(\vartheta), J(\vartheta))$, the set of discontinuities of ψ is a null set under $\mathcal{L}(J(\vartheta)|P_0)$. If we consider $\widehat{K}_n = \psi(\widehat{J}_n)$, the continuous mapping theorem allows us to extend the above convergence to

(+) $\quad \mathcal{L}(\widehat{K}_n, J_n^{-1}(\vartheta), \widehat{J}_n, J_n(\vartheta) \mid P_{n,\vartheta})$
$\quad \longrightarrow \mathcal{L}(J^{-1}(\vartheta), J^{-1}(\vartheta), J(\vartheta), J(\vartheta) \mid P_0)$

weakly in $\mathbb{R}^{d\times d}\times\mathbb{R}^{d\times d}\times\mathbb{R}^{d\times d}\times\mathbb{R}^{d\times d}$ as $n \to \infty$; recall from Definition 7.1 that $J_n(\vartheta)$ takes values $P_{n,\vartheta}$-almost surely in D^+ for all $n \geq 1$. In particular, (+) shows that both sequences $J_n^{-1}(\vartheta)$ under $P_{n,\vartheta}$ and $\widehat{K}_n$ under $P_{n,\vartheta}$ are tight as $n \to \infty$. Applying the continuous mapping theorem to differences or products of components in (+) we see that $\widehat{K}_n - J_n^{-1}(\vartheta)$ under $P_{n,\vartheta}$ converges weakly in $\mathbb{R}^{d\times d}$ as $n \to \infty$ to the matrix having all entries equal to 0, and that $\widehat{K}_n J_n(\vartheta)$ under $P_{n,\vartheta}$ converges weakly in $\mathbb{R}^{d\times d}$ to the identity matrix I_d. Weak convergence to a constant limit being equivalent to convergence in probability, assertion (ii) follows. □

7.18 Proposition. The discretisation G_n of Proposition 7.16 allows to define a *score with estimated parameter*

$$\widehat{S}_n := S_n(G_n) = \sum_{\xi \in \mathcal{S} \cap \Theta} \left[S_n(\xi) \cdot 1_{\{G_n = \xi\}}\right], \quad n \geq 1$$

which satisfies for every $\vartheta \in \Theta$

$$\widehat{S}_n = S_n(\vartheta) - J_n(\vartheta)\left[\delta_n^{-1}(\vartheta)(G_n - \vartheta)\right] + o_{P_{n,\vartheta}}(1), \quad n \to \infty.$$

In particular, $\mathcal{L}(\widehat{S}_n | P_{n,\vartheta})$ is tight as $n \to \infty$ for every $\vartheta \in \Theta$.

Proof. Again $\widehat{S}_n = S_n(G_n)$ is well defined. Fix $\vartheta \in \Theta$ and write

$$\widehat{S}_n = S_n(G_n) = S_n(\vartheta + \delta_n(\vartheta)\check{U}_n(\vartheta)), \quad \check{U}_n(\vartheta) := \delta_n^{-1}(\vartheta)(G_n - \vartheta)$$

where G_n is $\mathcal{S} \cap \Theta$-valued and where $\mathcal{L}(\check{U}_n(\vartheta)|P_{n,\vartheta})$ is tight as $n \to \infty$. Then we write

$$P_{n,\vartheta}\left(\left|S_n(G_n) - \{S_n(\vartheta) - J_n(\vartheta)\check{U}_n(\vartheta)\}\right| > \varepsilon\right) \leq P_{n,\vartheta}\left(|\check{U}_n(\vartheta)| > c\right)$$
$$+ P_{n,\vartheta}\left(\sup_{\xi \in \mathcal{S} \cap \Theta, |\xi - \vartheta| \leq c\delta_n(\vartheta)} |S_n(\xi) - \{S_n(\vartheta) - J_n(\vartheta)[\delta_n^{-1}(\vartheta)(\xi - \vartheta)]\}| > \varepsilon\right)$$

and apply part (C.i) of Assumption 7.14. □

Now we resume: Under Assumptions 7.14, with preliminary estimator sequence $(T_n)_n$ and its discretisation $(G_n)_n$ as in Proposition 7.16, with local scale with estimated parameter D_n, information with estimated parameter $\widehat{J}_n$ and score with estimated parameter $\widehat{S}_n$ as defined in Propositions 7.15, 7.17 and 7.18, with 'inverse' $\widehat{K}_n$ for $\widehat{J}_n$ as in Proposition 7.17, we have

7.19 Theorem. (a) With these assumptions and notations, the *one-step modification*

$$\widetilde{T}_n := G_n + D_n \widehat{K}_n \widehat{S}_n , \quad n \geq 1$$

yields an estimator sequence $(\widetilde{T}_n)_n$ for the unknown parameter in $(\mathcal{E}_n)_n$ which has the property

$$\delta_n^{-1}(\vartheta)(\widetilde{T}_n - \vartheta) = Z_n(\vartheta) + o_{(P_{n,\vartheta})}(1) , \quad n \to \infty$$

for every $\vartheta \in \Theta$.

(b) In the sense of Corollary 7.7, for every $\vartheta \in \Theta$, $(\widetilde{T}_n)_n$ works over shrinking neighbourhoods of ϑ as well as the maximum likelihood estimator in the limit model $\mathcal{E}_\infty(\vartheta) = \mathcal{E}(S(\vartheta), J(\vartheta))$.

(c) If LAMN or LAN holds at ϑ, the sequence $(\widetilde{T}_n)_n$ is regular and efficient at ϑ in the sense of the Convolution Theorem 7.10, and attains the local asymptotic minimax bound at ϑ according to the Local Asymptotic Minimax Theorem 7.12.

Proof. Only (a) requires a proof. Fix $\vartheta \in \Theta$. Then from Propositions 7.15 and 7.17

$$\delta_n^{-1}(\vartheta)(\widetilde{T}_n - \vartheta) = \delta_n^{-1}(\vartheta)(G_n - \vartheta) + \frac{D_n}{\delta_n(\vartheta)} \widehat{K}_n \widehat{S}_n$$

$$= \delta_n^{-1}(\vartheta)(G_n - \vartheta) + \left[1 + o_{P_{n,\vartheta}}(1)\right]\left[J_n^{-1}(\vartheta) + o_{P_{n,\vartheta}}(1)\right] \widehat{S}_n$$

$$= \delta_n^{-1}(\vartheta)(G_n - \vartheta) + J_n^{-1}(\vartheta) \widehat{S}_n + o_{P_{n,\vartheta}}(1)$$

which according to Proposition 7.18 equals

$$\delta_n^{-1}(\vartheta)(G_n - \vartheta)$$
$$+ J_n^{-1}(\vartheta)\left[S_n(\vartheta) - J_n(\vartheta)\left[\delta_n^{-1}(\vartheta)(G_n - \vartheta)\right] + o_{P_{n,\vartheta}}(1)\right] + o_{P_{n,\vartheta}}(1)$$

where terms $\delta_n^{-1}(\vartheta)(G_n - \vartheta)$ cancel out and the last line simplifies to

$$J_n^{-1}(\vartheta) S_n(\vartheta) + o_{P_{n,\vartheta}}(1) = Z_n(\vartheta) + o_{P_{n,\vartheta}}(1) , \quad n \to \infty.$$

This is the assertion. □

7.4 The Case of i.i.d. Observations

To establish LAN with local scale $\delta_n = 1/\sqrt{n}$ in statistical models for i.i.d. observations, we use in most cases Le Cam's Second Lemma 4.11 and 4.10. We recall this important special case here, and discuss some examples. During this section,

$$\mathcal{E} = (\Omega, \mathcal{A}, \mathcal{P} = \{P_\xi : \xi \in \Theta\})$$

is an experiment, $\Theta \subset \mathbb{R}^d$ is open, and $\mathcal{E}_n$ is the n-fold product experiment with canonical variable $(X_1, \ldots, X_n)$. We write $P_{n,\xi} = \otimes_{i=1}^n P_\xi$ for the laws in $\mathcal{E}_n$ and

$$\Lambda_{n,\xi}^{h/0} = \Lambda_n^{(\xi+h/\sqrt{n})/\xi} = \log \frac{d\, P_{n,\xi+h/\sqrt{n}}}{d\, P_{n,\xi}}$$

for $\xi \in \Theta$ and $h \in \Theta_{\xi,n}$, the set of all $h \in \mathbb{R}^d$ such that $\xi + h/\sqrt{n}$ belongs to Θ. We shall assume

(∗) $\begin{cases} \text{there is an open set } \Theta_0 \subset \mathbb{R}^d \text{ contained in } \Theta \text{ such} \\ \text{that the following holds: for every } \vartheta \in \Theta_0, \text{ the experiment} \\ \mathcal{E} \text{ is } L^2\text{-differentiable at } \xi = \vartheta \text{ with derivative } V_\vartheta. \end{cases}$

Recall from Assumptions 4.10, Corollary 4.5 and Definition 4.2 that $\vartheta \in \Theta_0$ implies that V_ϑ is centred and belongs to $L^2(P_\vartheta)$, write $J_\vartheta = E_\vartheta(V_\vartheta V_\vartheta^T)$ for the Fisher information in the sense of Definition 4.6. Then Le Cam's Second Lemma 4.11 yields a quadratic expansion of log-likelihood ratios

$$\Lambda_n^{(\vartheta+h_n/\sqrt{n})} = h_n^T S_n(\vartheta) - \frac{1}{2} h_n^T J_\vartheta h_n + o_{P_{n,\vartheta}}(1)$$

as $n \to \infty$, for arbitrary bounded sequences $(h_n)_n$ in $\mathbb{R}^d$, at every reference point $\vartheta \in \Theta_0$, with

$$S_n(\vartheta) = \frac{1}{\sqrt{n}} \sum_{j=1}^n V_\vartheta(X_i)$$

such that $\mathcal{L}(S_n(\vartheta) \mid P_{n,\vartheta}) \longrightarrow \mathcal{N}(0, J_\vartheta), \quad n \to \infty.$

In terms of Definitions 5.2 and 7.1 we can rephrase Le Cam's Second Lemma 4.11 as follows:

7.20 Theorem. For n-fold independent replication of an experiment $\mathcal{E}$ satisfying (∗) as $n \to \infty$, LAN holds at every $\vartheta \in \Theta_0$ with local scale $\delta_n(\vartheta) = n^{-1/2}$; the limit experiment $\mathcal{E}_\infty(\vartheta)$ is the Gaussian shift $\mathcal{E}(J_\vartheta)$.

We present some examples of i.i.d. models for which Le Cam's second lemma establishes LAN at all parameter values. The aim is to specify efficient estimator sequences, either by checking directly the coupling condition in Theorem 7.11, or by one-step modification according to Section 7.3. The first example is very close to Example 7.13'.

7.21 Example. Put $\Theta := (-\frac{1}{\pi}, +\frac{1}{\pi})$ and $P_0 := \mathcal{R}(I)$, the uniform distribution on $I := (-\pi, +\pi)$. In the set of all probability measures on $(\mathbb{R}, \mathcal{B}(\mathbb{R}))$, define a one-parametric path $\mathcal{E} = \{P_\xi : \xi \in \Theta\}$ in direction g through P_0 by

$$g(x) := \pi \sin(x),$$

$$P_\xi(dx) := (1 + \xi g(x)) P_0(dx) = \frac{1}{2\pi}(1 + \xi \pi \sin(x)) \, dx, \quad x \in I$$

as in Examples 1.3 and 4.3, where the parameterisation is motivated by

$$F_\xi(0) = \frac{1}{2} - \xi, \quad \xi \in \Theta$$

with $F_\xi(\cdot)$ the distribution function corresponding to P_ξ

$$F_\xi(x) = \frac{1}{2\pi}([x + \pi] - \xi \pi [\cos(x) + 1]) \quad \text{when} \quad x \in I.$$

(1) According to Example 4.3, the model $\mathcal{E}$ is L^2-differentiable at $\xi = \vartheta$ with derivative

$$V_\vartheta(x) = \frac{g}{1 + \vartheta g}(x) = \frac{\pi \sin(x)}{1 + \vartheta \pi \sin(x)}, \quad x \in I$$

at every reference point $\vartheta \in \Theta$. As resumed in Theorem 7.20, Le Cam's second lemma yields LAN at ϑ for every $\vartheta \in \Theta$, with local scale $\delta_n(\vartheta) = 1/\sqrt{n}$ and score

$$S_n(\vartheta)(X_1, \ldots, X_n) = \frac{1}{\sqrt{n}} \sum_{i=1}^n V_\vartheta(X_i)$$

at ϑ, and with finite Fisher information in the sense of Definition 4.6

$$J_\vartheta = E_\vartheta(V_\vartheta^2) < \infty, \quad \vartheta \in \Theta.$$

In this model, we prefer to keep the observed information

$$J_n(\vartheta) = \frac{1}{n} \sum_{i=1}^n V_\vartheta^2(X_i)$$

in the quadratic expansion of log-likelihood ratios in the local model at ϑ, and write

(+) $\quad \Lambda_n^{(\vartheta + h_n/\sqrt{n})} = h_n^\top S_n(\vartheta) - \frac{1}{2} h_n^\top J_n(\vartheta) h_n + o_{P_{n,\vartheta}}(1)$

as $n \to \infty$, for arbitrary bounded sequences $(h_n)_n$ in $\mathbb{R}^d$.

(2) We show that the set of Assumptions 7.14 is satisfied. A preliminary estimator
$$T_n := \frac{1}{2} - \widehat{F}_n(0)$$
for the unknown parameter ϑ in $\mathcal{E}_n$ is at hand, clearly $\sqrt{n}$-consistent as $n \to \infty$. From step (1), parts (A) and (B) of Assumptions 7.14 are granted; part (C.ii) holds by continuity of $J_n(\vartheta)$ in the parameter. Calculating
$$S_n(\xi) - \{S_n(\vartheta) - J_n(\vartheta)\left[\delta_n^{-1}(\vartheta)(\xi - \vartheta)\right]\}, \quad \text{with } J_n(\vartheta) \text{ as in } (+)$$
(the quantities arising in part (C.i) of Assumption 7.14), we find
$$S_n(\xi) - S_n(\vartheta) = -\sqrt{n}(\xi - \vartheta) \cdot \frac{1}{n} \sum_{i=1}^{n} V_\vartheta^2(X_i) \left[\frac{1 + \vartheta \pi \sin(X_i)}{1 + \xi \pi \sin(X_i)}\right]$$
from which we deduce easily that Assumption 7.14(C.i) is satisfied.

(3) Now the one-step modification according to Theorem 7.19 (there is no need for discretisation since score and observed information depend continuously on the parameter)
$$\widetilde{T}_n := T_n + \frac{1}{\sqrt{n}} [J_n(T_n)]^{-1} S_n(T_n), \quad n \geq 1$$
yields an estimator sequence $(\widetilde{T}_n)_n$ which is regular and efficient at every point $\vartheta \in \mathbb{R}$ (a simplified version of the proof of Theorem 7.19 is sufficient to check this). □

However, even if the set of assumptions 7.14 can be checked in a broad variety of statistical models, not all of the assumptions listed there are harmless.

7.22 Example. Let $\mathcal{E} = (\Omega, \mathcal{A}, \{P_\xi : \xi \in \mathbb{R}\})$ be the location model on $(\mathbb{R}, \mathcal{B}(\mathbb{R}))$ generated from the two-sided exponential distribution $P_0(dy) := \frac{1}{2} e^{-|y|} dy$; this example has already been considered in exercises 4.1''', 7.9' and 7.9''. We shall see that assumption 7.14 C) i) on the score with estimated parameter does not hold, hence one-step correction according to theorem 7.19 is not applicable. However, it is easy –for this model– to find optimal estimator sequences directly.

1) For every $\vartheta \in \Theta := \mathbb{R}$, we have L^2-differentiability at $\xi = \vartheta$ with derivative V_ϑ given by
$$V_\vartheta(x) = \operatorname{sgn}(x - \vartheta)$$
(cf. exercise 4.1'''). For all ϑ, put $\delta_n(\vartheta) = 1/\sqrt{n}$. Then Le Cam's second lemma (theorem 7.20 or theorem 4.11) establishes LAN at ϑ with score
$$S_n(\vartheta)(y_1, \ldots, y_n) = \frac{1}{\sqrt{n}} \sum_{i=1}^{n} V_\vartheta(y_i)$$
at ϑ, and with Fisher information $J_\vartheta = E_\vartheta(V_\vartheta^2) \equiv 1$ not depending on ϑ.

Section 7.4 The Case of i.i.d. Observations

2) We consider the set of assumptions 7.14. From 1), parts A) and B) of 7.14 are granted; 7.14 C) part ii) is trivial since Fisher information does not depend on the parameter. Calculate now

$$S_n(\xi) - \{S_n(\vartheta) - J_n(\vartheta)\left[\delta_n^{-1}(\vartheta)(\xi - \vartheta)\right]\}, \quad J_n(\vartheta) = J_\vartheta \equiv 1$$

in view of 7.14 C) part i). This takes a form $\frac{1}{\sqrt{n}} \sum_{i=1}^n Y_{n,i}$ where

$$Y_{n,i} := \begin{cases} 2\left[-1_{(\vartheta,\xi)}(X_i) + \frac{1}{2}(\xi - \vartheta)\right] & \text{for } \vartheta < \xi, \\ 2\left[1_{(\xi,\vartheta)}(X_i)) - \frac{1}{2}(\vartheta - \xi)\right] & \text{for } \xi < \vartheta. \end{cases}$$

Defining a function $\rho(r) := \int_0^r \frac{1}{2} e^{-y} \, dy$ for $r > 0$, we have

$$|E_\vartheta(Y_{n,i})| = |\vartheta - \xi| - 2\rho(|\vartheta - \xi|), \quad \text{Var}_\vartheta(Y_{n,i}) = 4[\rho(1-\rho)](|\vartheta - \xi|))$$

where $\rho(r) \sim \frac{1}{2} r$ as $r \downarrow 0$. We shall see that this structure violates assumption 7.14 C) i), thus one-step correction according to theorem 7.19 breaks down.

To see this, consider a constant sequence $(h_n)_n$ with $h_n := \frac{1}{2}$ for all n, and redefine the $Y_{n,i}$ above using $\xi_n = \vartheta + n^{-1/2} h$ in place of ξ: first, $E_\vartheta(Y_{n,i}) = o(n^{-1/2})$ as $n \to \infty$ and thus

$$\frac{1}{\sqrt{n}} \sum_{i=1}^n Y_{n,i} = \frac{1}{\sqrt{n}} \sum_{i=1}^n [Y_{n,i} - E_\vartheta(Y_{n,i})] + o_{P_{n,\vartheta}}(1),$$

second, $\text{Var}_\vartheta(Y_{n,i}) = O(n^{-1/2})$ as $n \to \infty$ and by choice of the particular sequence $(h_n)_n$

$$\frac{1}{\sqrt{n}} \sum_{i=1}^n Y_{n,i} = \sum_{i=1}^n \frac{Y_{n,i} - E_\vartheta(Y_{n,i})}{\sqrt{\text{Var}_\vartheta(Y_{n,1}) + \ldots + \text{Var}_\vartheta(Y_{n,i})}} + o_{P_{n,\vartheta}}(1)$$

The structure of the $Y_{n,i}$ implies that the Lindeberg condition holds, thus as $n \to \infty$

$$S_n(\xi_n) - \{S_n(\vartheta) - J_n(\vartheta)\left[\delta_n^{-1}(\vartheta)(\xi_n - \vartheta)\right]\} = \frac{1}{\sqrt{n}} \sum_{i=1}^n Y_{n,i} \longrightarrow \mathcal{N}(0,1)$$

weakly in $\mathbb{R}$. This is incompatible with 7.14 C) i).

3) In the location model generated by the two-sided exponential distribution, there is no need for one-step correction. Consider the median

$$T_n := \text{median}(X_1, \ldots, X_n)$$

which in our model is maximum likelihood in $\mathcal{E}_n$ for every n. Our model allows to apply a classical result on asymptotic normality of the median: from [128, p. 578]),

$$\mathcal{L}\left(\sqrt{n}\,(T_n - \vartheta) \mid P_{n,\vartheta}\right) \longrightarrow \mathcal{N}(0,1)$$

as $n \to \infty$. In all location models, the median is equivariant in $\mathcal{E}_n$ for all n. By this circonstance, the last convergence is already regularity in the sense of Hajek: for any $h \in \mathbb{R}$,

$$\mathcal{L}\left(\sqrt{n}\left(T_n - [\vartheta + n^{-1/2}h]\right) \mid P_{n,\vartheta+n^{-1/2}h}\right) \longrightarrow F := \mathcal{N}(0,1) .$$

Recall from 1) that Fisher information in $\mathcal{E}$ equals $J_\vartheta = 1$ for all ϑ. Thus F appearing here is the optimal limit distribution $F = \mathcal{N}(0, J_\vartheta^{-1})$ in Hajek's convolution theorem. By (7.10') and Theorem 7.11, the sequence $(T_n)_n$ is thus regular and efficient in the sense of the convolution theorem, at every reference point $\vartheta \in \Theta$. Theorem 7.11 establishes that the coupling condition

$$\sqrt{n}\,(T_n - \vartheta) = Z_n(\vartheta) + o_{(P_{n,\vartheta})}(1) \quad , \quad n \to \infty$$

holds, for every $\vartheta \in \mathbb{R}$. Asymptotically as $n \to \infty$, this links estimation errors of the median T_n to differences $\frac{1}{n}\sum_{i=1}^n \mathrm{sgn}(X_i - \vartheta)$ between the relative number of observations above and below ϑ. The coupling condition in turn implies that the estimator sequence $(T_n)_n$ attains the local asymptotic minimax bound 7.12. □

7.22' Exercise. Consider the location model $\mathcal{E} = (\Omega, \mathcal{A}, \{P_\xi : \xi \in \mathbb{R}\})$ generated from $P_0 = \mathcal{N}(0,1)$, show that L^2-differentiability holds with L^2-derivative $V_\vartheta = (\cdot - \vartheta)$ at every point $\vartheta \in \mathbb{R}$. Show that the set of Assumptions 7.14 is satisfied, with left-hand sides in Assumption 7.14(C) identical to zero. Construct an MDE sequence as in Chapter 2 for the unknown parameter, with tightness rate $\sqrt{n}$, and specify the one-step modification according to Theorem 7.19 which as $n \to \infty$ grants regularity and efficiency at all points $\vartheta \in \Theta$ (again, as in Example 7.21, there is no need for discretisation). Verify that this one-step modification directly replaces the preliminary estimator by the empirical mean, the MLE in this model. □

7.22'' Exercise. We continue with the location model generated from the two-sided exponential law, under all notations and assumptions of Exercises 7.9', 7.9'' and of Example 7.22. We focus on the Bayesians with 'uniform over $\mathbb{R}$ prior'

$$T_n^* = \frac{\int_{-\infty}^\infty \xi\, L_n^{\xi/0}\, d\xi}{\int_{-\infty}^\infty L_n^{\xi/0}\, d\xi} , \quad n \ge 1$$

and shall prove that $(T_n^*)_n$ is efficient in the sense of the Convolution Theorem 7.10 and of the Local Asymptotic Minimax Theorem 7.12.

(a) Fix any reference point $\vartheta \in \mathbb{R}$ and recall (Example 7.22, or Exercises 7.9'(a), or 4.1''' and Lemma 4.11) that LAN holds at ϑ in the form

$$\Lambda_{n,\vartheta}^{h_n/0} = h_n\, S_n(\vartheta) - \frac{1}{2} h_n^2 + o_{(P_{n,\vartheta})}(1), \quad S_n(\vartheta) := \frac{1}{\sqrt{n}} \sum_{i=1}^n \mathrm{sgn}(X_i - \vartheta)$$

as $n \to \infty$, for bounded sequences $(h_n)_n$ in $\mathbb{R}$. Also, we can write under $P_{n,\vartheta}$

$$\sqrt{n}\,(T_n^* - \vartheta) = \frac{\int_{-\infty}^\infty u\, L_{n,\vartheta}^{u/0}\, du}{\int_{-\infty}^\infty L_{n,\vartheta}^{u/0}\, du} .$$

Section 7.4 The Case of i.i.d. Observations

(b) Prove joint convergence

$$\left(S_n(\vartheta), \int_{-\infty}^{\infty} u \, L_{n,\vartheta}^{u/0} \, du, \int_{-\infty}^{\infty} L_{n,\vartheta}^{u/0} \, du \right) \quad \text{under } P_{n,\vartheta}$$

as $n \to \infty$ to a limit law

$$\left(S, \int_{-\infty}^{\infty} u \, e^{uS - \frac{1}{2}u^2} \, du, \int_{-\infty}^{\infty} e^{uS - \frac{1}{2}u^2} \, du \right)$$

where $S \sim \mathcal{N}(0, 1)$ generates the Gaussian limit experiment $\mathcal{E}(1) = \{\mathcal{N}(h, 1) : h \in \mathbb{R}\}$.

(Hint: finite-dimensional convergence of $(L_{n,\vartheta}^{u/0})_{u \in \mathbb{R}}$ allows to deal e.g. with integrals $\int_K u \, L_{n,\vartheta}^{u/0} \, du$ on compacts $K \subset \mathbb{R}$, as in Lemma 2.5, then give some bound for $\int_{K^c} u \, L_{n,\vartheta}^{u/0} \, du$ as $n \to \infty$).

(c) Let U^* denote the Bayesian with 'uniform over $\mathbb{R}$ prior' in the limit experiment $\mathcal{E}(1)$

$$U^* := \frac{\int_{-\infty}^{\infty} u \, e^{uS - \frac{1}{2}u^2} \, du}{\int_{-\infty}^{\infty} e^{uS - \frac{1}{2}u^2} \, du}$$

and recall from Exercise 5.4" that in a Gaussian shift experiment, U^* coincides with the central statistic Z.

(d) Use (a), (b) and (c) to prove that the coupling condition

$$\sqrt{n} \, (T_n^* - \vartheta) = Z_n(\vartheta) + o_{(P_{n,\vartheta})}(1) \quad \text{as } n \to \infty$$

holds. By Theorem 7.11, comparing to Theorem 7.10, (7.10') and Theorem 7.12, we have efficiency of $(T_n^*)_n$ at ϑ. □

Chapter 8

*Some Stochastic Process Examples for Local Asymptotics of Type LAN, LAMN and LAQ

Topics for Chapter 8:

8.1 *The Ornstein–Uhlenbeck Process with Unknown Parameter Observed over a Long Time Interval
The process and its long-time behaviour (8.1)–8.2
Statistical model, likelihoods and ML estimator 8.3
Local models at ϑ 8.3'
LAN in the ergodic case $\vartheta < 0$ 8.4
LAQ in the null recurrent case $\vartheta = 0$ 8.5
LAMN in the transient case $\vartheta > 0$ 8.6
Remark on non-finite second moments in the LAMN case 8.6'
Sequential observation schemes: LAN at all parameter values 8.7

8.2 *A Null Recurrent Diffusion Model
The process and its long-time behaviour (8.8)–8.9
Regular variation, i.i.d. cycles, and norming constants for invariant measure
    8.10–(8.11")
Statistical model, likelihoods and ML estimator 8.12
Convergence of martingales together with their angle brackets 8.13
LAMN for local models at every $\vartheta \in \Theta$ 8.14
Remarks on non-finite second moments 8.15
Random norming 8.16
One-step modification is possible 8.17

8.3 *Some Further Remarks
LAN or LAMN or LAQ arising in other stochastic process models
Example: a limit experiment with essentially different statistical properties

Exercises: 8.2', 8.6", 8.7'

We discuss in detail some examples for local asymptotics of type LAN, LAMN or LAQ in stochastic process models (hence the asterisk * in front of all sections). Martingale convergence and Harris recurrence (positive or null) will play an important role in our arguments and provide limit theorems which establish convergence of local models to a limit model. Background on these topics and some relevant apparatus are collected in an Appendix (Chapter 9) to which we refer frequently, so one might have a look to Chapter 9 first before reading the sections of the present chapter.

8.1 *Ornstein–Uhlenbeck Process with Unknown Parameter Observed over a Long Time Interval

We start in dimension $d = 1$ with the well-known example of Ornstein–Uhlenbeck processes depending on an unknown parameter, see [23] and [5, p. 4]). We have a probability space $(\Omega, \mathcal{A}, P)$ carrying a Brownian motion W, and consider the unique strong solution $X = (X_t)_{t \geq 0}$ to the Ornstein–Uhlenbeck SDE

(8.1) $$dX_t = \vartheta\, X_t\, dt + dW_t, \quad t \geq 0$$

for some value of the parameter $\vartheta \in \mathbb{R}$ and some starting point $x \in \mathbb{R}$. There is an explicit representation of the solution

$$X_t = e^{\vartheta t}\left(x + \int_0^t e^{-\vartheta s}\, dW_s\right), \quad t \geq 0$$

satisfying

$$\mathcal{L}\left(e^{\vartheta t}\int_0^t e^{-\vartheta s}\, dW_s\right) = \begin{cases} \mathcal{N}\left(0, \frac{1}{2\vartheta}\left(e^{2\vartheta t} - 1\right)\right) & \text{if } \vartheta \neq 0 \\ \mathcal{N}(0, t) & \text{if } \vartheta = 0, \end{cases}$$

and thus an explicit representation of the semigroup $(P_t(\cdot, \cdot))_{t \geq 0}$ of transition probabilities of X

$$P_t(x, dy) := P\left(X_{s+t} \in dy \mid X_s = x\right)$$
$$= \begin{cases} \mathcal{N}\left(e^{\vartheta t} x, \frac{1}{2\vartheta}\left(e^{2\vartheta t} - 1\right)\right)(dy) & \text{if } \vartheta \neq 0 \\ \mathcal{N}(x, t)(dy) & \text{if } \vartheta = 0 \end{cases}$$

where $x, y \in \mathbb{R}$ and $0 \leq s, t < \infty$.

8.2 Long-time Behaviour of the Process. Depending on the value of the parameter $\vartheta \in \Theta$, we have three different types of asymptotics for the solution $X = (X_t)_{t \geq 0}$ to equation (8.1).

(a) *Positive recurrence in the case where* $\vartheta < 0$: When $\vartheta < 0$, the process X is positive recurrent in the sense of Harris (cf. Definition 9.4) with invariant measure

$$\mu := \mathcal{N}\left(0, \frac{1}{2|\vartheta|}\right).$$

This follows from Proposition 9.12 in the Appendix where the function

$$S(x) = \int_0^x s(y)\, dy \quad \text{with} \quad s(y) = \exp\left(-\int_0^y 2\vartheta v\, dv\right) = e^{-\vartheta y^2}, \quad x, y \in \mathbb{R}$$

corresponding to the coefficients of equation (8.1) in the case where $\vartheta < 0$ is a bijection from $\mathbb{R}$ onto $\mathbb{R}$, and determines the invariant measure for the process (8.1) as $\frac{1}{s(x)} dx$, unique up to constant multiples. Normed to a probability measure this specifies μ as above.

Next, the Ratio Limit Theorem 9.6 yields for functions $f \in L^1(\mu)$

$$\lim_{t \to \infty} \frac{1}{t} \int_0^t f(\xi_s) \, ds = \mu(f) \quad \text{almost surely as } t \to \infty$$

for arbitrary choice of a starting point. We thus have strong laws of large numbers for a large class of additive functionals of X in the case where $\vartheta < 0$.

(b) *Null recurrence in the case where* $\vartheta = 0$: Here X is one-dimensional Brownian motion with starting point x, and thus null recurrent (cf. Definition 9.5') in the sense of Harris. The invariant measure is λ, the Lebesgue measure on $\mathbb{R}$.

(c) *Transience in the case where* $\vartheta > 0$: Here trajectories of X tend towards $+\infty$ or towards $-\infty$ exponentially fast. In particular, any given compact K in $\mathbb{R}$ will be left in finite time without return: thus X is transient in the case where $\vartheta > 0$. This is proved as follows. Write $\mathbb{F}$ for the (right-continuous) filtration generated by W. For fixed starting point $x \in \mathbb{R}$, consider the $(P, \mathbb{F})$-martingale

$$Y = (Y_t)_{t \geq 0}, \quad Y_t := e^{-\vartheta t} X_t = x + \int_0^t e^{-\vartheta s} \, dW_s, \quad t \geq 0.$$

Then $E((Y_t - x)^2) = \int_0^t e^{-2\vartheta s} \, ds$ and thus

$$\sup_{t \geq 0} E(Y_t^2) < \infty$$

in the case where $\vartheta > 0$: hence Y_t converges as $t \to \infty$ P-almost surely and in $L^2(\Omega, \mathcal{A}, P)$, and the limit is

$$Y_\infty = x + \int_0^\infty e^{-\vartheta s} \, dW_s \sim \mathcal{N}\left(x, \frac{1}{2\vartheta}\right).$$

But almost sure convergence of trajectories $t \to Y_t(\omega)$ signifies that for P-almost all $\omega \in \Omega$,

(×) $\qquad\qquad X_t(\omega) \sim Y_\infty(\omega) e^{\vartheta t} \quad \text{as} \quad t \to \infty.$

Asymptotics (×) can be transformed into a strong law of large numbers for some few additive functionals of X, in particular

$$\int_0^t X_s^2(\omega) \, ds \sim Y_\infty^2(\omega) \int_0^t e^{2\vartheta s} \, ds \sim Y_\infty^2(\omega) \frac{1}{2\vartheta} e^{2\vartheta t} \quad \text{as} \quad t \to \infty$$

for P-almost all $\omega \in \Omega$. $\qquad\square$

The following tool will be used several times in this chapter.

Section 8.1 Ornstein-Uhlenbeck Model

8.2' Lemma. On some space $(\Omega, \mathcal{A}, \mathbb{F}, P)$ with right-continuous filtration $\mathbb{F}$, consider a continuous $(P, \mathbb{F})$-semi-martingale X admitting a decomposition

$$X = X_0 + M + A$$

under P, where A is continuous $\mathbb{F}$-adapted with paths locally of bounded variation starting in $A_0 = 0$, and M a continuous local $(P, \mathbb{F})$-martingale starting in $M_0 = 0$. Let P' denote a second probability measure on $(\Omega, \mathcal{A}, \mathbb{F})$ such that X is a continuous $(P', \mathbb{F})$-semi-martingale with representation

$$X = X_0 + M' + A'$$

under P'. Let H be an $\mathbb{F}$-adapted process with left-continuous paths, locally bounded under both P and P'. Then we have the following: under the assumption $P \stackrel{loc}{\sim} P'$ relative to $\mathbb{F}$, any determination $(t, \omega) \to I(t, \omega)$ of the stochastic integral $\int H_s dX_s$ under P is also a determination of the stochastic integral $\int H_s dX_s$ under P'.

Proof. (1) For the process H, consider some localising sequence $(\tau_n)_n$ under P, and some localising sequence $(\tau'_n)_n$ under P'. For finite time horizon $N < \infty$ we have $P_N \sim P'_N$, hence events $\bigcap_n \{\tau_n \leq N\}$ and $\bigcap_n \{\tau'_n \leq N\}$ in $\mathcal{F}_N$ are null sets under both probability measures P and P'. This holds for all N, thus $(\tau_n \wedge \tau'_n)_n$ is a common localising sequence under both P and P'.

(2) Localising further, we may assume that H as well as M, $\langle M \rangle_P$, $\|A\|$ and M', $\langle M' \rangle_{P'}$, $\|A'\|$ are bounded (we write $\|A\|$ for the total variation process of A, and $\langle M \rangle_P$ for the angle bracket under P).

(3) Fix a version $(t, \omega) \to J(t, \omega)$ of the stochastic integral $\int H_s dX_s$ under P, and a version $(t, \omega) \to J'(t, \omega)$ of the stochastic integral $\int H_s dX_s$ under P'. For $n \geq 1$, define processes $(t, \omega) \to I^{(n)}(t, \omega)$

$$I_t^{(n)} = \sum_{k=0}^{\infty} H_{\frac{k}{2^n}} \left(X_{t \wedge \frac{k+1}{2^n}} - X_{t \wedge \frac{k}{2^n}} \right) = \int_0^t H_s^{(n)} dX_s,$$

$$H^{(n)} := \sum_{k=0}^{\infty} H_{\frac{k}{2^n}} 1_{]\!]\frac{k}{2^n}, \frac{k+1}{2^n}]\!]}$$

to be considered under both P and P'. Using [98, Thm. 18.4] with respect to the martingale parts of X, select a subsequence $(n_\ell)_\ell$ from $\mathbb{N}$ such that simultaneously as $\ell \to \infty$

$$\begin{cases} \text{for } P\text{-almost all } \omega: \text{ the paths } I^{(n_\ell)}(\cdot, \omega) \text{ converge uniformly on } [0, \infty) \\ \text{to the path } J(\cdot, \omega) \\ \text{for } P'\text{-almost all } \omega: \text{ the paths } I^{(n_\ell)}(\cdot, \omega) \text{ converge uniformly on } [0, \infty) \\ \text{to the path } J'(\cdot, \omega). \end{cases}$$

Since $P_N \sim P'_N$ for $N < \infty$, there is an event $A_N \in \mathcal{F}_N$ of full measure under both P and P' such that $\omega \in A_N$ implies $J(\cdot, \omega) = J'(\cdot, \omega)$ on $[0, N]$. As a consequence,

J and J' are indistinguishable processes for the probability measure P as well as for the probability measure P'. □

We turn to statistical models defined by observing a trajectory of an Ornstein–Uhlenbeck process continuously over a long time interval, under unknown parameter $\vartheta \in \mathbb{R}$.

8.3 Statistical Model. Consider the Ornstein–Uhlenbeck equation (8.1). Write $\Theta := \mathbb{R}$. Fix a starting point $x = x_0$ which does not depend on $\vartheta \in \Theta$. Let Q_ϑ denote the law of the solution to equation (8.1) under ϑ, on the canonical path space $(C, \mathcal{C}, \mathbb{G})$ or $(D, \mathcal{D}, \mathbb{G})$. Then as in Theorem 6.10 or Example 6.11, all laws Q_ϑ are locally equivalent relative to $\mathbb{G}$, and the density process of Q_ϑ with respect to Q_0 relative to $\mathbb{G}$ is

$$L_t^{\vartheta/0} = \exp\left\{ \vartheta \int_0^t \eta_s \, dm_s^{(0)} - \frac{1}{2}\vartheta^2 \int_0^t \eta_s^2 \, ds \right\}, \quad t \geq 0$$

with $m^{(0)}$ the Q_0-local martingale part of the canonical process η under Q_0.

(1) Fix a determination $(t, \omega) \to Y(t, \omega)$ of the stochastic integral

$$Y = \int \eta_s \, d\eta_s = \int \eta_s \, dm_s^{(0)} = \eta_0 \, m^{(0)} + \int m_s^{(0)} \, dm_s^{(0)} \quad \text{under } Q_0:$$

as in Example 6.11, using $\mathcal{L}([\eta - \eta_0], m^{(0)} \mid Q_0) = \mathcal{L}(B, B)$, there is an explicit representation

$$Y_t = \frac{1}{2}\left(\eta_t^2 - \eta_0^2 - t\right), \quad t \geq 0.$$

As a consequence, in statistical experiments $\mathcal{E}_t$ corresponding to observation of the canonical process η up to time $0 < t < \infty$, the likelihood function

$$\Theta \ni \vartheta \longrightarrow L_t^{\vartheta/0} = \exp\left\{ \vartheta Y_t - \frac{1}{2}\vartheta^2 \int_0^t \eta_s^2 \, ds \right\} \in (0, \infty)$$

and the maximum likelihood (ML) estimator

$$\widehat{\vartheta}_t := \frac{Y_t}{\int_0^t \eta_s^2 \, ds} = \frac{\frac{1}{2}\left(\eta_t^2 - \eta_0^2 - t\right)}{\int_0^t \eta_s^2 \, ds}$$

are expressed without reference to any particular probability measure.

(2) Simultaneously for all $\vartheta \in \Theta$, $(t, \omega) \to Y(t, \omega)$ provides a common determination for

$$\left(\int_0^t \eta_s \, d\eta_s\right)_{t \geq 0} = \left(\int_0^t \eta_s \, dm_s^{(\vartheta)} + \vartheta \int_0^t \eta_s^2 \, ds\right)_{t \geq 0} \quad \text{under } Q_\vartheta$$

by Lemma 8.2': η is a continuous semi-martingale under both Q_0 and Q_ϑ, and any determination $(t, \omega) \to Y(t, \omega)$ of $\int \eta_s d\eta_s$ under Q_0 is also a determination of $\int \eta_s d\eta_s$

Section 8.1 Ornstein-Uhlenbeck Model

under Q_ϑ. Obviously η has Q_ϑ-martingale part $m_t^{(\vartheta)} = \eta_t - \eta_0 - \vartheta \int_0^t \eta_s ds$. There are two statistical consequences:

(i) for every every $0 < t < \infty$, ML estimation errors under $\vartheta \in \Theta$ take the form of the ratio of a Q_ϑ-martingale divided by its angle bracket under Q_ϑ:

$$\widehat{\vartheta}_t - \vartheta = \frac{\int_0^t \eta_s \, dm_s^{(\vartheta)}}{\int_0^t \eta_s^2 \, ds} \quad \text{under } Q_\vartheta \, ;$$

(ii) for the density process $L^{\xi/\vartheta}$ of Q_ξ with respect to Q_ϑ relative to $\mathbb{G}$, the representation

$$L_t^{\xi/0}/L_t^{\vartheta/0} = \exp\left\{ (\xi - \vartheta) Y_t - \frac{1}{2}(\xi^2 - \vartheta^2) \int_0^t \eta_s^2 \, ds \right\}, \quad t \geq 0$$

coincides with the representation from Theorem 6.10 applied to Q_ξ and Q_ϑ:

(+) $\qquad L_t^{\xi/\vartheta} = \exp\left\{ (\xi - \vartheta) \int_0^t \eta_s \, dm_s^{(\vartheta)} - \frac{1}{2}(\xi - \vartheta)^2 \int_0^t \eta_s^2 \, ds \right\} .$

(3) The representation (+) allows to reparameterise the model $\mathcal{E}_t$ with respect to fixed reference points $\vartheta \in \Theta$, and makes quadratic models appear around ϑ. We will call

$$\left(\int_0^t \eta_s \, dm_s^{(\vartheta)} \right)_{t \geq 0} \quad \text{and} \quad \left(\int_0^t \eta_s^2 \, ds \right)_{t \geq 0} = \left\langle \int \eta \, dm^{(\vartheta)} \right\rangle \quad \text{under } Q_\vartheta$$

score martingale at ϑ and *information process at* ϑ. It is obvious from 8.2 that the law of the observed information depends on ϑ. Hence reparameterising as in (+) with respect to different reference points ϑ or ϑ' makes statistically different models appear. □

The last part of Model 8.3 allows to introduce local models at $\vartheta \in \mathbb{R}$.

8.3' Local Models at ϑ. (1) Localising around a fixed reference point $\vartheta \in \mathbb{R}$, write Q_ξ^n for Q_ξ restricted to $\mathcal{G}_n$. With suitable choice $\delta_n(\vartheta)$ of local scale to be specified below, consider local models

$$\mathcal{E}_{\vartheta,n} = \left(C, \mathcal{G}_n, \{ Q_{\vartheta + \delta_n(\vartheta)h}^n : h \in \mathbb{R} \} \right), \quad n \geq 1$$

when n tends to ∞. Then from (+), the log-likelihoods in $\mathcal{E}_{\vartheta,n}$ are

(◊) $\qquad \Lambda_{\vartheta,n}^{h/0} = \log L_n^{(\vartheta + \delta_n(\vartheta)h)/\vartheta}$

$= h \left[\delta_n(\vartheta) \int_0^n \eta_s \, dm_s^\vartheta \right] - \frac{1}{2} h^2 \left[\delta_n^2(\vartheta) \int_0^n \eta_s^2 \, ds \right], \quad h \in \mathbb{R} .$

By ($\diamond$) and in view of Definition 7.1, the problem of choice of local scale at ϑ turns out to be the problem of choice of norming constants for the score martingale: we need weak convergence as $n \to \infty$ of pairs

$$(\diamond\diamond) \quad (S_n(\vartheta), J_n(\vartheta)) := \left(\delta_n(\vartheta) \int_0^n \eta_s\, dm_s^\vartheta,\ \delta_n^2(\vartheta) \int_0^n \eta_s^2\, ds \right) \quad \text{under } Q_\vartheta$$

to some pair of limiting random variables

$$(S(\vartheta), J(\vartheta))$$

which generate as in Definition 6.1 and Remark 6.1" a quadratic limit experiment.
(2) From ($\diamond$), rescaled ML estimation errors at ϑ take the form

$$\delta_n^{-1}(\vartheta)\left(\widehat{\vartheta}_n - \vartheta\right) = \frac{\delta_n(\vartheta) \int_0^n \eta_s\, dm_s^\vartheta}{\delta_n^2(\vartheta) \int_0^n \eta_s^2\, ds} \quad \text{under } Q_\vartheta$$

and thus act in the local experiment $\mathcal{E}_{\vartheta,n}$ as estimators

$$(\diamond\diamond\diamond) \qquad \widehat{h}_n = \delta_n^{-1}(\vartheta)\left(\widehat{\vartheta}_n - \vartheta\right) = J_n^{-1}(\vartheta) S_n(\vartheta)$$

for the local parameter $h \in \mathbb{R}$. $\square$

Now we show that local asymptotic normality at ϑ holds in the case where $\vartheta < 0$, local asymptotic mixed normality in the case where $\vartheta > 0$, and local asymptotic quadraticity in the case where $\vartheta = 0$. We also specify local scale $(\delta_n(\vartheta))_n$.

8.4 LAN in the Positive Recurrent Case. By 8.2(a), in the case where $\vartheta < 0$, the canonical process η is positive recurrent under Q_ϑ with invariant probability $\mu^\vartheta = \mathcal{N}(0, \frac{1}{2|\vartheta|})$. For the information process in step (3) of Model 8.3 we thus have the following strong law of large numbers

$$\lim_{t\to\infty} \frac{1}{t}\int_0^t \eta_s^2\, ds = \int x^2\, \mu^\vartheta(dx) = \frac{1}{2|\vartheta|} =: \Lambda \quad Q_\vartheta\text{-almost surely.}$$

Correspondingly, if at stage n of the asymptotics we observe a trajectory over the time interval $[0, n]$, we take local scale $\delta_n(\vartheta)$ at ϑ such that

$$\delta_n(\vartheta) = n^{-1/2} \quad \text{for all values } \vartheta < 0 \text{ of the parameter.}$$

(1) Rescaling the score martingale in step (3) of Model 8.3 in space and time we put $\mathbb{G}^n := (\mathcal{G}_{tn})_{t\geq 0}$ and

$$M_\vartheta^n = (M_\vartheta^n(t))_{t\geq 0}, \quad M_\vartheta^n(t) := n^{-1/2} \int_0^{tn} \eta_s\, dm_s^{(\vartheta)}, \quad t \geq 0.$$

Section 8.1 Ornstein-Uhlenbeck Model

This yields a family $(M^n_\vartheta)_n$ of continuous $(Q_\vartheta, \mathbb{G}^n)$-martingales with angle brackets

(∗) $$\forall\, t \geq 0: \quad \langle M^n_\vartheta \rangle_t = \frac{1}{n} \int_0^{tn} \eta_s^2\, ds \;\longrightarrow\; t\, \frac{1}{2|\vartheta|} = t\, \Lambda$$

Q_ϑ-almost surely as $n \to \infty$.

From Jacod and Shiryaev [64, Cor. VIII.3.24], the martingale convergence theorem – we recall this in Appendix 9.1 below – establishes weak convergence in the Skorohod space D of càdlàg functions $[0, \infty) \to \mathbb{R}$ to standard Brownian motion with scaling factor $\Lambda^{1/2}$:

$$M^n_\vartheta \longrightarrow \Lambda^{1/2} B \quad \text{(weakly in } D \text{ under } Q_\vartheta, \text{ as } n \to \infty).$$

From this, for the particular time $t = 1$,

(∗∗) $$M^n_\vartheta(1) \longrightarrow \Lambda^{1/2} B_1 \quad \text{(weakly in } \mathbb{R} \text{ under } Q_\vartheta, \text{ as } n \to \infty)$$

since projection mappings $D \ni \alpha \to \alpha(t) \in \mathbb{R}$ are continuous at every $\alpha \in C$, cf. [64, VI.2.1]), and C has full measure under $\mathcal{L}\left(\Lambda^{1/2} B\right)$ in the space D.

(2) Combining (∗) and (∗∗) above with (◇) in 8.3', log-likelihoods in the local model $\mathcal{E}_{n,\vartheta}$ at ϑ are

$$\Lambda^{h/0}_{\vartheta,n} = \log L_n^{(\vartheta + n^{-1/2} h)/\vartheta} = h\, M^n_\vartheta(1) - \frac{1}{2} h^2 \langle M^n_\vartheta \rangle_1, \quad h \in \mathbb{R}$$

which gives for arbitrary bounded sequences $(h_n)_n$

(×) $$\begin{cases} \Lambda^{h_n/0}_{\vartheta,n} = h_n\, M^n_\vartheta(1) - \frac{1}{2} h_n^2\, \Lambda + o_{Q_\vartheta}(1) & \text{as } n \to \infty \\ \mathcal{L}\left(M^n_\vartheta(1) \mid Q_\vartheta\right) \longrightarrow \mathcal{N}(0, \Lambda) & \text{as } n \to \infty \\ \text{with } \Lambda = \frac{1}{2|\vartheta|}. & \end{cases}$$

This establishes LAN at parameter values $\vartheta < 0$, cf. Definition 7.1(c), and the limit experiment $\mathcal{E}_\infty(\vartheta)$ is the Gaussian shift $\mathcal{E}(\frac{1}{2|\vartheta|})$ in the notation of Definition 5.2.

(3) Once LAN is established, the assertion (◇◇◇) in 8.3' is the coupling condition of Theorem 7.11. From Hájek's Convolution Theorem 7.10 and the Local Asymptotic Minimax Theorem 7.12 we deduce the following properties for the ML estimator sequence $(\widehat{\vartheta}_n)_n$: at all parameter values $\vartheta < 0$, the maximum likelihood estimator sequence is regular and efficient for the unknown parameter, and attains the local asymptotic minimax bound. □

8.5 LAQ in the Null Recurrent Case. In the case where $\vartheta = 0$, the canonical process η under Q_0 is a Brownian motion with starting point x, cf. 8.2(b), and self-similarity properties of Brownian motion turn out to be the key to local asymptotics at $\vartheta = 0$, as pointed out by [23] or [38]. Writing B or $\widetilde{B}$ for standard Brownian motion, Ito formula and scaling properties give

$$\left(\int_0^t B_s \, dB_s \, , \, \int_0^t |B_s| \, ds \, , \, \int_0^t B_s^2 \, ds \right) = \left(\frac{1}{2}(B_t^2 - t) \, , \, \int_0^t |B_s| \, ds \, , \, \int_0^t B_s^2 \, ds \right)$$

$$\stackrel{d}{=} \left(\frac{1}{2}([\sqrt{t}\,\widetilde{B}_1]^2 - t) \, , \, \int_0^t \left|\sqrt{t}\,\widetilde{B}_{\frac{s}{t}}\right| ds \, , \, \int_0^t \left(\sqrt{t}\,\widetilde{B}_{\frac{s}{t}}\right)^2 ds \right)$$

$$\stackrel{d}{=} \left(t \int_0^1 B_s \, dB_s \, , \, t^{3/2} \int_0^1 |B_s| \, ds \, , \, t^2 \int_0^1 B_s^2 \, ds \right) .$$

From step (3) of Model 8.3, the score martingale at $\vartheta = 0$ is

$$\int \eta_s \, dm_s^{(0)} = \eta_0 \, m^{(0)} + \int (\eta_s - \eta_0) \, dm_s^{(0)} \quad \text{under } Q_0$$

where $\mathcal{L}([\eta - \eta_0], m^{(0)} \mid Q_0) = \mathcal{L}(B, B)$. Consequently, observing at stage n of the asymptotics the canonical process η up to time n, the right choice of local scale is

$$\delta_n(\vartheta) := \frac{1}{n} \quad \text{at parameter value } \vartheta = 0 \, ,$$

and rescaling of the score martingale works as follows: with $\mathbb{G}^n := (\mathcal{G}_{tn})_{t \geq 0}$,

$$M^n = (M^n(t))_{t \geq 0} \, , \quad M^n(t) := \frac{1}{n} \int_0^{tn} \eta_s \, dm_s^{(0)} \, , \quad t \geq 0$$

where in the case where $\vartheta = 0$ we suppress subscript $\vartheta = 0$ from our notation. Note the influence of the starting point x for equation (8.1) on the form of the score martingale.

(A) Consider first starting point $x = 0$ in equation (8.1) as a special case.

(1) Applying the above scaling properties, we have exact equality in law

(*) $\quad \mathcal{L}\left(M^n, \langle M^n \rangle \mid Q_0 \right) \stackrel{d}{=} \mathcal{L}\left(\int B_s \, dB_s \, , \, \int B_s^2 \, ds \right) \quad$ for all $n \geq 1$.

According to ($\diamond$) in 8.3', the log-likelihoods in the local model $\mathcal{E}_{n,0}$ at $\vartheta = 0$ are

(×) $\quad \Lambda_{0,n}^{h/0} = \log L_n^{(0+n^{-1}h)/0} = h \, M^n(1) - \frac{1}{2} h^2 \langle M^n \rangle_1 \, , \quad h \in \mathbb{R}$

for every $n \geq 1$. Thus, as a statistical experiment, local experiments $\mathcal{E}_{n,0} = \{Q_{0 + \frac{1}{n}h}^n : h \in \mathbb{R}\}$ at $\vartheta = 0$ coincide for all $n \geq 1$ with the experiment $\mathcal{E}_1 = \{Q_h^1 : h \in \mathbb{R}\}$ where an Ornstein–Uhlenbeck trajectory (having initial point $x = 0$) is observed over the time interval $[0, 1]$. Thus self-similarity creates a particular LAQ situation for which approximating local experiments and limit experiment coincide. We have seen in Example 6.11 that the limit experiment $\mathcal{E}_1$ is not mixed normal.

Section 8.1 Ornstein-Uhlenbeck Model

(2) We look to ML estimation in the special case of starting value $x = 0$ for equation (8.1). Combining the representation of rescaled ML estimation errors $(\diamond\diamond\diamond)$ in 8.3' with $(*)$ in step (1), the above scaling properties allow for equality in law which does not depend on $n \geq 1$

$$(\circ) \qquad \mathcal{L}\left(n\left(\widehat{\vartheta}_n - \left[0 + \frac{1}{n}h\right]\right) \mid Q_{[0+\frac{1}{n}h]}\right) = \mathcal{L}(\widehat{\vartheta}_1 - h \mid Q_h)$$

at every value $h \in \mathbb{R}$. To see this, write for functions $f \in \mathcal{C}_b(\mathbb{R})$

$$E_{Q_{[0+\frac{1}{n}h]}}\left(f\left(n\left(\widehat{\vartheta}_n - \left[0 + \frac{1}{n}h\right]\right)\right)\right)$$
$$= E_{Q_{[0+\frac{1}{n}h]}}\left(f\left(\widehat{h}_n - h\right)\right)$$
$$= E_{Q_0}\left(L_n^{[0+\frac{1}{n}h]/0} f\left(\widehat{h}_n - h\right)\right)$$
$$= E_{Q_0}\left(\exp\left\{h M^n(1) - \frac{1}{2}h^2 \langle M^n\rangle_1\right\} f\left(\frac{M^n(1)}{\langle M^n\rangle_1} - h\right)\right)$$

which by $(*)$ above is free of $n \geq 1$. We can rephrase $(\circ)$ as follows: observing over longer time intervals, we do not gain anything except scaling factors.

(B) Now we consider the general case of starting values $\eta_0 = x \neq 0$ for equation (8.1). In this case, by the above decomposition of the score martingale, M^n under Q_0 is of type

$$\frac{1}{n}\int_0^{tn}(x + B_s)\,dB_s = \frac{x}{n}B_{tn} + \frac{1}{n}\int_0^{tn} B_s\,dB_s\,,\qquad t \geq 0$$

for standard Brownian motion B. Decomposing M^n in this sense, we can control

$$\sup_{0\leq s \leq t}\left|M^n(s) - \frac{1}{n}\int_0^{sn}(\eta - \eta_0)_v\,dm_v^{(0)}\right|\quad \text{under } Q_0$$

in the same way as $\frac{x}{\sqrt{n}}\sup_{0\leq s\leq t}|B_s|$, for arbitrary n and t, and

$$\sup_{0\leq s\leq t}\left|\langle M^n\rangle(s) - \frac{1}{n^2}\int_0^{sn}(\eta - \eta_0)_v^2\,dv\right|\quad \text{under } Q_0$$

in the same way as $\frac{x^2 t}{n} + \frac{2|x|}{\sqrt{n}}\int_0^t|B_s|\,ds$, where we use the scaling property stated at the start.

(1) In the general situation (B), the previous equality in law $(*)$ of score and information is replaced by weak convergence of the pair (score martingale, information process) under the parameter value $\vartheta = 0$

$$(**)\qquad \mathcal{L}\left(M^n, \langle M^n\rangle \mid Q_0\right) \longrightarrow \mathcal{L}\left(\int B_s\,dB_s,\int B_s^2\,ds\right)$$
$$\text{weakly in } D(\mathbb{R}^2) \text{ as } n\to\infty.$$

Let us write $\widetilde{\mathcal{E}}_1$ for the limit experiment which appears in (A.1), in order to avoid confusion about the different starting values. In our case (B) of starting value $x \neq 0$ for equation (8.1), we combine ($\times$) for the likelihood ratios in the local models $\mathcal{E}_{n,0}$ at $\vartheta = 0$ (which is ($\diamond$) in 8.3')

$$\Lambda_{0,n}^{h/0} = \log L_n^{[0+\frac{1}{n}h]/0} = h\, M^n(1) - \frac{1}{2} h^2 \langle M^n \rangle_1, \quad h \in \mathbb{R}$$

with weak convergence (**) to establish LAQ at $\vartheta = 0$ with limit experiment $\widetilde{\mathcal{E}}_1$.

(2) Given LAQ at $\vartheta = 0$ with limit experiment $\widetilde{\mathcal{E}}_1$, write $\widetilde{Q}_h$ for the laws in $\widetilde{\mathcal{E}}_1$. Then Corollary 7.7 shows for ML estimation

(oo)
$$\sup_{|h| \leq C} \left| E_{Q_{[0+\frac{1}{n}h]}}\left(\ell\left(n\,(\widehat{\vartheta}_n - [0 + n^{-1}h])\right)\right) - E_{\widetilde{Q}_h}\left(\ell\left(\widehat{\vartheta}_1 - h\right)\right) \right| \longrightarrow 0$$

for arbitrary loss functions $\ell(\cdot)$ which are continuous and bounded and for arbitrary constants $C < \infty$. In case (B) of starting value $x \neq 0$ for equation (8.1), (oo) replaces equality of laws (o) which holds in case (A) above. Recall that (oo) merely states the following: the ML estimator $\widehat{h}_n$ for the local parameter h in the local model $\mathcal{E}_{n,\vartheta}$ at $\vartheta = 0$ works approximately as well as the ML estimator in the limit model $\widetilde{\mathcal{E}}_1$. In particular, (oo) is not an optimality criterion. □

8.6 LAMN in the Transient Case. By 8.2(c), in the case where $\vartheta > 0$, the canonical process η under Q_ϑ is transient, and we have the following asymptotics for the information process when $\vartheta > 0$:

$$e^{-2\vartheta t} \int_0^t \eta_s^2\, ds \longrightarrow Y_\infty^2(\vartheta)\, \frac{1}{2\vartheta} \quad Q_\vartheta\text{-almost surely as } t \to \infty.$$

Here $Y_\infty(\vartheta)$ is the $\mathcal{G}_\infty$-measurable limit variable

$$Y_\infty = Y_\infty(\vartheta) \sim \mathcal{N}\left(x, \frac{1}{2\vartheta}\right),$$

for the martingale Y of 8.2(c), with x the starting point for equation (8.1). Observing at stage n of the asymptotics a trajectory of η up to time n, we have to chose local scale as

$$\delta_n(\vartheta) = e^{-\vartheta n} \quad \text{at } \vartheta > 0.$$

Thus in the transient case, local scale depends on the value of the parameter. Space-time scaling of the score martingale is done as follows. With notation $f^+ = f \vee 0$ for the positive part of f, we put $\mathbb{G}^n := \left(\mathcal{G}_{(n+\log(t))^+}\right)_{t \geq 0}$ and

$$M_\vartheta^n = \left(M_\vartheta^n(t)\right)_{t \geq 0}, \quad M_\vartheta^n(t) := e^{-\vartheta n} \int_0^{(n+\log(t))^+} \eta_s\, dm_s^{(\vartheta)}, \quad t \geq 0$$

Section 8.1 Ornstein-Uhlenbeck Model

for $n \geq 1$. Then angle brackets of M_ϑ^n under Q_ϑ satisfy as $n \to \infty$

$$\langle M_\vartheta^n \rangle_t = e^{-2\vartheta n} \int_0^{(n+\log(t))^+} \eta_s^2 \, ds \quad \longrightarrow \quad Y_\infty^2(\vartheta) \frac{1}{2\vartheta} t^{2\vartheta} \quad Q_\vartheta\text{-almost surely}$$

for every $0 < t < \infty$ fixed. Writing

$$\varphi_\vartheta(t) := Y_\infty^2(\vartheta) \frac{1}{2\vartheta} t^{2\vartheta}, \quad t \geq 0,$$

we have a collection of $\mathcal{G}_\infty$-measurable random variables with the properties

$$\begin{cases} \varphi_\vartheta(0) \equiv 0, \; t \to \varphi_\vartheta(t) \text{ is continuous and strictly increasing, } \lim_{t \to \infty} \varphi_\vartheta(t) = +\infty; \\ \text{for every } 0 < t < \infty: \; \langle M_\vartheta^n \rangle_t \; \longrightarrow \; \varphi_\vartheta(t) \; Q_\vartheta\text{-almost surely as } n \to \infty. \end{cases}$$

All M_ϑ^n being continuous $(Q_\vartheta, \mathbb{G}^n)$-martingales, a martingale convergence theorem (from Jacod and Shiryaev [64, VIII.5.7 and VIII.5.42]) which we recall in Theorem 9.2 in the Appendix (the nesting condition there is satisfied for our choice of the $\mathbb{G}_n$) yields

$$M_\vartheta^n \quad \longrightarrow \quad B \circ \varphi_\vartheta \quad (\text{weak convergence in } D, \text{ under } Q_\vartheta, \text{ as } n \to \infty)$$

where standard Brownian motion B is independent from φ_ϑ. Again by continuity of all M_ϑ^n, we also have weak convergence of pairs under Q_ϑ as $n \to \infty$

$$\left(M_\vartheta^n, \langle M_\vartheta^n \rangle \right) \quad \longrightarrow \quad (B \circ \varphi_\vartheta, \varphi_\vartheta)$$

(cf. [64, VI.6.1]); we recall this in Theorem 9.3 in the Appendix) in the Skorohood space $D(\mathbb{R}^2)$ of càdlàg functions $[0, \infty) \to \mathbb{R}^2$. By continuity of projection mappings on a subset of $D(\mathbb{R}^2)$ of full measure, we end up with weak convergence

(*)
$$\left(M_\vartheta^n(1), \langle M_\vartheta^n \rangle_1 \right) \quad \longrightarrow \quad (B(\varphi_\vartheta(1)), \varphi_\vartheta(1))$$
$$(\text{weakly in } \mathbb{R}^2, \text{ under } Q_\vartheta, \text{ as } n \to \infty).$$

(1) Log-likelihood ratios in local models $\mathcal{E}_{\vartheta,n}$ at ϑ are

$$\Lambda_{0,n}^{h/0} = \log L_n^{[\vartheta + e^{-\vartheta n} h]/\vartheta} = h \, M_\vartheta^n(1) - \frac{1}{2} h^2 \langle M_\vartheta^n \rangle_1, \quad h \in \mathbb{R}$$

according to ($\diamond$) in 8.3'. Combining this with weak convergence (*), we have established LAMN at parameter values $\vartheta > 0$; the limit model $\mathcal{E}_\infty(\vartheta)$ is Brownian motion with unknown drift observed up to the independent random time $\varphi_\vartheta(1)$ as above. This type of limit experiment was studied in 6.16.

(2) According to step (2) in 8.3', rescaled ML estimator errors at ϑ

$$e^{\vartheta n} \left(\hat{\vartheta}_n - \vartheta \right) = \frac{M_\vartheta^n(1)}{\langle M_\vartheta^n \rangle_1} = Z_n(\vartheta) \quad \text{under } Q_\vartheta, \text{ for all } n \geq 1$$

coincide with the central sequence at ϑ and converge to the limit law

(×) $$\frac{B(\varphi_\vartheta(1))}{\varphi_\vartheta(1)} \sim \int \mathcal{L}(\varphi_\vartheta(1))(du) \, \mathcal{N}\left(0, \frac{1}{u}\right).$$

By local asymptotic mixed normality according to step (1), Theorems 7.11, 7.10 and 7.12 apply and show the following: at all parameter values $\vartheta > 0$, the ML estimator sequence $(\widehat{\vartheta}_n)_n$ is regular and efficient in the sense of Jeganathan's version of the Convolution Theorem 7.10, and attains the local asymptotic minimax bound of Theorem 7.12. □

8.6' Remark. Under assumptions and notations of 8.6, we comment on the limit law arising in (×). With x the starting point for equation (8.1), recall from 8.2(c) and the start of 8.6

$$Y_\infty(\vartheta) \sim \mathcal{N}\left(x, \frac{1}{2\vartheta}\right), \quad \varphi_\vartheta(1) = Y_\infty^2(\vartheta) \frac{1}{2\vartheta}$$

which gives (use e.g. [4, Sect. VII.1])

$$\mathcal{L}(\varphi_\vartheta(1)) = \begin{cases} \Gamma\left(\frac{1}{2}, 2\vartheta^2\right) & \text{in the case where } x = 0 \\ \Gamma\left(\frac{1}{2}, x\sqrt{2\vartheta^3}, 2\vartheta^2\right) & \text{in the case where } x \neq 0 \end{cases}$$

where notation $\Gamma(a, \lambda, p)$ is used for decentral Gamma laws ($a > 0, \lambda > 0, p > 0$)

$$\Gamma(a, \lambda, p) = \sum_{m=0}^{\infty} \frac{e^{-\lambda} \lambda^k}{k!} \Gamma(a+m, p) .$$

In the case where $\lambda = 0$ this reduces to the usual $\Gamma(a, p)$. It is easy to see that variance mixtures of type

$$\int \Gamma(a, p)(du) \, \mathcal{N}\left(0, \frac{1}{u}\right) \quad \text{where} \quad 0 < a < 1$$

do not admit finite second moments. $\Gamma(a, p)$ is the first contribution (summand $m = 0$) to $\Gamma(a, \lambda, p)$. Thus the limit law (×) which is best concentrated in the sense of Jeganathan's Convolution Theorem 7.10 and in the sense of the Local Asymptotic Minimax Theorem 7.12 in the transient case $\vartheta > 0$

$$\int \mathcal{L}(\varphi_\vartheta(1))(du) \, \mathcal{N}\left(0, \frac{1}{u}\right) = \int \mathcal{L}\left(Y_\infty^2(\vartheta) \frac{1}{2\vartheta}\right)(du) \, \mathcal{N}\left(0, \frac{1}{u}\right)$$

is of infinite variance, for all choices of a starting point $x \in \mathbb{R}$ for SDE (8.1). Recall in this context Remark 6.6": optimality criteria in mixed normal models are conditional on the observed information, never in terms of moments of the laws of rescaled estimation errors. □

Section 8.1 Ornstein-Uhlenbeck Model

Let us resume the tableau 8.4, 8.5 and 8.6 for convergence of local models when we observe an Ornstein–Uhlenbeck trajectory under unknown parameter over a long time interval: optimality results are available in restriction to submodels where the process either is positive recurrent or is transient; except in the positive recurrent case, the rates of convergence and the limit experiments are different at different values of the unknown parameter.

For practical purposes, one might wish to have limit distributions of homogeneous and easily tractable structure which hold over the full range of parameter values. Depending on the statistical model, one may try either random norming of estimation errors or sequential observation schemes.

8.6'' Exercise (Random norming). In the Ornstein–Uhlenbeck model, the information process $t \to \int_0^t \eta_s^2 ds$ is observable, its definition does not involve the unknown parameter. Thus we may consider random norming for ML estimation errors using the observed information.

(a) Consider first the positive recurrent cases $\vartheta < 0$ in 8.4 and the transient cases $\vartheta > 0$ in 8.6. Using the structure of the limit laws for the pairs

$$\left(M_\vartheta^n(1), \langle M_\vartheta^n \rangle_1 \right) \quad \text{under } Q_\vartheta \text{ as } n \to \infty$$

from (∗)+(∗∗) in 8.4 and (∗) in 8.6 combined with the representation

$$\widehat{\vartheta}_n - \vartheta = \frac{\int_0^n \eta_s \, dm_s^{(\vartheta)}}{\int_0^n \eta_s^2 \, ds}, \quad n \in \mathbb{N}, \quad \text{under } Q_\vartheta$$

of ML estimation errors we can write under Q_ϑ

$$\sqrt{\int_0^n \eta_s^2 \, ds} \left(\widehat{\vartheta}_n - \vartheta \right) = \frac{M_\vartheta^n(1)}{\left(\langle M_\vartheta^n \rangle_1 \right)^{\frac{1}{2}}} \longrightarrow \mathcal{N}(0, 1), \quad n \to \infty$$

to get a unified result which covers the cases $\vartheta \neq 0$.

(b) However, random norming as in the last line is not helpful when $\vartheta = 0$. This has been noted by Feigin [23]. Using the scaling properties in 8.5, we find in the case where $\vartheta = 0$ that the weak limit of the laws

$$\mathcal{L}\left(\sqrt{\int_0^n \eta_s^2 \, ds} \left(\widehat{\vartheta}_n - 0 \right) \mid Q_0 \right) \quad \text{as } n \to \infty$$

is the law $\mathcal{L}(\frac{\frac{1}{2}(B_1^2-1)}{(\int_0^1 B_s^2 \, ds)^{1/2}})$. Feigin notes simply that this law lacks symmetry around 0

$$P \left(\frac{\frac{1}{2}(B_1^2 - 1)}{(\int_0^1 B_s^2 \, ds)^{1/2}} \leq 0 \right) = P \left(B_1^2 \leq 1 \right) = P \left(B_1 \in [-1, 1] \right) \approx 0.68 \neq \frac{1}{2}$$

and thus cannot be a normal law. Hence there is no unified result extending (a) to cover all cases $\vartheta \in \mathbb{R}$. □

In our model, the information process can be calculated from the observation without knowing the unknown parameter, cf. step (3) in Model 8.3. This allows to define a time change and thus a sequential observation scheme by stopping when the observed information hits a prescribed level.

8.7 LAN at all Points $\vartheta \in \Theta$ by Transformation of Time. For the Ornstein–Uhlenbeck Model 8.3 with information process $\int \eta_s^2 ds$, define for every $n \in \mathbb{N}$ a random time change

$$\tau(n,u) := \inf\left\{ t > 0 : \int_0^t \eta_s^2 ds > un \right\}, \quad u \geq 0.$$

Define score martingale and information process scaled and time-changed by $u \to \tau(n,u)$:

$$\widetilde{M}_\vartheta^n(u) := \frac{1}{\sqrt{n}} \int_0^{\tau(n,u)} \eta_s \, dm_s^{(\vartheta)}, \quad \widetilde{\mathbb{G}}^n := \left(\mathcal{G}_{\tau(n,u)} \right)_{u \geq 0},$$

$$\left\langle \widetilde{M}_\vartheta^n \right\rangle_u = \frac{1}{n} \int_0^{\tau(n,u)} \eta_s^2 \, ds = u.$$

Then P. Lévy's characterisation theorem (cf. [61, p. 74]) shows for all $n \in \mathbb{N}$ and for all $\vartheta \in \Theta$ that $\widetilde{M}_\vartheta^n = \left(\widetilde{M}_\vartheta^n(u) \right)_{u \geq 0}$ is a $(\widetilde{\mathbb{G}}^n, Q_\vartheta)$-standard Brownian motion.

(1) If at stage n of the asymptotics we observe a trajectory of the canonical process η up to the random time $\tau(n,1)$, and write $\widetilde{\mathcal{E}}_{n,\vartheta}$ for the local model at ϑ with local scale $1/\sqrt{n}$:

$$\widetilde{\mathcal{E}}_{n,\vartheta} = \left(C, \mathcal{G}_{\tau(n,1)}, \{ Q_{\vartheta + n^{-1/2}h} \mid \mathcal{G}_{\tau(n,1)} : h \in \mathbb{R} \} \right),$$

then log-likelihoods in the local model at ϑ are

$$\widetilde{\Lambda}_{\vartheta,n}^{h/0} := \log L_{\tau(n,1)}^{(\vartheta + n^{-1/2}h)/\vartheta} = h \widetilde{M}_\vartheta^n(1) - \frac{1}{2} h^2 \left\langle \widetilde{M}_\vartheta^n \right\rangle_1 = h \widetilde{M}_\vartheta^n(1) - \frac{1}{2} h^2, \quad h \in \mathbb{R}$$

where we have for all $n \in \mathbb{N}$ and all $\vartheta \in \Theta$

$$\mathcal{L}\left(\widetilde{M}_\vartheta^n(1) \mid Q_\vartheta \right) = \mathcal{N}(0, 1).$$

According to Definition 7.1(c), this is LAN at all parameter values $\vartheta \in \mathbb{R}$, and even more than that: not only for all values of the parameter $\vartheta \in \Theta$ are limit experiments $\widetilde{\mathcal{E}}_\infty(\vartheta)$ given by the same Gaussian shift $\mathcal{E}(1)$, in the notation of Definition 5.2, but also all local experiments $\widetilde{\mathcal{E}}_{n,\vartheta}$ at all levels $n \geq 1$ of the asymptotics coincide with $\mathcal{E}(1)$. Writing

$$\widetilde{\vartheta}_n = \widehat{\vartheta}_{\tau(n,1)} = \frac{\eta_{\tau(n,1)}^2 - \eta_0^2 - \tau(n,1)}{2 \int_0^{\tau(n,1)} \eta_s^2 \, ds} = \frac{\eta_{\tau(n,1)}^2 - \eta_0^2 - \tau(n,1)}{2n}$$

for the maximum likelihood estimator when we observe up to the stopping time $\tau(n, 1)$, cf. step (1) of Model 8.3, we have at all parameter values $\vartheta \in \Theta$ the following properties: the ML sequence is regular and efficient for the unknown parameter at ϑ (Theorems 7.11 and 7.10), and attains the local asymptotic minimax bound at ϑ (Remark 7.13). This sequential observation scheme allows for a unified treatment over the whole range of parameter values $\vartheta \in \mathbb{R}$ (in fact, everything thus reduces to an elementary normal distribution model $\{\mathcal{N}(h, 1) : h \in \mathbb{R}\}$). □

8.7' Exercise. We compare the observation schemes used in 8.7 and in 8.6 in the transient case $\vartheta > 0$. Consider only the starting point $x = 0$ for equation (8.1). Define

(∗) $\qquad \sigma_\vartheta(u) := \inf\{t > 0 : \varphi_\vartheta(t) > u\} = (u \, [\varphi_\vartheta(1)]^{-1})^{\frac{1}{2\vartheta}}$

for $0 < u < \infty$, with $\sigma_\vartheta(0) \equiv 0$, and recall from 8.6 that

$$\varphi_\vartheta(1) = Y_\infty^2(\vartheta) \frac{1}{2\vartheta} \sim \Gamma\left(\frac{1}{2}, 2\vartheta^2\right).$$

(a) Deduce from (∗) that for all parameter values $\vartheta > 0$

$$E\left(|\log \sigma_\vartheta(u)|\right) < \infty.$$

(b) Consider in the case where $\vartheta > 0$ the time change $u \to \tau(m, u)$ in 8.7 and prove that for $0 < u < \infty$ fixed,

$$\tau(m, u) - \frac{1}{2\vartheta} \log(m) \quad \text{under } Q_\vartheta$$

converges weakly as $m \to \infty$ to

$$\log(\sigma_\vartheta(u)).$$

Hint: Observe that the definition of $\langle M_\vartheta^n \rangle_t$ in 8.6 and $\tau(n, u)$ in 8.7 allows us to replace $n \in \mathbb{N}$ by arbitrary $v \in (0, \infty)$. Using this, we can write

$$P(\sigma_\vartheta(u) > t) = P(\varphi_\vartheta(t) < u) = \lim_{n \to \infty} Q_\vartheta\left(\langle M_\vartheta^n \rangle_t < u\right)$$
$$= \lim_{n \to \infty} Q_\vartheta\left(\tau(e^{2\vartheta n}, u) > [n + \log(t)]^+\right)$$

for fixed values of u and t in $(0, \infty)$, where the last limit is equal to

$$\lim_{m \to \infty} Q_\vartheta\left(\tau(m, u) > \frac{1}{2\vartheta} \log(m) + \log(t)\right) = \lim_{m \to \infty} Q_\vartheta\left(e^{\tau(m, u) - \frac{1}{2\vartheta} \log(m)} > t\right). \quad \square$$

8.2 *A Null Recurrent Diffusion Model

We discuss a statistical model where the diffusion process under observation is recurrent null for all values of the parameter. Our presentation follows Höpfner and Kutoyants [54] and makes use of the limit theorems in Höpfner and Löcherbach [53] and of a result by Khasminskii [73].

In this section, we have a probability space $(\Omega, \mathcal{A}, P)$ carrying a Brownian motion W; for some constant $\sigma > 0$ we consider the unique strong solution to equation

(8.8) $$dX_t = \vartheta \frac{X_t}{1 + X_t^2} dt + \sigma \, dW_t, \quad X_0 \equiv x$$

depending on a parameter ϑ which ranges over the parameter space

(8.8') $$\Theta := \left(-\frac{1}{2}\sigma^2, \frac{1}{2}\sigma^2\right).$$

We shall refer frequently to the Appendix (Chapter 9).

8.9 Long Time Behaviour of the Process. Under $\vartheta \in \Theta$ where the parameter space Θ is defined by equation (8.8'), the process X in equation (8.8) is recurrent null in the sense of Harris with invariant measure

(8.9') $$m(dx) = \frac{1}{\sigma^2}\sqrt{1 + y^2}^{-\frac{2\vartheta}{\sigma^2}} dx, \quad x \in \mathbb{R}.$$

We prove this as follows. Fix $\vartheta \in \Theta$, write $b(v) = \vartheta \frac{v}{1+v^2}$ for the drift coefficient in equation (8.8), and consider the mapping $S : \mathbb{R} \to \mathbb{R}$ defined by

$$S(x) := \int_0^x s(y) \, dy \quad \text{where} \quad s(y) := \exp\left(-\int_0^y \frac{2b}{\sigma^2}(v) \, dv\right), \quad x, y \in \mathbb{R}$$

as in Proposition 9.12 in the Appendix; we have $\int_0^y \frac{2b}{\sigma^2}(v) \, dv = \frac{\vartheta}{\sigma^2} \ln(1 + y^2)$ and thus

$$s(y) = (1 + y^2)^{-\frac{\vartheta}{\sigma^2}} = \sqrt{1 + y^2}^{-\frac{2\vartheta}{\sigma^2}},$$

$$S(x) \sim \text{sign}(x) \frac{1}{1 - \frac{2\vartheta}{\sigma^2}} |x|^{1 - \frac{2\vartheta}{\sigma^2}} \quad \text{as } x \to \pm\infty.$$

Since $\left|\frac{2\vartheta}{\sigma^2}\right| < 1$ by equation (8.8'), the function $S(\cdot)$ is a bijection onto $\mathbb{R}$: thus Proposition 9.12 shows that X under ϑ is Harris with invariant measure

$$m(dx) = \frac{1}{\sigma^2}\frac{1}{s(x)} dx \quad \text{on } (\mathbb{R}, \mathcal{B}(\mathbb{R}))$$

which gives equation (8.9'). We have null recurrence since m has infinite total mass. □

We remark that Θ defined by equation (8.8') is the maximal open interval in $\mathbb{R}$ such that null recurrence holds for all parameter values. For the next result, recall from Remark 6.17 the definition of a Mittag–Leffler process $V^{(\alpha)}$ of index $0 < \alpha < 1$: to the stable increasing process $S^{(\alpha)}$ of index $0 < \alpha < 1$, the process with independent and stationary increments having Laplace transforms

$$E\left(e^{-\lambda (S^{(\alpha)}_{t_2} - S^{(\alpha)}_{t_1})}\right) = e^{-(t_2 - t_1)\lambda^\alpha}, \quad \lambda \geq 0, \quad 0 \leq t_1 < t_2 < \infty$$

and starting from $S_0^{(\alpha)} \equiv 0$, we associate the process of level crossing times $V^{(\alpha)}$. Paths of $V^{(\alpha)}$ are continuous and non-decreasing, with $V_0^{(\alpha)} = 0$ and $\lim_{t \to \infty} V_t^{(\alpha)} = \infty$. Part (a) of the next result is a consequence of two results due to Khasminskii [73] which we recall in Proposition 9.14 of the Appendix. Based on (a), part (b) then is a well-known and classical statement, see Feller [24, p. 448] or Bingham, Goldie and Teugels [12, p. 349], on domains of attraction of one-sided stable laws. For regularly varying functions see [12]. X being a one-dimensional process with continuous trajectories, there are many possibilities to define a sequence of renewal times $(R_n)_{n \geq 1}$ which decompose the trajectory of X into i.i.d. excursions $(X \mathbf{1}_{[[R_n, R_{n+1}]]})_{n \geq 1}$ away from 0; a particular choice is considered below.

8.10 Regular Variation. Fix $\vartheta \in \Theta$, consider the function $S(\cdot)$ and the measure $m(dx)$ of 8.9 (both depending on ϑ), and define

(8.10′) $$\alpha = \alpha(\vartheta) = \frac{1}{2}\left(1 - \frac{2\vartheta}{\sigma^2}\right) \in (0, 1).$$

(a) The sequence of renewal times $(R_n)_{n \geq 1}$ defined by

$$R_n := \inf\{t > S_n : X_t < 0\}, \quad S_n := \inf\{t > R_{n-1} : X_t > S^{-1}(1)\},$$
$$n \geq 1, \quad R_0 \equiv 0$$

has the following properties (i) and (ii) under ϑ:

(i) $$P(R_{n+1} - R_n > t) \sim \left(\frac{1}{2\sigma^2}\right)^\alpha \frac{4}{\Gamma(\alpha)} t^{-\alpha} \quad \text{as } t \to \infty,$$

(ii) $$E\left(\int_{R_n}^{R_{n+1}} f(X_s)\, ds\right) = 2m(f), \quad f \in L^1(m).$$

(b) For any norming function $a(\cdot)$ with the property

(8.10″) $$a(t) \sim \frac{1}{\Gamma(1-\alpha)\, P(R_2 - R_1 > t)} \quad \text{as } t \to \infty$$

and for any function $f \in L^1(m)$ we have weak convergence

(8.10‴) $$\frac{1}{a(n)} \int_0^{\cdot n} f(X_s)\, ds \longrightarrow 2m(f) V^{(\alpha)}$$
(weakly in D, under ϑ, as $n \to \infty$)

where $V^{(\alpha)}$ is the Mittag–Leffler process of index α, $0 < \alpha < 1$. In particular, the norming function in equation (8.10″) varies regularly at ∞ with index $\alpha = \alpha(\vartheta)$, and all objects above depend on $\vartheta \in \Theta$.

Proof. (1) From Khasminskii's results [73] which we recall in Proposition 9.14 of the Appendix, we deduce the assertions (a.i) and (a.ii):

Fix $\vartheta \in \Theta$. We have the functions $S(\cdot)$, $s(\cdot)$ defined in 8.9 which depend on ϑ. $S(\cdot)$ is a bijection onto $\mathbb{R}$. According to step (3) in the proof of Proposition 9.12, the process $\widetilde{X} := S(X)$

$$d\widetilde{X}_t = \widetilde{\sigma}(\widetilde{X}_t)\, dW_t \quad \text{where} \quad \widetilde{\sigma} = (s \cdot \sigma) \circ S^{-1}$$

is a diffusion without drift, is Harris recurrent, and has invariant measure

$$\widetilde{m}(d\widetilde{x}) = \frac{1}{\widetilde{\sigma}^2(\widetilde{x})}\, d\widetilde{x} = \frac{1}{\sigma^2\,[s \circ S^{-1}]^2(\widetilde{x})}\, d\widetilde{x} \quad \text{on } (\mathbb{R}, \mathcal{B}(\mathbb{R})).$$

Write for short $\gamma := \frac{2\vartheta}{\sigma^2}$: by choice of the parameter space Θ in equation (8.8'), γ belongs to $(-1, 1)$. From 8.9 we have the following asymptotics:

$$s(y) \sim |y|^{-\gamma}, \quad y \to \pm\infty$$

$$S(x) \sim \text{sign}(x)\frac{1}{1-\gamma}|x|^{1-\gamma}, \quad x \to \pm\infty$$

$$S^{-1}(z) \sim \text{sign}(z)((1-\gamma)|z|)^{\frac{1}{1-\gamma}}, \quad z \to \pm\infty$$

$$[s \circ S^{-1}](v) \sim ((1-\gamma)|v|)^{\frac{-\gamma}{1-\gamma}}, \quad v \to \pm\infty.$$

By the properties of $S(\cdot)$, the sequence $(R_n)_{n \geq 1}$ defined in (a) can be written in the form

(+)
$$R_n := \inf\{t > S_n : \widetilde{X}_t < 0\}, \quad S_n := \inf\{t > R_{n-1} : \widetilde{X}_t > 1\},$$
$$n \geq 1, \quad R_0 \equiv 0$$

with respect to $\widetilde{X} = S(X)$. As a consequence of (+), the stopping times $(R_n)_{n \geq 1}$ induce the particular decomposition of trajectories of $\widetilde{X}$ into i.i.d. excursions $(\widetilde{X}\, 1_{[\![R_n, R_{n+1}]\!]})_{n \geq 1}$ away from 0 to which Proposition 9.14 applies. With respect to $(R_n)_{n \geq 1}$ and $\widetilde{X}$ we have

$$\frac{2}{\widetilde{\sigma}^2(v)} = \frac{2}{\sigma^2\,[s \circ S^{-1}]^2}(v) \sim \frac{2}{\sigma^2}\,((1-\gamma)|v|)^{\frac{2\gamma}{1-\gamma}} = \left[\frac{2}{\sigma^2}(1-\gamma)^{\frac{2\gamma}{1-\gamma}}\right]|v|^{\frac{2\gamma}{1-\gamma}},$$
$$v \to \pm\infty$$

which shows that the condition in Proposition 9.14(b) is satisfied:

$$\lim_{x \to \pm\infty} \frac{1}{x}\int_0^x |v|^{-\beta}\,\frac{2}{\widetilde{\sigma}^2(v)}\, dv = A_{\pm},$$

$$\beta := \frac{2\gamma}{1-\gamma} > -1, \quad A_+ = A_- := \frac{2}{\sigma^2}(1-\gamma)^{\frac{2\gamma}{1-\gamma}}.$$

Section 8.2 *A Null Recurrent Diffusion Model

Recall that γ, β and $A_\pm$ depend on ϑ. With $\alpha = \alpha(\vartheta)$ defined as in equation (8.10')

$$\alpha := \frac{1}{\beta+2} = \frac{1}{2}(1-\gamma) = \frac{1}{2}\left(1 - \frac{2\vartheta}{\sigma^2}\right) \in (0,1)$$

we have $\beta = \frac{1-2\alpha}{\alpha}$ and rewrite the constants $A_\pm$ in terms of $\alpha = \alpha(\vartheta)$ as

$$A_\pm = \frac{2}{\sigma^2}(1-\gamma)^\beta = \frac{2}{\sigma^2}(2\alpha)^{\frac{1-2\alpha}{\alpha}}.$$

From this, Proposition 9.14(b) gives us

$$P(R_{n+1} - R_n > t) \sim \frac{\alpha^{2\alpha}([A_+]^\alpha + [A_-]^\alpha)}{\Gamma(1+\alpha)} t^{-\alpha} = \left(\frac{1}{2\sigma^2}\right)^\alpha \frac{4}{\Gamma(\alpha)} t^{-\alpha}$$

as $t \to \infty$. This proves part (a.i) of the assertion. Next, we apply Proposition 9.14(a) to the process $\widetilde{X}$: the i.i.d. excursions defined by $(R_n)_{n\geq 1}$ correspond to the norming constant

$$E\left(\int_{R_n}^{R_{n+1}} \widetilde{f}(\widetilde{X}_s)\,ds\right) = 2\,\widetilde{m}(\widetilde{f}), \quad \widetilde{f} \geq 0 \text{ measurable}$$

for the invariant measure $\widetilde{m}$. If we write $\widetilde{f} := f \circ S^{-1}$ and apply formula (∞) in the proof of Proposition 9.12

$$m(f) = \widetilde{m}\left(f \circ S^{-1}\right)$$

we obtain

$$E\left(\int_{R_n}^{R_{n+1}} f(X_s)\,ds\right) = 2\,m(f), \quad f \geq 0 \text{ measurable}$$

which proves part (a.ii) of the assertion. Thus, 8.10(a) is now proved.

(2) We prove 8.10(b). Under $\vartheta \in \Theta$, for α given by equation (8.10') and for the renewal times $(R_n)_{n\geq 1}$ considered in (a), select a strictly increasing continuous norming function $a(\cdot)$ such that

$$a(t) \sim \frac{1}{\Gamma(1-\alpha)\,P(R_2 - R_1 > t)} \quad \text{as } t \to \infty.$$

In particular, $a(\cdot)$ varies regularly at ∞ with index α. Fix an asymptotic inverse $b(\cdot)$ to $a(\cdot)$ which is strictly increasing and continuous. All this depends on ϑ, and (a.i) above implies

(8.11) $$b(t) \sim \frac{1}{2\sigma^2}\left[\frac{4\Gamma(1-\alpha)}{\Gamma(\alpha)}\right]^{\frac{1}{\alpha}} t^{\frac{1}{\alpha}} \quad \text{as } t \to \infty.$$

From $a(b(n)) \sim n$ as $n \to \infty$, a well-known result on convergence of sums of i.i.d. variables to one-sided stable laws (cf. [24, p. 448], [12, p. 349]) gives weak convergence

$$\frac{R_n}{b(n)} \longrightarrow S_1^{(\alpha)} \quad \text{(weakly in } \mathbb{R}, \text{ under } \vartheta, \text{ as } n \to \infty\text{)}$$

where $S_1^{(\alpha)}$ follows the one-sided stable law on $(0, \infty)$ with Laplace transform $\lambda \to \exp(-\lambda^\alpha)$, $\lambda \geq 0$. Regular variation of $b(\cdot)$ at ∞ with index $\frac{1}{\alpha}$ implies that $b(tn) \sim t^{\frac{1}{\alpha}} b(n)$ as $n \to \infty$. Thus, scaling properties and independence of increments in the one-sided stable process S^α of index $0 < \alpha < 1$ show that the last convergence extends to finite-dimensional convergence

(8.11') $$\frac{R_{[\cdot n]}}{b(n)} \xrightarrow{f.d.} S^{(\alpha)} \quad \text{as } n \to \infty.$$

Associate a counting process

$$N = (N_t)_{t \geq 0}, \quad N_t := \max\{j \in \mathbb{N} : R_j \leq t\},$$

to the renewal times $(R_n)_n$ and write for arbitrary $0 < t_1 < \cdots < t_m < \infty$, $A_i \in \mathcal{B}(\mathbb{R})$ and $x_i > 0$

$$P\left(\frac{N_{t_i b(n)}}{n} < \frac{[x_i n]}{n}, 1 \leq i \leq m\right) = P\left(\frac{R_{[x_i n]}}{b(n)} > t_i, 1 \leq i \leq m\right).$$

Then (8.11') gives finite dimensional convergence under ϑ as $n \to \infty$

$$\frac{N_{\cdot b(n)}}{n} \xrightarrow{f.d.} V^{(\alpha)}$$

to the process inverse $V^{(\alpha)}$ of $S^{(\alpha)}$. On the left-hand side of the last convergence, we may replace n by $a(n)$ and make use of $b(a(n)) \sim n$ as $n \to \infty$. Then the last convergence takes the form

(8.11'') $$\frac{N_{\cdot n}}{a(n)} \xrightarrow{f.d.} V^{(\alpha)} \quad \text{as } n \to \infty.$$

By (a.ii) we have for functions $f \geq 0$ belonging to $L^1(m)$

$$E\left(\int_{R_n}^{R_{n+1}} f(X_s)\, ds\right) = 2m(f).$$

The counting process N increasing by 1 on $]]R_n, R_{n+1}]]$, we have almost sure convergence under ϑ

$$\lim_{t \to \infty} \frac{1}{N_t} \int_0^t f(X_s)\, ds = \lim_{n \to \infty} \frac{1}{n} \int_0^{R_n} f(X_s)\, ds = 2m(f)$$

from the classical strong law of large numbers with respect to the i.i.d. excursions $X\,1_{[[R_j, R_{j+1}]]}$, $j \geq 1$. Together with (8.11'') we arrive at

$$\frac{1}{a(n)} \int_0^{\cdot n} f(X_s)\, ds \xrightarrow{f.d.} 2m(f)\, V^{(\alpha)} \quad \text{as } n \to \infty$$

Section 8.2 *A Null Recurrent Diffusion Model

for functions $f \geq 0$ belonging to $L^1(m)$. All processes in the last convergence are increasing processes, and the limit process is continuous. In this case, according to Jacod and Shiryaev (1987, VI.3.37), finite dimensional convergence and weak convergence in D are equivalent, and part (b) of 8.10 is proved. □

We turn to statistical models defined by observation of a trajectory of the process (8.8) under unknown $\vartheta \in \Theta$ over a long time interval, with parameter space Θ defined by equation (8.8').

8.12 Statistical Model. For Θ given by equation (8.8') and for some starting point $x_0 \in \mathbb{R}$ which does not depend on $\vartheta \in \Theta$, let Q_ϑ denote the law of the solution to (8.8) under ϑ, on the canonical path space $(C, \mathcal{C}, \mathbb{G})$ or $(D, \mathcal{D}, \mathbb{G})$. Applying Theorem 6.10, all laws Q_ϑ are locally equivalent relative to $\mathbb{G}$, and the density process of Q_ϑ with respect to Q_0 relative to $\mathbb{G}$ is

$$L_t^{\vartheta/0} = \exp\left\{ \vartheta \int_0^t \gamma(\eta_s)\, dm_s^{(0)} - \frac{1}{2}\vartheta^2 \int_0^t \gamma^2(\eta_s)\, \sigma^2 ds \right\}, \quad t \geq 0$$

with $m^{(0)}$ the Q_0-local martingale part of the canonical process η under Q_0, and

(8.12') $$\gamma(x) := \frac{1}{\sigma^2} \frac{x}{1+x^2}, \quad x \in \mathbb{R}.$$

(1) Fix a determination $(t, \omega) \to Y(t, \omega)$ of the stochastic integral

$$\int \gamma(\eta_s)\, d\eta_s = \int \gamma(\eta_s)\, dm_s^{(0)} \quad \text{under } Q_0$$

where $\mathcal{L}(\eta - \eta_0 | Q_0) = \mathcal{L}(m^{(0)} | Q_0) = \mathcal{L}(\sigma B)$. Thus, in statistical experiments $\mathcal{E}_t$ corresponding to observation of the canonical process η up to time $0 < t < \infty$, both likelihood function

$$\Theta \ni \vartheta \longrightarrow L_t^{\vartheta/0} = \exp\left\{ \vartheta Y_t - \frac{1}{2}\vartheta^2 \int_0^t \gamma^2(\eta_s)\, \sigma^2 ds \right\} \in (0, \infty)$$

and maximum likelihood (ML) estimator

$$\widehat{\vartheta}_t := \frac{Y_t}{\int_0^t \gamma^2(\eta_s)\, \sigma^2 ds}$$

are expressed without reference to a particular probability measure.

(2) Simultaneously for all $\vartheta \in \Theta$, $(t, \omega) \to Y(t, \omega)$ provides a common determination for

$$\left(\int_0^t \gamma(\eta_s)\, d\eta_s \right)_{t \geq 0} = \left(\int_0^t \gamma(\eta_s)\, dm_s^{(\vartheta)} + \vartheta \int_0^t \gamma^2(\eta_s)\, \sigma^2 ds \right)_{t \geq 0} \quad \text{under } Q_\vartheta :$$

any determination $(t, \omega) \to Y(t, \omega)$ of the stochastic integral $\int \gamma(\eta_s) \, d\eta_s$ under Q_0 is also a determination of the stochastic integral $\int \gamma(\eta_s) \, d\eta_s$ under Q_ϑ, by Lemma 8.2', and the Q_ϑ-martingale part of η equals $m_t^{(\vartheta)} = \eta_t - \eta_0 - \vartheta \int_0^t \gamma(\eta_s) \sigma^2 ds$. This allows us to write ML estimation errors under $\vartheta \in \Theta$ in the form

$$\hat{\vartheta}_t - \vartheta = \frac{\int_0^t \gamma(\eta_s) \, dm_s^{(\vartheta)}}{\left\langle \int \gamma(\eta_s) \, dm_s^{(\vartheta)} \right\rangle_t} \quad \text{under } Q_\vartheta$$

and allows us to write density processes $L^{\xi/\vartheta}$ of Q_ξ with respect to Q_ϑ relative to $\mathbb{G}$ either as

$$L_t^{\xi/0} / L_t^{\vartheta/0} = \exp\left\{ (\xi - \vartheta) Y_t - \frac{1}{2} (\xi^2 - \vartheta^2) \int_0^t \gamma^2(\eta_s) \sigma^2 ds \right\}, \quad t \geq 0$$

or equivalently – coinciding with Theorem 6.10 applied to Q_ξ and Q_ϑ – as

$$(+) \quad L_t^{\xi/\vartheta} = \exp\left\{ (\xi - \vartheta) \int_0^t \gamma(\eta_s) \, dm_s^{(\vartheta)} - \frac{1}{2} (\xi - \vartheta)^2 \int_0^t \gamma^2(\eta_s) \sigma^2 ds \right\}.$$

(3) The representation (+) allows to reparameterise the model $\mathcal{E}_t$ with respect to fixed reference points ϑ such that the model around ϑ is quadratic in $(\xi - \vartheta)$. We call

$$\left(\int_0^t \gamma(\eta_s) \, dm_s^{(\vartheta)} \right)_{t \geq 0}$$

and

$$\left(\int_0^t \gamma^2(\eta_s) \sigma^2 ds \right)_{t \geq 0} = \left\langle \int \gamma(\eta) \, dm^{(\vartheta)} \right\rangle \quad \text{under } Q_\vartheta$$

score martingale at ϑ and *information process at* ϑ. It is clear from (8.10''') that the law of the observed information depends on ϑ. Hence reparameterising as in (+) with respect to a reference point ϑ or with respect to $\vartheta' \neq \vartheta$ yields statistical models which are different. □

From now on we shall keep trace of the parameter ϑ in some more notations: instead of m as in (8.9') we write

$$\mu^\vartheta(dx) = \frac{1}{\sigma^2} \sqrt{1 + y^2}^{\frac{2\vartheta}{\sigma^2}} dx, \quad x \in \mathbb{R}$$

for the invariant measure of the canonical process η under Q_ϑ; instead of $a(\cdot)$ we write $a_\vartheta(\cdot)$ for the norming function in (8.10'') which is regularly varying at ∞ with index

$$\alpha(\vartheta) = \frac{1}{2}\left(1 - \frac{2\vartheta}{\sigma^2}\right) \in (0, 1)$$

Section 8.2 *A Null Recurrent Diffusion Model

as in (8.10′). By choice of Θ in (8.8′), there is a one-to-one correspondence between indices $0 < \alpha < 1$ and parameter values $\vartheta \in \Theta$. Given the representation of log-likelihoods in Model 8.12, the following proposition allows to prove convergence of local models at ϑ, or convergence of rescaled estimation errors at ϑ for a broad class of estimators. It is obtained as a special case of Theorems 9.8 and 9.10 and of Corollary 9.10' (from Höpfner and Löcherbach [53]) in the Appendix.

8.13 Proposition. For functions $f \in L^2(\mu^\vartheta)$ and with respect to $\mathbb{G}^n = (\mathcal{G}_{tn})_{t \geq 0}$, consider locally square integrable local Q_ϑ-martingales

$$M^n = (M^n_t)_{t \geq 0}, \quad M^n_t := \frac{1}{\sqrt{n^{\alpha(\vartheta)}}} \int_0^{tn} f(\eta_s)\, dm_s^{(\vartheta)}$$

with $m^{(\vartheta)}$ the Q_ϑ-martingale part of the canonical process η. Under Q_ϑ, we have weak convergence

$$\left(M^n, \langle M^n \rangle \right) \longrightarrow \left((\Lambda(\vartheta))^{1/2}\, B \circ V^{(\alpha(\vartheta))},\; \Lambda(\vartheta)\, V^{(\alpha(\vartheta))} \right)$$

in $D(\mathbb{R} \times \mathbb{R})$ as $n \to \infty$: in this limit, Brownian motion B and Mittag–Leffler process $V^{(\alpha(\vartheta))}$ of index $0 < \alpha(\vartheta) < 1$ are independent, and the constant is

(8.13′) $$\Lambda(\vartheta) := \left(2\sigma^2\right)^{1+\alpha(\vartheta)} \frac{\Gamma(\alpha(\vartheta))}{4\,\Gamma(1-\alpha(\vartheta))}\, \mu^\vartheta(f^2).$$

Proof. (1) Fix $\vartheta \in \Theta$. For functions $f \geq 0$ in $L^1(\mu^\vartheta)$, we have from 8.10(b) weak convergence in D, under Q_ϑ, of integrable additive functionals:

$$\frac{1}{a_\vartheta(n)} \int_0^n f(\eta_s)\, ds \longrightarrow 2\,\mu^\vartheta(f)\, V^{\alpha(\vartheta)}, \quad n \to \infty.$$

We rephrase this as weak convergence on D, under Q_ϑ, of

$$\frac{1}{n^{\alpha(\vartheta)}} \int_0^n f(\eta_s)\, ds \longrightarrow C(\vartheta)\, \mu^\vartheta(f)\, V^{\alpha(\vartheta)}, \quad n \to \infty$$

for a constant $C(\vartheta)$ which according to 8.10 is given by

(8.13″) $$C(\vartheta) := 2 \cdot \left[\left(\frac{1}{2\sigma^2}\right)^{\alpha(\vartheta)} \frac{4\,\Gamma(1-\alpha(\vartheta))}{\Gamma(\alpha(\vartheta))} \right]^{-1}.$$

(2) Step (1) allows to apply Theorem 9.8(a) from the Appendix: we have the regular variation condition (a.i) there with $\alpha = \alpha(\vartheta)$, $m = \mu^{(\vartheta)}$ and $\ell(\cdot) \equiv 1/C(\vartheta)$ on the right-hand side.

(3) With the same notations, we apply Theorem 9.10 from the Appendix: for locally square integrable local martingales $\widetilde{M}$ satisfying the conditions made there, this gives weak convergence of

$$\frac{1}{\sqrt{n^{\alpha(\vartheta)} C(\vartheta)}} \left(\widetilde{M}_{tn}\right)_{t \geq 0} = \frac{1}{\sqrt{n^{\alpha(\vartheta)}/\ell(n)}} \left(\widetilde{M}_{tn}\right)_{t \geq 0} \quad \text{under } Q_\vartheta$$

in D as $n \to \infty$ to
$$(\widetilde{\Lambda}(\vartheta))^{1/2} B \circ V^{(\alpha(\vartheta))}$$
where Brownian motion B and Mittag–Leffler process $V^{(\alpha(\vartheta))}$ are independent, with constant

(8.13''') $$\widetilde{\Lambda}(\vartheta) := E_{\mu^\vartheta}(\langle \widetilde{M} \rangle_1).$$

(4) We put together steps (1)–(3) above. For $g \in L^2(\mu^\vartheta)$ consider the local $(Q_\vartheta, \mathbb{G})$-martingale M
$$M_t := \int_0^t g(\eta_s)\, dm_s^{(\vartheta)}, \quad t \geq 0$$
where $m^{(\vartheta)}$ is the Q_ϑ-local martingale part of the canonical process η. Then M is a martingale additive functional as defined in Definition 9.9 and satisfies
$$E_{\mu^\vartheta}(\langle M \rangle_1) = E_{\mu^\vartheta}\left(\int_0^1 g^2(\eta_s)\sigma^2\, ds\right) = \sigma^2 \mu^\vartheta(g^2) < \infty.$$

Thus, as a consequence of step (3),
$$M^n := \frac{1}{\sqrt{n^{\alpha(\vartheta)}}}(M_{tn})_{t \geq 0}$$
under Q_ϑ converges weakly in D as $n \to \infty$ to
$$(\Lambda(\vartheta))^{1/2} B \circ V^{(\alpha(\vartheta))}$$
where combining (8.13'') and (8.13''') we have
$$\Lambda(\vartheta) = C(\vartheta)\widetilde{\Lambda}(\vartheta) = C(\vartheta)\sigma^2 \mu^\vartheta(g^2) = (2\sigma^2)^{1+\alpha(\vartheta)} \frac{\Gamma(\alpha(\vartheta))}{4\,\Gamma(1-\alpha(\vartheta))} \mu^\vartheta(g^2).$$

This constant $\Lambda(\vartheta)$ appears in (8.13').

(5) The martingales in step (4) are continuous. Thus Corollary 9.10' in the Appendix extends the result of step (4) to weak convergence of martingales together with their angle brackets, and concludes the proof of Proposition 8.13. □

With Model 8.12 and Proposition 8.13 we have all elements which we need to prove LAMN at arbitrary reference points $\vartheta \in \Theta$. Recall that the parameter space Θ is defined by (8.8'), and that the starting point for equation (8.8) is fixed and does not depend on ϑ. Combining the representation of likelihoods with respect to ϑ in formula (+) of Model 8.12 with Proposition 8.13 we get the following:

8.14 LAMN at Every Reference Point $\vartheta \in \Theta$. Consider the statistical model $\mathcal{E}_n$ which corresponds to observation of a trajectory of the solution to equation (8.8) continuously over the time interval $[0, n]$, under unknown $\vartheta \in \Theta$.

Section 8.2 *A Null Recurrent Diffusion Model

(a) For every $\vartheta \in \Theta$, we have LAMN at ϑ with local scale

$$\delta(n) := \frac{1}{\sqrt{n^{\alpha(\vartheta)}}} \quad \text{for } \alpha(\vartheta) \in (0,1) \text{ defined by (8.10')}:$$

for every n, log-likelihoods in the local model $\mathcal{E}_{\vartheta,n}$ are quadratic in the local parameter

$$\Lambda_{\vartheta,n}^{h/0} = \Lambda_n^{(\vartheta + \delta_n(\vartheta)h)/\vartheta} = h\, S_n(\vartheta) - \frac{1}{2} h^2\, J_n(\vartheta), \quad h \in \mathbb{R}$$

and we have weak convergence as $n \to \infty$ of score and information at ϑ

$$(S_n(\vartheta), J_n(\vartheta) \mid Q_\vartheta) \longrightarrow \left(\Lambda^{1/2}(\vartheta)\, B\left(V_1^{(\alpha(\vartheta))}\right),\, \Lambda(\vartheta)\, V_1^{(\alpha(\vartheta))} \right) =: (S, J)$$

with $\Lambda(\vartheta)$ given by (8.13'), and with Mittag–Leffler process $V^{\alpha(\vartheta)}$ independent from B. For the limit experiment $\mathcal{E}(S,J)$ at ϑ see Construction 6.16 and Example 6.18.

(b) For every $\vartheta \in \Theta$, rescaled ML estimation errors at ϑ coincide with the central sequence at ϑ. Hence ML estimators are regular and efficient in the sense of Jeganathan's version 7.10(b) of the convolution theorem, and attain the local asymptotic minimax bound of Theorem 7.12.

Proof. (1) Localising around a fixed reference point $\vartheta \in \mathbb{R}$, write Q_ξ^n for Q_ξ restricted to $\mathcal{G}_n$. With

$$\delta_n(\vartheta) := \frac{1}{\sqrt{n^{\alpha(\vartheta)}}} \quad \text{with} \quad \alpha(\vartheta) = \frac{1}{2}\left(1 - \frac{2\vartheta}{\sigma^2}\right) \in (0,1)$$

according to 8.10, Model 8.12 and Proposition 8.13 we consider local models at ϑ

$$\mathcal{E}_{\vartheta,n} = \left(C, \mathcal{G}_n, \left\{ Q_{\vartheta + \delta_n(\vartheta)h}^n : h \in \mathbb{R} \right\} \right), \quad n \geq 1$$

when n tends to ∞. According to (+) in step (2) of Model 8.12, log-likelihoods in $\mathcal{E}_{\vartheta,n}$ are

$$(\diamond) \quad \Lambda_{\vartheta,n}^{h/0} = h \left[\delta_n(\vartheta) \int_0^n \gamma(\eta_s)\, dm_s^\vartheta \right] - \frac{1}{2} h^2 \left[\delta_n^2(\vartheta) \int_0^n \gamma^2(\eta_s)\, \sigma^2 ds \right], \quad h \in \mathbb{R}.$$

Now we can apply Proposition 8.13: note that for every $\vartheta \in \Theta$, $\gamma(\cdot)$ belongs to $L^2(\mu^\vartheta)$ since

$$\gamma^2(x) = O\left(x^{-2}\right) \text{ as } |x| \to \infty, \quad d\mu^\vartheta(x) = O\left(|x|^{\frac{2\vartheta}{\sigma^2}}\right) dx \text{ as } |x| \to \infty$$

where $\left|\frac{2\vartheta}{\sigma^2}\right| < 1$. Proposition 8.13 yields weak convergence as $n \to \infty$ of pairs

$$(\diamond\diamond) \quad (S_n(\vartheta), J_n(\vartheta)) := \left(\frac{1}{\sqrt{n^{\alpha(\vartheta)}}} \int_0^n \gamma(\eta_s)\, dm_s^\vartheta,\, \frac{1}{n^{\alpha(\vartheta)}} \int_0^n \gamma^2(\eta_s)\, \sigma^2 ds \right)$$

under Q_ϑ

to the pair of limiting random variables

(8.14′) $\quad (S(\vartheta), J(\vartheta)) = \left(\Lambda^{1/2}(\vartheta) B(V_1^{(\alpha(\vartheta))}), \Lambda(\vartheta) V_1^{(\alpha(\vartheta))} \right)$

where Brownian motion B and Mittag–Leffler process $V^{\alpha(\vartheta)}$ are independent, and

$$\Lambda(\vartheta) = (2\sigma^2)^{1+\alpha(\vartheta)} \frac{\Gamma(\alpha(\vartheta))}{4\Gamma(1-\alpha(\vartheta))} \mu^\vartheta(\gamma^2)$$

according to (8.13′), with $\gamma(\cdot)$ from (8.12′). We have proved in Construction 6.16 and in Example 6.18 that the pair of random variables (8.14′) indeed generates a mixed normal experiment.

(2) Combining this with step (2) of Model 8.12, we see that rescaled ML estimation errors at ϑ coincide with the central sequence at ϑ:

(◇ ◇ ◇) $\quad \sqrt{n^{\alpha(\vartheta)}} (\widehat{\vartheta}_n - \vartheta) = J_n^{-1}(\vartheta) S_n(\vartheta) = Z_n(\vartheta), \quad n \geq 1$.

In particular, the coupling condition of Theorem 7.11 holds. By LAMN at ϑ, by the Convolution Theorem 7.10 and the Local Asymptotic Minimax Theorem 7.12, we see that the ML estimator sequence is regular and efficient at ϑ, and attains the local asymptotic minimax bound as in Theorem 7.12(b). □

8.15 Remark. According to (8.14′) and to Remark 6.6', the limit law for (◇ ◇ ◇) as $n \to \infty$ has the form

$$\mathcal{L}(Z(\vartheta)|P_0) = \int \mathcal{L}(\Lambda(\vartheta) V_1^{\alpha(\vartheta)})(du) \mathcal{N}\left(0, \frac{1}{u}\right).$$

As mentioned in Example 6.18, this law – the best concentrated limit distribution for rescaled estimation errors at ϑ, in the sense of the Convolution Theorem 6.6 or of the Local Asymptotic Minimax Theorem 6.8 – does not admit finite second moments (see Exercise 6.18′). □

8.16 Remark. Our model presents different speeds of convergence at different parameter values, different limit experiments at different parameter values, but has an information process

$$\int_0^t \gamma^2(\eta_s) \sigma^2 ds = \frac{1}{\sigma^2} \int_0^t \left[\frac{\eta_s}{1+\eta_s^2}\right]^2 ds, \quad t \geq 0$$

which can be calculated from the observation $(\eta_t)_{t \geq 0}$ without knowing the unknown parameter $\vartheta \in \Theta$. This fact allows for random norming using the observed information: directly from LAMN at ϑ in 8.14, combining (◇), (◇◇) and (◇ ◇ ◇) there, we can write

$$\sqrt{\int_0^n \left[\frac{\eta_s}{1+\eta_s^2}\right]^2 ds} \, (\widehat{\vartheta}_n - \vartheta) \quad \longrightarrow \quad \mathcal{N}(0, \sigma^2)$$

Section 8.2 *A Null Recurrent Diffusion Model

for all values of $\vartheta \in \Theta$. The last representation allows for practical work, e.g. to fix confidence intervals for the unknown parameter determined from an asymptotically efficient estimator sequence, and overcomes the handicap caused by non-finiteness of second moments in Remark 8.15. □

We conclude the discussion of the statistical model defined by (8.8) and (8.8') by pointing out that one-step correction is possible, and that we may start from any preliminary estimator sequence $(T_n)_n$ for the unknown parameter in $(\mathcal{E}_n)_n$ whose estimation errors at ϑ are tight at rate $\sqrt{n^{\alpha(\vartheta)}}$ as $n \to \infty$ for every $\vartheta \in \Theta$, and modify $(T_n)_n$ according to Theorem 7.19 in order to obtain a sequence of estimators $(\widetilde{T}_n)_n$ which satisfies the coupling condition of Theorem 7.11

$$\sqrt{n^{\alpha(\vartheta)}}\,(\widetilde{T}_n - \vartheta) = Z_n(\vartheta) + o_{(Q_\vartheta)}(1) \quad \text{as } n \to \infty$$

at every $\vartheta \in \Theta$. Again there is no need for discretisation as in Proposition 7.16 since local scale, observed information and score depend continuously on ϑ. To do this, it is sufficient to check the set of Conditions 7.14 for our model.

8.17 Proposition. With the above notations, the sequence of models $(\mathcal{E}_n)_n$ satisfies all assumptions stated in 7.14. For $(T_n)_n$ as above, one-step modification

$$\widetilde{T}_n := T_n + \frac{1}{\sqrt{n^{\alpha(T_n)}}} \frac{1}{J_n(T_n)} S_n(T_n) = \widehat{\vartheta}_n$$

directly leads to the ML estimator $\widehat{\vartheta}_n$.

Proof. Let Y denote a common determination – as in step (2) of Model 8.12, by Lemma 8.2' – of the stochastic integral $\int \gamma(\eta_s)\,d\eta_s$ under all laws Q_ξ, $\xi \in \Theta$. We check the set of conditions in 7.14.

First, it is obvious that the model allows for a broad collection of preliminary estimator sequences. As an example, select $A \in \mathcal{B}(\mathbb{R})$ with $\lambda(A) > 0$, write $\check{\gamma}(x) := \gamma(x)1_A(x)$, choose a common determination $\check{Y}$ for the stochastic integral $\int \check{\gamma}(\eta_s)\,d\eta_s$ under all laws Q_ξ, $\xi \in \Theta$, and put

$$\check{T}_n := \frac{\check{Y}_n}{\int_0^n \check{\gamma}^2(\eta_s)\,\sigma^2 ds}, \quad n \geq 1.$$

Then Proposition 8.13 establishes weak convergence of $\mathcal{L}(\sqrt{n^{\alpha(\vartheta)}}(\check{T}_n - \vartheta) \mid Q_\vartheta)$ as $n \to \infty$ under all values of the parameter $\vartheta \in \Theta$. Even if such estimators can be arbitrarily bad, depending on the choice of A, they converge at the right speed at all points of the model: this establishes Assumption 7.14(D).

In the local model $\mathcal{E}_{\vartheta,n}$ at ϑ

$$S_n(\vartheta) = \frac{1}{\sqrt{n\alpha(\vartheta)}} \int_0^n \gamma(\eta_s) \, dm_s^{(\vartheta)} = \frac{1}{\sqrt{n\alpha(\vartheta)}} \left[Y_n - \vartheta \int_0^n \gamma^2(\eta_s) \sigma^2 ds \right]$$

is a version of the score, according to 8.14 and (+) in Model 8.12, and the information

$$J_n(\vartheta) = \frac{1}{n\alpha(\vartheta)} \int_0^n \gamma^2(\eta_s) \sigma^2 ds$$

depends on ϑ only through local scale. In particular,

(+) $\qquad J_n(\xi) - J_n(\vartheta) = \left(\dfrac{n\alpha(\vartheta)}{n\alpha(\xi)} - 1 \right) J_n(\vartheta) ;$

writing down $S_n(\xi)$ and transforming $\int \gamma(\eta_s) \, dm^{(\xi)}$ into $\int \gamma(\eta_s) \, dm^{(\vartheta)}$ plus correction term we get

(++) $\qquad S_n(\xi) - S_n(\vartheta) = \left(\dfrac{\sqrt{n\alpha(\vartheta)}}{\sqrt{n\alpha(\xi)}} - 1 \right) S_n(\vartheta) - \dfrac{\sqrt{n\alpha(\vartheta)}}{\sqrt{n\alpha(\xi)}} \left[\sqrt{n\alpha(\vartheta)}(\xi - \vartheta) \right] J_n(\vartheta) .$

From (+) and (++) it is obvious that parts (B) and (C) of Assumptions 7.14 simultaneously will follow from

(*) $\qquad \sup\limits_{|h| \le c} \left| \dfrac{n\alpha(\vartheta + h/\sqrt{n\alpha(\vartheta)})}{n\alpha(\vartheta)} - 1 \right| \to 0 \quad \text{as } n \to \infty$

for arbitrary values of a constant $0 < c < \infty$. By (8.10'), $\alpha(\xi)$ differs from $\alpha(\vartheta)$ by $[-\frac{1}{\sigma^2}](\xi - \vartheta)$. Hence, to check (*) it is sufficient to cancel out $n\alpha(\vartheta)$ in the ratio in (*) and then take logarithms (note that this argument exploits again $0 < \alpha(\vartheta) < 1$ for every $\vartheta \in \Theta$). Now parts (B) and (C) of Assumptions 7.14 are established. This finishes the proof, and we write down the one-step modification

$$\widetilde{T}_n := T_n + \frac{1}{\sqrt{n\alpha(T_n)}} \frac{1}{J_n(T_n)} S_n(T_n) = T_n + \frac{Y_n - T_n \int_0^n \gamma^2(\eta_s) \sigma^2 ds}{\int_0^n \gamma^2(\eta_s) \sigma^2 ds} = \widehat{\vartheta}_n$$

according to (a simplified version of) Theorem 7.19. □

8.3 *Some Further Remarks

We point out several (out of many more) references on LAN, LAMN, LAQ in stochastic process models of different types. Some of these papers prove LAN or LAMN or LAQ in a particular stochastic process model, others establish properties of estimators

Section 8.3 *Some Further Remarks

at ϑ which indicate that the underlying statistical model should be LAMN at ϑ or LAQ at ϑ.

Cox–Ingersoll–Ross process models are treated in Overbeck [103], Overbeck and Ryden [104] and Ben Alaya and Kebaier [8]. For ergodic diffusions with state space $\mathbb{R}$

$$dX_t = b(\vartheta, X_t)\, dt + \sigma(X_t)\, dW_t, \quad t \geq 0$$

where the drift depends on a d-dimensional parameter $\vartheta \in \Theta$, the books by Kutoyants [78, 80] present a large variety of interesting models and examples.

For LAN in parametric models for non-time-homogeneous diffusions – where it is assumed that the drift is periodic in time and that in a suitable sense 'periodic ergodicity' holds – see Höpfner and Kutoyants [55]. An intricate tableau of LAN, LAMN or LAQ properties arising in a setting of delay equations can be found in Gushchin and Küchler [39].

Markov step process models with transition intensities depending on an unknown parameter $\vartheta \in \Theta$ are considered in Höpfner, Jacod and Ladelli [51] and in Höpfner [46, 48, 49]. For general semi-martingale models with jump parts and with diffusive parts see Sørensen [119] and Luschgy [95–97]. For branching diffusions – some finite random number of particles diffusing in space, branching with random number of offspring at rates which depend on the position and on the whole configuration, whereas immigrants are arriving according to Poisson random measure – LAN or LAMN has been proved by Löcherbach [90, 91]; ergodicity properties and invariant measures for such processes have been investigated by Hammer [44].

Diffusion process models where the underlying process is not observed continuously in time but at discrete time steps $t_i = i\Delta$ with $0 \leq i \leq \frac{T}{\Delta}$ (either with T fixed or with T increasing to ∞), and where asymptotics are in Δ tending to 0, have received a lot of interest: see e.g. Yoshida [129], Genon-Catalot and Jacod [25], Kessler [70], or Shimizu and Yoshida [118]. Proofs establishing LAMN (in the case where T is fixed) or LAN (in the case where $T \uparrow \infty$ provided the process is ergodic) have been given by Dohnal [22] and by Gobet [30, 31]. The book [71] edited by Kessler, Lindner and Sørensen gives a broad recent overview on discretely observed semi-martingale models.

If the underlying process is a point process, e.g. Poisson random measure $\mu(dx)$ on some space $(E, \mathcal{E})$ with deterministic intensity $\nu_\vartheta(dx)$ depending on some unknown parameter $\vartheta \in \Theta$, one may consider sequences of windows $K_n \subset E$ which increase to E as $n \to \infty$, and identify the experiment $\mathcal{E}_n$ at stage n of the asymptotics with observation of the point process in restriction to the window K_n, cf. Kutoyants [79] or Höpfner and Jacod [50].

If our Chapters 7 and 8 deal with local asymptotics of type LAN, LAMN or LAQ, let us mention one example for a limit experiment of different type, inducing statistical properties via local asymptotics which are radically different from what we have discussed above. Ibragimov and Khasminskii [60] put forward a limit experiment where the likelihood ratios are

(IK) $$L^{h/0} = e^{\widetilde{W}_u - \frac{1}{2}|u|}, \quad h \in \mathbb{R}$$

with two-sided Brownian motion $(\widetilde{W}_u)_{u \in \mathbb{R}}$, and studied convergence to this limit experiment in 'signal in white noise' models. Obviously (IK) is not a quadratic experiment since the parameter plays the role of time: in this sense, the experiment (IK) is linked to the Gaussian shift limit experiment as we pointed out in Example 1.16'. Dachian [17] associated to the limit experiment (IK) approximating experiments where likelihood ratios have – separately on the positive and on the negative branch – the form of the trajectory of a particular Poisson process with suitable linear terms subtracted. In the limit model (IK), Rubin and Song [114] could calculate the risk of the Bayesian u^* – for quadratic loss, and with 'uniform prior over the real line' – and could show that quadratic risk of u^* is by some factor smaller than quadratic risk of the maximum likelihood estimator $\widehat{u}$. Recent investigations by Dachian show quantitatively how this feature carries over to the approximating experiments which he considers: there is a large domain of parameter values in the approximating experiments where rescaled estimation errors of an analogously defined u_n^* outperform those of $\widehat{u}_n$ under quadratic risk. However, not much seems to be known on comparison of $\widehat{u}$ and u^* using other loss functions than the quadratic, a fortiori not under a broader class of loss functions, e.g. subconvex and bounded. There is nothing like a central statistic in the sense of Definition 6.1 for the experiment (IK) where the only sufficient statistic is the whole two-sided Brownian path, hence nothing like a central sequence in the sense of Definition 7.1'(c) in the approximating experiments. Still, both estimators $\widehat{u}$ and u^* in the experiment (IK) are equivariant in the sense of Definition 5.4: in Höpfner and Kutoyants [56, 57] we have studied local asymptotics with limit experiment (IK) – in a context of diffusions carrying a deterministic discontinuous periodic signal in their drift, and being observed continuously over a long time interval – using some of the techniques of Chapter 7; the main results of Chapter 7 have no counterpart here. The limit experiment (IK) is of importance in a broad variety of contexts, e.g. Golubev [32], Pflug [106], Küchler and Kutoyants [76] and the references therein.

Chapter 9

*Appendix

Topics:

9.1 *Convergence of Martingales
Convergence to a Gaussian martingale 9.1
Convergence to a conditionally Gaussian martingale 9.2
Convergence of pairs (martingale, angle bracket) 9.3

9.2 *Harris Recurrent Markov Processes
Harris recurrent Markov processes 9.4
Invariant measure 9.5–9.5'
Additive functionals 9.5"
Ratio limit theorem 9.6
Tightness rates under null recurrence (9.7)
Regular variation condition, Mittag-Leffler processes, weak convergence of integrable additive functionals 9.8
Martingale additive functionals 9.9–9.9'
Regular variation condition: weak convergence of martingale additive functionals together with their angle brackets 9.10–9.10'

9.3 *Checking the Harris Condition
A variant of the Harris condition 9.11
From grid chains via segment chains to continuous time processes 9.11'

9.4 *One-dimensional Diffusions
Harris properties for diffusions with values in $\mathbb{R}$ 9.12
Some examples 9.12'
Diffusions taking values in some open interval I in $\mathbb{R}$ 9.13
Some examples 9.13'–9.13"
Exact constants for i.i.d. cycles in null recurrent diffusions without drift 9.14

This Appendix collects facts of different nature which we quote in the stochastic process sections of this book. In most cases, they are stated without proof, and we indicate references. An asterisk * in front of this chapter (and in front of all its sections) indicates that the reader should be acquainted with basic properties of stochastic processes in continuous time, with semi-martingales and stochastic differential equations. Our principal references are the following. The book by Métivier [98] represents a well-written source for the theory of stochastic processes; a useful overview appears in the appendix sections of Bremaud [14]. A detailed treatment can be found in Dellacherie and Meyer [20]. For stochastic differential equations, we refer to Karatzas and Shreve [69] and Ikeda and Watanabe [61]. For semi-martingales and their (weak) convergence, see Jacod and Shiryaev [64].

9.1 *Convergence of Martingales

All filtrations which appear in this section are right-continuous; all processes below have càdlàg paths. $(D, \mathcal{D}, \mathbb{G})$ denotes the Skorohod space of d-dimensional càdlàg functions (see [64, Chap. VI]). A d-dimensional locally square integrable local martingale $M = (M_t)_{t \geq 0}$ starting from $M_0 = 0$ is called a continuous Gaussian martingale if there are no jumps and if the angle bracket process $\langle M \rangle$ is continuous and deterministic: in this case, M has independent increments, and all finite dimensional distributions are Gaussian laws. We quote the following from [64, Coroll. VIII.3.24]:

9.1 Theorem. For $n \geq 1$, consider d-dimensional locally square integrable local martingales

$$M^{(n)} = \left(M_t^{(n)}\right)_{t \geq 0} \quad \text{defined on } (\Omega^{(n)}, \mathcal{A}^{(n)}, \mathbb{F}^{(n)}, P^{(n)})$$

starting from $M_0^{(n)} = 0$. Let $\mu^{(n)}(ds, dy)$ denote the point process of jumps of $M^{(n)}$ and $\nu^{(n)}(ds, dy)$ its $(P^{(n)}, \mathbb{F}^{(n)})$-compensator, for (s, y) in $(0, \infty) \times (\mathbb{R}^d \setminus \{0\})$. Assume a Lindeberg condition

for all $0 < t < \infty$ and all $\varepsilon > 0$,

$$\int_0^t \int_{\{|y| > \varepsilon\}} |y|^2 \, \nu^{(n)}(ds, dy) = o_{P^{(n)}}(1) \quad \text{as } n \to \infty.$$

Let M' denote a continuous Gaussian martingale

$$M' = \left(M_t'\right)_{t \geq 0} \quad \text{defined on } (\Omega', \mathcal{A}', \mathbb{F}', P')$$

and write $C := \langle M' \rangle$ for the deterministic angle bracket. Then stochastic convergence

for every $0 < t < \infty$ fixed, $\quad \langle M^{(n)} \rangle_t = C_t + o_{P^{(n)}}(1) \quad \text{as } n \to \infty$

implies weak convergence of martingales

$$Q^{(n)} := \mathcal{L}(M^{(n)} \mid P^{(n)}) \quad \longrightarrow \quad Q' := \mathcal{L}(M' \mid P')$$

in the Skorohod space D as $n \to \infty$.

Section 9.1 *Convergence of Martingales

Next we fix one probability space $(\Omega^{(n)}, \mathcal{A}^{(n)}, P^{(n)}) = (\Omega, \mathcal{A}, P)$ for all n, equipped with a filtration $\mathbb{F}$, and assume that $M^{(n)}$ and $\mathbb{F}^{(n)}$ as above are derived from the same locally square integrable local $(P, \mathbb{F})$-martingale $M = (M_t)_{t \geq 0}$ through space-time rescaling (such as e.g. $\mathcal{F}_t^{(n)} := \mathcal{F}_{tn}$ and $M_t^{(n)} := n^{-1/2} M_{tn}$). We need the following nesting condition (cf. [64, VIII.5.37]):

(+) $\quad \begin{cases} \text{there is a sequence of positive real numbers } \alpha_n \downarrow 0 \text{ such that} \\ \mathcal{F}_{\alpha_n}^{(n)} \text{ is contained in } \mathcal{F}_{\alpha_{n+1}}^{(n+1)} \text{ for all } n \geq 1, \text{ and} \\ \sigma\left(\bigcup_n \mathcal{F}_{\alpha_n}^{(n)}\right) = \sigma\left(\bigcup_t \mathcal{F}_t\right) =: \mathcal{F}_\infty . \end{cases}$

On $(\Omega, \mathcal{A}, P)$, we also need a collection of $\mathcal{F}_\infty$-measurable random variables $\Phi = (\Phi_t)_{t \geq 0}$ such that

(++) $\quad \begin{aligned} &\text{paths } t \to \Phi_t \text{ are continuous and strictly increasing with } \Phi_0 = 0 \\ &\text{and } \lim_{t \to \infty} \Phi_t = \infty, \ P\text{-a.s.} \end{aligned}$

(such as e.g. $\Phi_t = t \zeta$ for some $\mathcal{F}_\infty$-measurable random variable $\zeta > 0$). On some other probability space $(\Omega', \mathcal{A}', \mathbb{F}', P')$, consider a continuous Gaussian martingale M' with deterministic angle bracket $\langle M' \rangle = C$. We define M' subject to independent time change $t \to \Phi_t$

$$M' \circ \Phi = (M'_{\Phi_t})_{t \geq 0}$$

as follows. Let $K'(\cdot, \cdot)$ denote a transition probability from $(\Omega, \mathcal{F}_\infty)$ to $(D, \mathcal{D})$ such that for the first argument $\omega \in \Omega$ fixed, the canonical process η on $(D, \mathcal{D})$ under $K'(\omega, \cdot)$ is a continuous Gaussian $\mathbb{G}$-martingale with angle bracket

$$t \to (C \circ \Phi)(\omega, t) = C(\Phi_t(\omega)).$$

Lifting Φ and η to

$$\left(\Omega \times D, \mathcal{A} \otimes \mathcal{D}, (\mathcal{F}_\infty \otimes \mathcal{G}_t)_{t \geq 0}, (PK')(d\omega, df) := P(d\omega)K'(\omega, df)\right)$$

the pair $(\Phi, M' \circ \Phi)$ is well defined on this space (cf. [64, p. 471]). By this construction, $M' \circ \Phi$ is a conditionally Gaussian martingale. We quote the following result from Jacod and Shiryaev [64, VIII.5.7 and VIII.5.42]:

9.2 Theorem. For $n \geq 1$, for filtrations $\mathbb{F}^{(n)}$ in $\mathcal{A}$ such that the nesting condition (+) above holds, consider d-dimensional locally square integrable local martingales

$$M^{(n)} = \left(M_t^{(n)}\right)_{t \geq 0} \quad \text{on } (\Omega, \mathcal{A}, \mathbb{F}^{(n)}, P)$$

starting from $M_0^{(n)} = 0$ and satisfying a Lindeberg condition

for all $0 < t < \infty$ and all $\varepsilon > 0$,

$$\int_0^t \int_{\{|y| > \varepsilon\}} |y|^2 \, \nu^{(n)}(ds, dy) = o_P(1) \quad \text{as } n \to \infty.$$

As above, for a collection of $\mathcal{F}_\infty$-measurable variables $\Phi = (\Phi_t)_{t\geq 0}$ satisfying (++) and for a continuous Gaussian martingale M' with deterministic angle bracket $\langle M'\rangle = C$, we have a (continuous) conditionally Gaussian martingale $M' \circ \Phi$ living on $(\Omega \times D, \mathcal{A} \otimes \mathcal{D}, (\mathcal{F}_\infty \otimes \mathcal{G}_t)_{t\geq 0}, PK')$. Then stochastic convergence

$$\text{for every } 0 < t < \infty \text{ fixed,} \quad \langle M^{(n)}\rangle_t = (C \circ \Phi)_t + o_P(1) \quad \text{as } n \to \infty$$

implies weak convergence of martingales

$$\mathcal{L}(M^{(n)} \mid P) \quad \longrightarrow \quad \mathcal{L}(M' \circ \Phi \mid PK')$$

in the Skorohod space D as $n \to \infty$.

Finally, consider locally square integrable local martingales $M^{(n)}$

$$M^{(n)} = M^{(n,c)} + \int_0^{\cdot} \int_{\mathbb{R}^d \setminus \{0\}} y\, (\mu^{(n)} - \nu^{(n)})(ds, dy)$$

where $M^{(n,c)}$ is the continuous local martingale part. Writing $M^{(n,i)}$, $1 \leq i \leq d$, for the components of $M^{(n)}$ and $[M^{(n)}]$ for the quadratic covariation process, we obtain for $0 < t < \infty$ and $i, j = 1, \ldots, d$

$$\sup_{s \leq t} \left| [M^{(n,i)}, M^{(n,j)}]_s - \langle M^{(n,i)}, M^{(n,j)}\rangle_s \right| = o_P(1) \quad \text{as } n \to \infty$$

provided the sequence $(M^{(n)})_n$ satisfies the Lindeberg condition. In this situation, since Jacod and Shiryaev [64, VI.6.1] show that weak convergence of $M^{(n)}$ as $n \to \infty$ in D to a continuous limit martingale implies weak convergence of pairs $(M^{(n)}, [M^{(n)}])$ in the Skorohod space of càdlàg functions $[0, \infty) \to \mathbb{R}^d \times \mathbb{R}^{d\times d}$, we also have weak convergence of pairs $(M^{(n)}, \langle M^{(n)}\rangle)$. Thus the following result is contained in [64, VI.6.1]:

9.3 Theorem. For $n \geq 1$, consider d-dimensional locally square integrable local martingales

$$M^{(n)} = \left(M_t^{(n)}\right)_{t\geq 0} \quad \text{defined on } (\Omega^{(n)}, \mathcal{A}^{(n)}, \mathbb{F}^{(n)}, P^{(n)})$$

with $M_0^{(n)} = 0$. Assume the Lindeberg condition of Theorem 9.1. Let $\widetilde{M}$ denote a continuous local martingale

$$\widetilde{M} = \left(\widetilde{M}_t\right)_{t\geq 0} \quad \text{defined on } (\widetilde{\Omega}, \widetilde{\mathcal{A}}, \widetilde{\mathbb{F}}, \widetilde{P})$$

starting from $\widetilde{M}_0 = 0$. Then weak convergence in D

$$\mathcal{L}(M^{(n)} \mid P^{(n)}) \quad \longrightarrow \quad \mathcal{L}(\widetilde{M} \mid \widetilde{P}), \quad n \to \infty$$

implies weak convergence

$$\mathcal{L}\left(M^{(n)},\langle M^{(n)}\rangle \mid P^{(n)}\right) \longrightarrow \mathcal{L}\left(\widetilde{M},\langle\widetilde{M}\rangle \mid \widetilde{P}\right), \quad n \to \infty$$

in the Skorohod space $D(\mathbb{R}^d \times \mathbb{R}^{d\times d})$ of càdlàg functions $[0,\infty) \to \mathbb{R}^d \times \mathbb{R}^{d\times d}$.

9.2 *Harris Recurrent Markov Processes

For Harris recurrence of Markov chains, we refer to Revuz [111] and Nummelin [101]. For Harris recurrence of continuous time Markov processes, our main reference is Azema, Duflo and Revuz [2]. On some underlying probability space, we consider a time homogeneous strong Markov process $X = (X_t)_{t\geq 0}$ taking values in $\mathbb{R}^d$, with càdlàg paths, having infinite life time, and its semigroup

$$(P_t(\cdot,\cdot))_{t\geq 0} :$$
$$P_t(x,dy) = P(X_{s+t} \in dy \mid X_s = x), \quad x,y \in \mathbb{R}^d, \ s,t \in [0,\infty).$$

We write U^1 for the potential kernel $U^1(x,dy) = \int_0^\infty e^{-t} P_t(x,dy)\,dt$ which corresponds to observation of X after an independent exponential time. For σ-finite measures m on $(\mathbb{R}^d, \mathcal{B}(\mathbb{R}^d))$, measurable functions $f : \mathbb{R}^d \to [0,\infty]$ and $t \geq 0$ write

$$P_t f(x) = \int P_t(x,dy)\, f(y),$$
$$mP_t(dy) = \int m(dx)\, P_t(x,dy),$$
$$E_m f = \int m(dx)\, E_x(f).$$

A σ-finite measure m on $(\mathbb{R}^d, \mathcal{B}(\mathbb{R}^d))$ with the property

$$m P_t = m \quad \text{for all} \quad t \geq 0$$

is termed invariant for X. On the canonical path space $(D, \mathcal{D}, \mathbb{G})$ for $\mathbb{R}^d$-valued càdlàg processes, write Q_x for the law of the process X starting from $x \in \mathbb{R}^d$. Let $\eta = (\eta_t)_{t\geq 0}$ denote the canonical process on $(D, \mathcal{D}, \mathbb{G})$, and $(\theta_t)_{t\geq 0}$ the collection of shift operators on $(D, \mathcal{D})$: $\theta_t(\alpha) := (\alpha(t+s))_{s\geq 0}$ for $\alpha \in D$. Systematically, we speak of 'properties of the process X' when we mean 'properties of the semigroup $(P_t(\cdot,\cdot))_{t\geq 0}$': these in turn will be formulated as properties of the canonical process η in the system $(D, \mathcal{D}, \mathbb{G}, (\theta_t)_{t\geq 0}, (Q_x)_{x\in\mathbb{R}^d})$.

9.4 Definition. The process $X = (X_t)_{t\geq 0}$ is called *recurrent in the sense of Harris* (or *Harris* for short) if there is a σ-finite measure ϕ on $(\mathbb{R}^d, \mathcal{B}(\mathbb{R}^d))$ such that the following holds:

($\diamond$)
$$A \in \mathcal{B}(\mathbb{R}^d), \phi(A) > 0 \implies$$
$$\int_0^\infty 1_A(\eta_s)\, ds = +\infty \quad Q_x\text{-almost surely, for every } x \in \mathbb{R}^d.$$

In Definition 9.4, a process X satisfying ($\diamond$) will accumulate infinite occupation time in sets of positive ϕ-measure, independently of the starting point. The following is from [2, 2.4–2.5]:

9.5 Theorem. If X is recurrent in the sense of Harris,

(a) there is a unique (up to multiplicative constants) invariant measure m on $(\mathbb{R}^d, \mathcal{B}(\mathbb{R}^d))$,

(b) condition ($\diamond$) of Definition 9.4 holds with m in place of ϕ,

(c) any measure ϕ in Definition 9.4 for which condition ($\diamond$) holds satisfies $\phi \ll m$ and $\phi U^1 \sim m$.

9.5' Definition. A Harris process X is called *positive recurrent* if the invariant measure m is of finite total mass on $\mathbb{R}^d$; X is called *null recurrent* otherwise.

9.5" Definition. An *additive functional of* X is an adapted process $A = (A_t)_{t \geq 0}$ on $(D, \mathcal{D}, \mathbb{G})$, all paths càdlàg non-decreasing and starting from $A_0 = 0$, such that
$$A_t - A_s = A_{t-s} \circ \theta_s \quad \text{for all } 0 \leq s < t < \infty.$$
If X is Harris with invariant measure m, A is termed *integrable* if $E_m(A_1) < \infty$.

From [2, 3.1], we quote Theorem 9.6(a.i) below; the following assertions of Theorem 9.6(a.ii) and 9.6(b) are immediate but important consequences.

9.6 Ratio Limit Theorem. (a) If X is recurrent in the sense of Harris, with invariant measure m,

(i) we have for integrable additive functionals A, B such that $0 < E_m(B_1) < \infty$
$$\text{for every } x \in \mathbb{R}^d: \quad \frac{A_t}{B_t} \longrightarrow \frac{E_m(A_1)}{E_m(B_1)} \quad Q_x\text{-almost surely as } t \to \infty;$$

(ii) we have for functions $f, g \in L^1(\mathbb{R}^d, \mathcal{B}(\mathbb{R}^d), m)$ such that $m(g) \neq 0$
$$\text{for every } x \in \mathbb{R}^d: \quad \frac{\int_0^t f(\eta_s)\, ds}{\int_0^t g(\eta_s)\, ds} \longrightarrow \frac{m(f)}{m(g)} \quad Q_x\text{-almost surely as } t \to \infty.$$

(b) If X is positive recurrent, with $g \equiv 1$ and norming factor such that $m(\mathbb{R}^d) = 1$,

for every $x \in \mathbb{R}^d$: $\quad \dfrac{1}{t}\displaystyle\int_0^t f(\eta_s)\,ds \longrightarrow m(f) \quad Q_x$-almost surely as $t \to \infty$

for arbitrary functions $f \in L^1(\mathbb{R}^d, \mathcal{B}(\mathbb{R}^d), m)$.

Note that integrability of additive functionals as required in Theorem 9.6(a) turns out to be a restrictive condition in null recurrent cases where the invariant measure m has infinite total mass on $\mathbb{R}^d$.

Whereas Theorem 9.6(b) provides a strong law of large numbers for suitably normed integrable additive functionals of a Harris process X which is positive recurrent, not even convergence in law holds true when X is null recurrent unless additional conditions are imposed. In the general null recurrent case, all that remains granted is existence of a tightness rate for the class of integrable additive functionals, i.e. of some deterministic norming function $t \to v(t)$ such that

(9.7) $\quad\displaystyle\lim_{M\to\infty} \liminf_{t\to\infty} Q_x\left(\dfrac{1}{M} < \dfrac{1}{v(t)}\int_0^t f(\eta_s)\,ds < M\right) = 1$

for all functions $f \in L^1(\mathbb{R}^d, \mathcal{B}(\mathbb{R}^d), m)$ with $m(f) \neq 0$, independently of the choice of the starting point $x \in \mathbb{R}^d$. This has been proved by Loukianova and Loukianov [94] for the special case of one-dimensional diffusions, using local time. For general Harris processes, the result is due to Löcherbach and Loukianova [93], using a new Nummelin-like splitting technique.

In the case of null recurrence, weak convergence of integrable additive functionals requires an extra condition on the semigroup of X, of regular variation type. This goes back to Darling and Kac [18] (see [12, Chap. 8.11]). Below, we give the regular variation condition in a form due to Touati [123]. Then suitably normed integrable additive functionals of X converge weakly to a Mittag–Leffler process $V^{(\alpha)}$ of index $0 < \alpha \leq 1$, the process inverse of the stable increasing process $S^{(\alpha)}$ of index $0 < \alpha \leq 1$. When $0 < \alpha < 1$, we refer to the definition given in Remark 6.17: the process $S^{(\alpha)}$ starting from $S_0^{(\alpha)} = 0$ has independent and stationary increments with Laplace transform

$$E\left(e^{-\lambda(S_{t_2}^{(\alpha)} - S_{t_1}^{(\alpha)})}\right) = e^{-(t_2 - t_1)\lambda^\alpha}, \quad \lambda \geq 0, \quad 0 \leq t_1 < t_2 < \infty,$$

$V^{(\alpha)}$ is the process of level crossing times of $S^{(\alpha)}$

$$V_t^{(\alpha)} = \inf\left\{v > 0: S_v^{(\alpha)} > t\right\}, \quad t \geq 0,$$

and paths of $V^{(\alpha)}$ are continuous and non-decreasing with $V_0^{(\alpha)} \equiv 0$ and $\displaystyle\lim_{t\to\infty} V_t^{(\alpha)} = \infty$. We extend this definition to the case $\alpha = 1$ by

$$S^{(1)} = \mathrm{id} = V^{(1)}, \text{ i.e. } S_t^{(1)} \equiv t \equiv V_t^{(1)}, \quad t \geq 0.$$

The last definition is needed in view of null recurrent processes where suitably normed integrable additive functionals converge weakly to $V^{(1)} = \mathrm{id}$. As an example, we might have a recurrent atom in the state space of X such that the distribution of the time between successive visits in the atom is 'relatively stable' in the sense of [12, Chap. 8.8 combined with p. 359]; at every visit, the process might spend an exponential time in the atom; then the occupation time A_t of the atom up to time t defines an additive functional $A = (A_t)_{t\geq 0}$ of X for which suitable norming functions vary regularly at ∞ with index $\alpha = 1$. Index $\alpha = 1$ is also needed for the strong law of large numbers in Theorem 9.6(b) in positive recurrent Harris processes. We quote the following from [53, Thm. 3.15].

9.8 Theorem. Consider a Harris process X with invariant measure m.

(a) For $0 < \alpha \leq 1$ and $\ell(\cdot)$ varying slowly at ∞, the following assertions (i) and (ii) are equivalent:

(i) for every function $g : \mathbb{R}^d \to [0, \infty)$ which is $\mathcal{B}(\mathbb{R}^d)$-measurable and satisfies $0 < m(g) < \infty$, one has

$$(R_{1/t}g)(x) := E_x \left(\int_0^\infty e^{-\frac{1}{t}s} g(\eta_s)\, ds \right) \sim \frac{t^\alpha}{\ell(t)} m(g) \quad \text{as } t \to \infty$$

for m-almost all $x \in \mathbb{R}^d$, the exceptional null set depending on g;

(ii) for every $f : \mathbb{R}^d \to [0, \infty)$ which is $\mathcal{B}(\mathbb{R}^d)$-measurable with $0 < m(f) < \infty$, for every $x \in \mathbb{R}^d$,

$$\left(\frac{\ell(n)}{n^\alpha} \int_0^{tn} f(\eta_s)\, ds \right)_{t \geq 0} \quad \text{under } Q_x$$

converges weakly in D as $n \to \infty$ to

$$m(f)\, V^{(\alpha)}$$

where $V^{(\alpha)}$ is the Mittag–Leffler process of index α.

(b) The cases in (a) are the only ones where for functions f as in (a.ii) and for suitable choice of a norming function $v(\cdot)$, weak convergence in D of

$$\left(\frac{1}{v(n)} \int_0^{tn} f(\eta_s)\, ds \right)_{t \geq 0} \quad \text{under } Q_x$$

to a continuous non-decreasing limit process V (starting at $V_0 = 0$, and such that $\mathcal{L}(V_1)$ is not Dirac measure ϵ_0 at 0) is available.

In statistical models for Markov processes under time-continuous observation, we have to consider martingales when the starting point $x \in \mathbb{R}^d$ for the process is arbitrary but fixed. As an example, in the context of Chapter 8, the score martingale is a locally square integrable local Q_x-martingale $M = (M_t)_{t \geq 0}$ on $(D, \mathcal{D}, \mathbb{G})$. We need limit theorems for the pair $(M, \langle M \rangle)$ under Q_x to prove convergence of local models. Under suitable assumptions, Theorem 9.8 above can be used to settle the problem for the angle bracket $\langle M \rangle$ under Q_x. Thus we need weak convergence for the score martingale M under Q_x. From this – thanks to a result in [64], which was recalled in Theorem 9.3 above – we can pass to joint convergence of the pair $(M, \langle M \rangle)$ under Q_x.

9.9 Definition. On the path space $(D, \mathcal{D}, \mathbb{G})$, for given $x \in \mathbb{R}^d$, consider a locally square integrable local Q_x-martingale $M = (M_t)_{t \geq 0}$ together with its quadratic variation $[M]$ and its angle bracket $\langle M \rangle$ under Q_x; assume in addition that $\langle M \rangle$ under Q_x is locally bounded. We call M a *martingale additive functional* if the following holds:

(i) $[M]$ and $\langle M \rangle$ under Q_x admit versions which are additive functionals of η;

(ii) for every choice of $0 \leq s < t < \infty$ and $y \in \mathbb{R}^d$, one has $M_t - M_s = M_{t-s} \circ \theta_s$ Q_y-almost surely.

9.9' Example. (a) On $(D, \mathcal{D}, \mathbb{G})$, for starting point $x \in \mathbb{R}$, write Q_x for the law of a diffusion
$$dX_t = b(X_t)\,dt + \sigma(X_t)\,dW_t$$
and $m^{(x)}$ for the local martingale part of the canonical process η on $(D, \mathcal{D}, \mathbb{G})$ under Q_x. Write $\mathcal{L}$ for the Markov generator of X and consider some $\mathcal{C}^2$ function $F : \mathbb{R} \to \mathbb{R}$ with derivative f. Then
$$M_t = \left(\int_0^t f(\eta_s)\,dm_s^{(x)} \right)_{t \geq 0}, \quad t \geq 0$$
is a locally square integrable local Q_x-martingale admitting a version $(t, \omega) \to Y(t, \omega)$
$$Y_t := \left(F(\eta_t) - F(\eta_0) - \int_0^t \mathcal{L}F(\eta_s)\,ds \right), \quad t \geq 0$$
by Ito formula. This version satisfies $Y_t - Y_s = Y_{t-s} \circ \theta_s$ for $0 \leq s < t < \infty$. Quadratic variation and angle bracket of M under Q_x
$$[M]_t = \langle M \rangle_t = \int_0^t f^2(\eta_s)\sigma^2(\eta_s)\,ds, \quad t \geq 0$$
are additive functionals. Hence M is a martingale additive functional as defined in Definition 9.9.

(b) Prepare a Poisson process $N = (N_t)_{t \geq 0}$ with parameter $\lambda > 0$ independent of Brownian motion W. Under parameter values $\mu > 0$, $\sigma > 0$, $\varepsilon > 0$, for starting point $x \in \mathbb{R}$, write Q_x for the law on $(D, \mathcal{D}, \mathbb{G})$ of the solution to
$$dX_t = -\mu\,1_{\{X_{t-}>0\}}\,dt + \varepsilon\,1_{\{X_{t-}<0\}}\,dN_t + \sigma\,dW_t$$

driven by the pair (N, W). Let M denote the local martingale part – sum of a continuous and a purely discontinuous martingale – of the canonical process η under Q_x. It admits a version $(t, \omega) \to Y(t, \omega)$

$$Y_t := \eta_t - \eta_0 - \int_0^t \left(\varepsilon\lambda\, 1_{(-\infty,0)}(\eta_s) - \mu\, 1_{(0,\infty)}(\eta_s) \right) ds \, , \quad t \geq 0$$

which satisfies $Y_t - Y_s = Y_{t-s} \circ \theta_s$ for $0 \leq s < t < \infty$. The quadratic variation of M under Q_x is

$$\langle M \rangle_t = \sigma^2 t + \varepsilon^2 \lambda \int_0^t 1_{(-\infty,0)}(\eta_s)\, ds \, .$$

The angle bracket of M under Q_x admits a version (with notation $\Delta\eta_s = \eta_s - \eta_{s-}$)

$$[M]_t = \sigma^2 t + \varepsilon^2 \sum_{0 < s \leq t} 1_{\{\Delta\eta_s = \varepsilon\}}$$

since the sum process on the right-hand side is indistinguishable from $(\sum_{0 < s \leq t} (\Delta\eta_s)^2 1_{\{|\Delta\eta_s| > 0\}})_{t \geq 0}$ under Q_x (see also [64, p. 55]). All requirements of Definition 9.9 are satisfied: thus M is a martingale additive functional. □

We quote the following from Höpfner and Löcherbach [53, Thm. 3.16].

9.10 Theorem. Consider a Harris process X, $\mathbb{R}^d$-valued, with invariant measure m. Assume that for X there are $0 < \alpha \leq 1$ and $\ell(\cdot)$ varying slowly at ∞ such that

condition (i) of Theorem 9.8(a) is satisfied with α and $\ell(\cdot)$.

For given $x \in \mathbb{R}^d$, consider a locally square integrable local Q_x-martingale $M = (M_t)_{t \geq 0}$ on $(D, \mathcal{D}, \mathbb{G})$. Assume that $E_m(\langle M \rangle_1)$ is finite, and that M is a martingale additive functional as defined in Definition 9.9. Then as $n \to \infty$

$$M^n := \frac{1}{\sqrt{n^\alpha / \ell(n)}} (M_{tn})_{t \geq 0} \quad \text{under } Q_x$$

converges weakly in D to the limit martingale M

$$M = \Lambda^{1/2}\, B \circ V^{(\alpha)}, \quad \Lambda := E_m(\langle M \rangle_1)$$

where Brownian motion B and Mittag–Leffler process $V^{(\alpha)}$ are independent.

Recall from Remark 6.17 that $\mathcal{L}(V_t^{(\alpha)})$ admits finite moments of arbitrary order. The proof in [53] is via Nummelin splitting [102] along independent exponential waiting times in a sequence of strong Markov processes approaching (an extension of) η: we construct renewal times and independent life cycles between successive renewal times in the approximating processes and apply limit theorems for sums of i.i.d. random variables from Resnick and Greenwood [110]. The theorem contains the positive recurrent cases where $\alpha = 1$, $\ell(\cdot) \equiv 1$ and $V^{(1)} = \mathrm{id}$. Note that Theorem 9.10 is not a martingale convergence theorem in the sense of Theorem 9.2: there is no stochastic convergence of angle brackets $\langle M^n \rangle_t$, only convergence in law.

Via Theorem 9.3, the following is a direct consequence of Theorems 9.8 and 9.10.

9.10' Corollary. With notations of Theorem 9.10, and under all assumptions made there, consider
$$M^n := \frac{1}{\sqrt{n^\alpha/\ell(n)}} (M_{tn})_{t \geq 0} \quad \text{under } Q_x$$

as $n \to \infty$. With $\nu^n(ds, dy)$ the $(Q_x, (\mathcal{G}_{tn})_{t \geq 0})$-compensator of the point process of jumps of M^n, assume in addition a Lindeberg condition

for all $0 < t < \infty$ and all $\varepsilon > 0$,
$$\int_0^t \int_{\{|y| > \varepsilon\}} |y|^2 \nu^n(ds, dy) = o_{Q_x}(1) \quad \text{as } n \to \infty.$$

Then the assertion of Theorem 9.10 strengthens to weak convergence of the pairs
$$(M^n, \langle M^n \rangle) \quad \text{under } Q_x$$

in $D(\mathbb{R}^d \times \mathbb{R}^{d \times d})$ as $n \to \infty$ to the pair
$$(\Lambda^{1/2} B \circ V^{(\alpha)}, \Lambda V^{(\alpha)}).$$

9.3 *Checking the Harris Condition

We continue under the basic assumptions and notations stated at the start of Section 9.2, and present conditions which imply that a càdlàg time-continuous strong Markov process is Harris. The first one is from [112, pp. 394–395].

9.11 Proposition. If an invariant measure m for $X = (X_t)_{t \geq 0}$ is known, the condition

(o) $\quad A \in \mathcal{B}(\mathbb{R}^d), \ m(A) > 0 \implies$
$\quad\quad\quad \limsup_{t \to \infty} 1_A(\eta_t) = 1 \ Q_x$-almost surely, for every $x \in \mathbb{R}^d$

implies Harris recurrence.

Proof. We use arguments of Revuz [111, pp. 94–95] and Revuz and Yor [112, p. 395]. Since m in condition (o) is invariant, we consider in (o) sets $A \in \mathcal{B}(\mathbb{R}^d)$ with

$E_m(\int_0^t 1_A(\eta_s)\,ds) = tm(A) > 0$. As a consequence, we can specify some $\varepsilon > 0$ and some $\delta > 0$ such that

(oo)
$$m\left(\{x \in \mathbb{R}^d : Q_x(\tau < \infty) > \delta\}\right) > 0$$
where $\tau := \tau_A = \inf\{t > 0 : \int_0^t 1_A(\eta_s)\,ds > \varepsilon\}$.

(1) Write $f : \mathbb{R}^d \to [0,1]$ for the $\mathcal{B}(\mathbb{R}^d)$-measurable function $v \to Q_v(\tau < \infty)$. Then
$$f(\eta_t) = E_\bullet\left(1_{\{t+\tau\circ\theta_t < \infty\}} \mid \mathcal{G}_t\right), \quad t \geq 0$$
('dot' indicating a common version of the conditional expectation under P_x for all x in virtue of the Markov property) and from this
$$P_t f(v) = E_v(f(\eta_t)) = Q_v(t + \tau \circ \theta_t < \infty) \leq Q_v(\tau < \infty) = f(v), \quad v \in \mathbb{R}^d$$
which establishes the inequality
$$P_t f \leq f \quad \text{for} \quad 0 < t < \infty.$$

Define a process $N = (N_t)_{t \geq 0}$ on $(D, \mathcal{D})$ by $N_t := f(\eta_t)$, $t \geq 0$. By $P_t f \leq f$, N is a non-negative $(\mathbb{G}, Q_y)$-supermartingale, for every value of a starting point $y \in \mathbb{R}^d$. Hence, for every value of $y \in \mathbb{R}^d$, there is a limit variable $N_\infty^{(y)}$ under Q_y such that
$$N_t \longrightarrow N_\infty^{(y)} \quad Q_y\text{-almost surely as } t \to \infty.$$

On the one hand, this implies $f(y) \geq E_y(N_\infty^{(y)})$ by Fatou, for every $y \in \mathbb{R}^d$. On the other hand, the set $\{x : f(x) > \delta\}$ has positive m-measure by (oo), and thus will be visited infinitely often, almost surely, by the process η under Q_y as a consequence of (o), for every value of $y \in \mathbb{R}^d$: hence we must have $N_\infty^{(y)} \geq \delta$ Q_y-almost surely, for every $y \in \mathbb{R}^d$. Both assertions combined give $f(y) \geq \delta$ for all $y \in \mathbb{R}^d$. Thus from definition of N, the supermartingale N together with its limit variables $N_\infty^{(y)}$ under Q_y for starting points $y \in \mathbb{R}^d$ is separated away from 0.

(2) Consider the event
$$R := R_A = \left\{\int_0^\infty 1_A(\eta_s)\,ds = \infty\right\} = \bigcap_t \{t + \tau \circ \theta_t < \infty\} \in \mathcal{G}_\infty.$$

For starting point $x \in \mathbb{R}^d$ fixed and for arbitrarily large $k < \infty$, introduce uniformly integrable $(\mathbb{G}, Q_x)$-martingales $N^{(1,x)}$, $N^{(2,x,k)}$:
$$N_t^{(1,x)} := E_x(1_R \mid \mathcal{G}_t), \quad N_t^{(2,x,k)} := E_x\left(1_R + 1_{R^c \cap \{k+\tau\circ\theta_k < \infty\}} \mid \mathcal{G}_t\right),$$
$$t \geq 0.$$

By definition of the supermartingale N in step (1) and since the events $\{t+\tau\circ\theta_t < \infty\}$ are decreasing to R as $t \to \infty$, we can compare conditional expectations, and obtain

Section 9.3 *Checking the Harris Condition   255

Q_x-almost surely

$$1_R = \lim_{t \to \infty} N_t^{(1,x)} \leq \lim_{t \to \infty} N_t = N_\infty^{(x)} \leq \lim_{t \to \infty} N_t^{(2,x,k)} = 1_R + 1_{R^c \cap \{k + \tau \circ \theta_k < \infty\}}.$$

Letting k tend to ∞ we arrive at

$$1_R = N_\infty^{(x)} \quad Q_x\text{-almost surely}$$

where $x \in \mathbb{R}^d$ is arbitrary. We have seen in step (1) that all limit variables $N_\infty^{(x)}$ under Q_x are separated from 0 by $\delta > 0$. Hence the event R is of full measure under Q_x for every $x \in \mathbb{R}^d$. $\square$

Revuz and Yor [112, pp. 394–395] take (∘) in Proposition 9.11 as a definition for Harris recurrence of $(X_t)_{t \geq 0}$. From the very beginning, we have to know an invariant measure. The following sufficient criterion for Harris recurrence of the process $(X_t)_{t \geq 0}$ avoids this: for some $0 < T < \infty$ which is deterministic we might be able to establish that $(X_{kT})_{k \in \mathbb{N}_0}$ is a Harris chain.

9.11' Proposition. Assume that for some step size $T \in (0, \infty)$ and for some σ-finite measure $\widehat{\phi}$ on $(\mathbb{R}^d, \mathcal{B}(\mathbb{R}^d))$ we have

(∘∘)
$$A \in \mathcal{B}(\mathbb{R}^d), \widehat{\phi}(A) > 0 \implies \sum_{k=1}^\infty 1_A(\eta_{kT}) = +\infty \quad Q_x\text{-almost surely, for every } x \in \mathbb{R}^d.$$

Then the process $X = (X_t)_{t \geq 0}$ is Harris recurrent. In fact, condition (∘∘) implies the following stronger statement: path segments over $[kt, (k+1)T]$ taken from the path of X form a Harris chain

$$\bigl((X_{kT+v})_{0 \leq v \leq T}\bigr)_{k \in \mathbb{N}_0}$$

taking values in the Skorohod space $(D([0, T]), \mathcal{D}([0, T]))$ of càdlàg functions $[0, T] \to \mathbb{R}^d$.

Proof. In this proof, we write for short $(E, \mathcal{E})$ instead of $(\mathbb{R}^d, \mathcal{B}(\mathbb{R}^d))$, everything remaining valid when $(E, \mathcal{E})$ is a Polish space and $(D, \mathcal{D}, \mathbb{G})$ the space of càdlàg functions $[0, \infty) \to E$ with canonical process $\eta = (\eta_t)_{t \geq 0}$.

(1) By condition (∘∘), the grid chain $\widehat{\eta} := (\eta_{kT})_{k \in \mathbb{N}_0}$ is a Harris chain, has a unique invariant measure $\widehat{m}$ (unique up to multiplication with a constant), and (∘∘) holds with $\widehat{m}$ in place of $\widehat{\phi}$ (this is from Harris [45], see also [101, p. 43] and [111, p. 92]).

(2) We show that when (∘∘) holds, T-segments in the path of η form a Harris chain.

(a) As a consequence of step (1), pairs $(\eta_{kT}, \eta_{(k+1)T})_{k \in \mathbb{N}_0}$ form a Harris chain on $(E \times E, \mathcal{E} \otimes \mathcal{E})$ whose invariant measure is $\widehat{m}(dy_1) P_T(y_1, dy_2)$.

(b) The next argument is as in [56, Sect. 2]. Write $(\pi_t)_{t\geq 0}$ for the process of coordinate projections on $(D([0,T]), \mathcal{D}([0,T]))$, and let $\overline{m}$ denote the unique measure on $\mathcal{D}([0,T])$ specified by the following set of finite dimensional distributions:

$$\begin{cases} \overline{m}\left(\pi_{t_j} \in A_j, 0 \leq j \leq l\right) = \\ \quad \int_{E^{l+1}} \widehat{m}(dy_0) 1_{A_0}(y_0) \prod_{j=1}^{l} P_{t_j - t_{j-1}}(y_{j-1}, dy_j) 1_{A_j}(y_j) \\ \text{with } 0 = t_0 < t_1 < \cdots < t_l = T \text{ arbitrary, } l \geq 1, \text{ and } A_j \in \mathcal{E} \text{ for } 0 \leq j \leq l. \end{cases}$$

Since $D([0,T])$ is a Polish space, $\overline{m}$ admits the following disintegration: there is a regular version of the conditional law under $\overline{m}$ of the process $(\pi_t)_{0 \leq t \leq T}$ given the pair (π_0, π_T), i.e. a transition probability $K(\cdot, \cdot)$ from $E \times E$ to $\mathcal{D}([0,T])$ such that

$$(\times) \quad \overline{m}(F) = \int_{E \times E} \widehat{m}(dy_1) P_T(y_1, dy_2) \, K((y_1, y_2), F), \quad F \in \mathcal{D}([0,T]).$$

Here the measure $m(dy_1) P_T(y_1, dy_2)$ is as in (a), and the laws $K((y_1, y_2), \cdot)$ on $\mathcal{D}([0,T])$ correspond to bridges from state y_1 at time 0 to state y_2 at time T.

(c) Consider a set $F \in \mathcal{D}([0,T])$ with $\overline{m}(F) > 0$. Then by ($\times$), there is some $\varepsilon > 0$ such that

$$B_F := \{(y_1, y_2) : K((y_1, y_2), F) > \varepsilon\}$$

has strictly positive $\widehat{m}(dy_1) P_T(y_1, dy_2)$-measure.

As a consequence of the Harris property in (a), the chain of pairs $(\eta_{kT}, \eta_{(k+1)T})_{k \in \mathbb{N}_0}$ will visit infinitely often the set B_F, independently of the choice of a starting point. Then by definition of B_F combined with ($\times$), the kernel K pasting in bridges, the segment chain will visit F infinitely often: we obtain

$$F \in \mathcal{D}([0,T]), \, \overline{m}(F) > 0 : \quad \sum_{k=0}^{\infty} 1_F\left((\eta_{kT+v})_{0 \leq v \leq T}\right) = \infty$$

almost surely, independently of the choice of a starting point. Thus the segment chain

$$(+) \qquad \overline{\eta} = (\overline{\eta}_k)_{k \in \mathbb{N}_0}, \quad \overline{\eta}_k := (\eta_{kT+v})_{0 \leq v \leq T}$$

is a Harris chain taking values in $(D([0,T]), \mathcal{D}([0,T]))$. It admits a unique invariant measure. From the definition of $\overline{m}$ we see that the measure $\overline{m}$ is invariant for the segment chain. Hence $\overline{\eta}$ is Harris with invariant measure $\overline{m}$.

(3) We show that $(\eta_t)_{t \geq 0}$ is a Harris process. Introduce a σ-finite measure ϕ on $(E, \mathcal{E})$:

$$\phi(A) := \int_0^T ds \, [\widehat{m} P_s](A), \quad A \in \mathcal{E}.$$

Fix some set $B \in \mathcal{E}$ with $\widehat{m}(B) = 1$ (recall that we are free to multiply $\widehat{m}$ with positive constants). Together with the $\mathcal{D}([0,T])$-measurable function

$$G : D([0,T]) \to \mathbb{R} \quad \text{given by} \quad \alpha \longrightarrow 1_{\{\pi_0 \in B\}}(\alpha), \quad \alpha \in D([0,T])$$

Section 9.3 *Checking the Harris Condition   257

consider a family of $\mathcal{D}([0, T])$-measurable functions indexed by $A \in \mathcal{E}$,

$$F_A : D([0, T]) \to \mathbb{R} \quad \text{given by} \quad \alpha \longrightarrow \int_0^T ds\, 1_A(\alpha(s)), \quad \alpha \in D([0, T]).$$

Note that $\overline{m}(F_A) = \phi(A)$ for $A \in \mathcal{E}$, and $\overline{m}(G) = \widehat{m}(B) = 1$. The ratio limit theorem for the segment chain (+) yields the convergence

$$(++) \quad \lim_{m\to\infty} \frac{\int_0^{mT} 1_A(\eta_s)\,ds}{\sum_{k=0}^{m-1} 1_B(\eta_{kT})} = \lim_{m\to\infty} \frac{\sum_{k=0}^{m-1} F_A(\overline{\eta}_k)}{\sum_{m=0}^{m-1} G(\overline{\eta}_k)} = \frac{\overline{m}(F_A)}{\overline{m}(G)} = \phi(A)$$

Q_x-almost surely, for every choice of a starting point $x \in E$. From $\widehat{m}(B) > 0$ and the Harris property of the grid chain $\widehat{\eta} = (\eta_{kT})_{k\in\mathbb{N}_0}$ in (1), denominators in (++) tend to ∞ Q_x-almost surely, for every choice of a starting point $x \in E$. Hence, for sets $A \in \mathcal{E}$ with $\phi(A) > 0$, also numerators in (++) increase to ∞ Q_x-almost surely, for every choice of a starting point $x \in E$:

$$A \in \mathcal{E}, \; \phi(A) > 0 : \quad \int_0^\infty 1_A(\eta_s)\,ds = \infty \quad Q_x\text{-almost surely, for every } x \in E.$$

This is property $(\diamond)$ in Definition 9.4: hence the continuous-time process $\eta = (\eta_t)_{t\geq 0}$ is Harris. By Theorem 9.5, η admits a unique (up to constant multiples) invariant measure m on $(E, \mathcal{E})$; it remains to identify m.

(4) We show that the three measures m, $\widehat{m}$, ϕ can be identified. Select some set $C \in \mathcal{E}$ with the property $\overline{m}(F_C) = \phi(C) = 1$. Then similarly to (++), for every $A \in \mathcal{E}$, the ratio limit theorem for the segment chain (+) yields

$$\lim_{t\to\infty} \frac{\int_0^t 1_A(\eta_s)\,ds}{\int_0^t 1_C(\eta_s)\,ds} = \frac{\overline{m}(F_A)}{\overline{m}(F_C)} = \phi(A) \quad Q_x\text{-almost surely, for every } x \in E.$$

Thus the measures m and ϕ coincide, up to some constant multiple, by Theorem 9.6 and the Harris property of $(\eta_t)_{t\geq 0}$ in step (3). Next, the measure $\phi = \int_0^T ds\,[\widehat{m}\,P_s]$ is by definition invariant for the grid chain $(\eta_{kT})_{k\in\mathbb{N}_0}$. Thus the Harris property in step (1) shows that $\widehat{m}$ and ϕ coincide up to constant multiples. This concludes the proof of the proposition. □

It follows from Azema, Duflo and Revuz [2] that the Harris property of $(X_t)_{t\geq 0}$ in continuous time is equivalent to the Harris property of the chain $(X_{S_n})_{n\in\mathbb{N}_0}$ which corresponds to observation of the continuous-time process after independent exponential waiting times (i.e. $S_n := \tau_1 + \cdots + \tau_n$ where $(\tau_j)_j$ are i.i.d. exponentially distributed and independent of X).

9.4 *One-dimensional Diffusions

For one-dimensional diffusions, it is an easy task to check Harris recurrence, in contrast to higher dimensions. We consider first diffusions with state space $(-\infty, \infty)$ and then – see [69, Sect. 5.5] – diffusions taking values in open intervals $I \subset \mathbb{R}$. The following criterion is widely known; we take it from from Khasminskii [73, 1st. ed., Chap. III.8, Ex. 2 on p. 105].

9.12 Proposition. In dimension $d = 1$, consider a continuous semi-martingale
$$dX_t = b(X_t)\,dt + \sigma(X_t)\,dW_t$$
where W is one-dimensional Brownian motion, and where $b(\cdot)$ and $\sigma(\cdot) > 0$ are continuous on $\mathbb{R}$.

(a) If the (strictly increasing) mapping $S: \mathbb{R} \to \mathbb{R}$ defined by

$$(*) \quad S(x) := \int_0^x s(y)\,dy \quad \text{where} \quad s(y) := \exp\left(-\int_0^y \frac{2b}{\sigma^2}(v)\,dv\right), \quad x, y \in \mathbb{R}$$

is a bijection onto $\mathbb{R}$, then the process $X = (X_t)_{t \geq 0}$ is recurrent in the sense of Harris, and the invariant measure m is given by

$$m(dx) = \frac{1}{s(x)}\frac{1}{\sigma^2(x)}\,dx = \frac{1}{\sigma^2(x)} \exp\left(\int_0^x \frac{2b}{\sigma^2}(v)\,dv\right)\,dx \quad \text{on } (\mathbb{R}, \mathcal{B}(\mathbb{R})).$$

The mapping $S(\cdot)$ defined by $(*)$ is called the *scale function*.

(b) In the special case $b(\cdot) \equiv 0$, we have $S(\cdot) = \text{id}$ on $\mathbb{R}$: thus a diffusion X without drift – for $\sigma(\cdot)$ continuous and strictly positive on $\mathbb{R}$ – is Harris with invariant measure $\frac{1}{\sigma^2(x)}\,dx$.

Proof. Let Q_x on $(D, \mathcal{D}, \mathbb{G})$ denote the law of X with starting point x. We give a proof in three steps, from Brownian motion via the driftless case to continuous semi-martingales as above.

(1) Special case $\sigma(\cdot) \equiv 1$ and $b(\cdot) \equiv 0$. Here X is one-dimensional Brownian motion starting from x, thus Lebesgue measure λ on $\mathbb{R}$ is an invariant measure since

$$\lambda P_t(f) = \int dx \int dy \, \frac{1}{\sqrt{2\pi t}} e^{-\frac{1}{2}\frac{(y-x)^2}{t}} f(y) = \int dy\, f(y) = \lambda(f).$$

For any Borel set A of positive Lebesgue measure and for any starting point x for X, select n large enough for $\lambda(A \cap B_{n-1}(0)) > 0$ and $x \in B_{n-1}(0)$, and define $\mathbb{G}$-stopping times

$$S_m := \inf\{t > T_{m-1} : \eta_t > n\}, \quad T_m := \inf\{t > S_m : \eta_t < -n\},$$
$$m \geq 0, \quad T_0 \equiv 0.$$

Section 9.4 *One-dimensional Diffusions

By the law of the iterated logarithm, these are finite Q_x-almost surely, and we have $T_m \uparrow \infty$ Q_x-almost surely as $m \to \infty$. Hence A is visited infinitely often by the canonical process η under Q_x, and Proposition 9.11 establishes Harris recurrence. By Theorem 9.5, $m = \lambda$ is the unique invariant measure, and we have null recurrence since m has infinite total mass on $\mathbb{R}$.

(2) *Special case* $b(\cdot) \equiv 0$. Here X is a diffusion without drift. On $(D, \mathcal{D}, \mathbb{G})$, let A denote the additive functional

$$A_t = \int_0^t \sigma^2(\eta_s)\, ds, \quad t \geq 0$$

and define $M := \eta_t - \eta_0$, $t \geq 0$. Then M is a continuous local martingale under Q_x with angle bracket $\langle M \rangle = A$. A result by Lepingle [85] states that Q_x-almost surely

on the event $R^c := \left\{ \lim_{t \to \infty} \langle M \rangle_t < \infty \right\}$, $\lim_{t \to \infty} M_t$ exists in $\mathbb{R}$.

Since $\sigma(\cdot)$ is continuous and strictly positive on $\mathbb{R}$, we infer that the event $R^c = \{\int_0^\infty \sigma^2(\eta_s)\, ds < \infty\}$ is a null set under Q_x. As a consequence, A is (strictly) increasing to ∞ Q_x-almost surely. Define

$$\tau(u) := \inf\{t > 0 : A_t > u\}, \quad 0 < u < \infty, \quad \tau_0 \equiv 0.$$

Then $\tau(u)$ tends to ∞ Q_x-almost surely as $u \to \infty$. As a consequence, the characterisation theorem of P. Lévy (cf. [61, p. 85]) shows that

$$B := \left(M_{\tau(u)}\right)_{u \geq 0} := \left(\eta_{\tau(u)} - \eta_0\right)_{u \geq 0}$$

is a standard Brownian motion. For $f \geq 0$ measurable, the time change gives

(+) $$\int_0^{\tau(v)} f(\eta_s)\,\sigma^2(\eta_s)\,ds = \int_0^{\tau(v)} f(\eta_s)\, dA_s$$
$$= \int_0^{\tau(v)} f(\eta_0 + B_{A_s})\, dA_s$$
$$= \int_0^v f(\eta_0 + B_r)\, dr.$$

By assumption on $\sigma(\cdot)$, we may replace measurable functions $f \geq 0$ in (+) by $\frac{f}{\sigma^2}$ and obtain

(++) $$\int_0^{\tau(v)} f(\eta_s)\, ds = \int_0^v \left(\frac{f}{\sigma^2}\right)(\eta_0 + B_r)\, dr, \quad 0 < v < \infty.$$

By step (1), one-dimensional Brownian motion is a Harris process. Letting v tend to ∞ in (++) and exploiting again continuity and strict positivity of $\sigma(\cdot)$, we obtain from (++)

$$A \in \mathcal{B}(\mathbb{R}), \; \lambda(A) > 0 \implies \int_0^\infty 1_A(\eta_s)\, ds = \infty \quad Q_x\text{-almost surely}.$$

This holds for an arbitrary choice of a starting point $x \in \mathbb{R}$. We thus have condition ($\diamond$) in Definition 9.4 with $\phi := \lambda$, thus the driftless diffusion X is Harris recurrent. It remains to determine the invariant measure m, unique by Theorem 9.5 up to constant multiples. We shall show
$$m(dx) = \frac{1}{\sigma^2(x)} dx .$$

Select $g \geq 0$ measurable with $m(g) = 1$. Then for measurable functions $f \geq 0$, the Ratio Limit Theorem 9.6 combined with (++) gives

$$m(f) = \lim_{t \to \infty} \frac{\int_0^t f(\eta_s) ds}{\int_0^t g(\eta_s) ds} = \lim_{v \to \infty} \frac{\int_0^{\tau(v)} f(\eta_s) ds}{\int_0^{\tau(v)} g(\eta_s) ds} = \lim_{v \to \infty} \frac{\int_0^v \frac{f}{\sigma^2}(\eta_0 + B_s) ds}{\int_0^v \frac{g}{\sigma^2}(\eta_0 + B_s) ds} .$$

By step (1) above, B is a Harris process whose invariant measure λ is translation invariant. Thus the ratio limit theorem for B applied to the last right-hand side shows

$$m(f) = \frac{\lambda\left(\frac{f}{\sigma^2}\right)}{\lambda\left(\frac{g}{\sigma^2}\right)} \quad \text{for all } f \geq 0 \text{ measurable} .$$

As a consequence we obtain $0 < \lambda\left(\frac{g}{\sigma^2}\right) < \infty$ (by definition of an invariant measure, m is σ-finite and non-null); then, up to multiplication with a constant, we identify m on $(\mathbb{R}, \mathcal{B}(\mathbb{R}))$ as $m(dx) = \frac{1}{\sigma^2(x)} dx$. This proves part (b) of the proposition.

(3) We consider a semi-martingale X admitting the above representation, and assume that $S : \mathbb{R} \to \mathbb{R}$ defined by (*) is a bijection onto $\mathbb{R}$. Let $\mathcal{L}$ denote the Markov generator of X acting on $\mathcal{C}^2$ functions

$$\mathcal{L}f(x) = b(x) f'(x) + \frac{1}{2}\sigma^2(x) f''(x), \quad x \in \mathbb{R} .$$

It follows from the definition in (*) that $\mathcal{L}S \equiv 0$. Thus $S \circ X$ is a local martingale by Ito formula. More precisely,

$$\widetilde{X} := S \circ X = (S(X_t))_{t \geq 0}$$

is a diffusion without drift as in step (2) solving the equation

$$d\widetilde{X}_t = \widetilde{\sigma}(\widetilde{X}_t) dW_t \quad \text{with} \quad \widetilde{\sigma} := (s \cdot \sigma) \circ S^{-1} .$$

It has been shown in step (2) that $\widetilde{X}$ is a Harris process with invariant measure $\widetilde{m}$ given by

$$\widetilde{m}(d\widetilde{x}) = \frac{1}{\widetilde{\sigma}^2(\widetilde{x})} d\widetilde{x} = \frac{1}{(s \cdot \sigma)^2(S^{-1}(\widetilde{x}))} d\widetilde{x} , \quad \widetilde{x} \in \mathbb{R}$$

and thus

($\diamond$) $\quad \widetilde{f}$ non-negative, $\mathcal{B}(\mathbb{R})$-measurable, $\widetilde{m}(\widetilde{f}) > 0 : \int_0^\infty \widetilde{f}(\widetilde{X}_t) dt = \infty$

almost surely, independently of the choice of a starting point for $\widetilde{X}$.

Section 9.4 *One-dimensional Diffusions

If we define a measure m on $(\mathbb{R}, \mathcal{B}(\mathbb{R}))$ as the image of $\widetilde{m}$ under the mapping $\widetilde{x} \to S^{-1}(\widetilde{x})$, then

$$(\text{oo}) \qquad m(f) = \int \widetilde{m}(d\widetilde{x}) \, f(S^{-1}(\widetilde{x})) = \widetilde{m}\left(f \circ S^{-1}\right)$$

and the transformation formula gives

$$m(dx) = \frac{1}{\widetilde{\sigma}^2(S(x))} s(x) \, dx = \frac{1}{\sigma^2(x)} \frac{1}{s(x)} \, dx, \quad x \in \mathbb{R}.$$

Now, for $f \geq 0$ measurable, we combine (o) and (oo) with the obvious representation

$$\int_0^t f(X_v) \, dv = \int_0^t \widetilde{f}(\widetilde{X}_v) \, dv \quad \text{for } \widetilde{f} := f \circ S^{-1}$$

in order to finish the proof. First, in this combination, the Harris property of $\widetilde{X}$ gives

$$f \text{ non-negative, } \mathcal{B}(\mathbb{R})\text{-measurable, } m(f) > 0: \quad \int_0^\infty f(X_t) \, dt = \infty$$

almost surely, for all choices of a starting point for X. This is property ($\diamond$) in Definition 9.4 with $\phi = m$. Thus X is a Harris process. Second, for $g \geq 0$ as in step 2 above and $\widetilde{g} := g \circ S^{-1}$ which implies $m(g) = \widetilde{m}(\widetilde{g}) = 1$, the ratio limit theorem

$$\lim_{t \to \infty} \frac{\int_0^t f(X_s) \, ds}{\int_0^t g(X_s) \, ds} = \lim_{t \to \infty} \frac{\int_0^t \widetilde{f}(\widetilde{X}_s) \, ds}{\int_0^t \widetilde{g}(\widetilde{X}_s) \, ds} = \widetilde{m}\left(f \circ S^{-1}\right) = m(f)$$

identifies m as the invariant measure for X. We have proved part (a) of the proposition. $\square$

9.12' Examples. The following examples (a)–(c) are applications of Proposition 9.12.

(a) The Ornstein–Uhlenbeck process $dX_t = \vartheta X_t \, dt + \sigma \, dW_t$ with parameters $\vartheta < 0$ and $\sigma > 0$ is positive Harris recurrent with invariant probability $m = \mathcal{N}(0, \frac{\sigma^2}{2|\vartheta|})$.

(b) The diffusion $dX_t = (\alpha - \beta X_t) \, dt + \sqrt{1 + X_t^2}^\gamma \, dW_t$ with parameters $\alpha \in \mathbb{R}$, $\beta > 0$, $0 < \gamma < 1$ is positive Harris recurrent. We have

$$-\int_0^x \frac{2b}{\sigma^2}(v) \, dv \sim \frac{\beta}{1-\gamma} |x|^{2(1-\gamma)} \quad \text{as } x \to +\infty \text{ or } x \to -\infty$$

with notations of ($*$) in Proposition 9.12. The invariant measure m of X admits finite moments of arbitrary order.

(c) The diffusion $dX_t = \vartheta \frac{X_t}{1+X_t^2} \, dt + \gamma \, dW_t$ with parameters $\gamma > 0$ and $\vartheta \leq \frac{\gamma^2}{2}$ is recurrent in the sense of Harris. Normed as in Proposition 9.12, the invariant measure is $m(dx) = \frac{1}{\gamma^2}\sqrt{1+x^2}^{\frac{2\vartheta}{\gamma^2}}\,dx$. Null recurrence holds if and only if $|\vartheta| \leq \frac{\gamma^2}{2}$.

The following is a variant of Proposition 9.12 for diffusions taking values in open intervals $I \subset \mathbb{R}$.

9.13 Proposition. In dimension $d = 1$, consider an open interval $I \subset \mathbb{R}$, write $\mathcal{B}(I)$ for the Borel-σ-field. Consider a continuous semi-martingale X taking values in I and admitting a representation

$$dX_t = b(X_t)\,dt + \sigma(X_t)\,dW_t, \quad t \geq 0.$$

We assume that $b(\cdot)$ and $\sigma(\cdot) > 0$ are continuous functions $I \to \mathbb{R}$. Fix any point $x_0 \in I$. If the mapping $S : I \to \mathbb{R}$ defined by

(*) $\qquad S(x) := \int_{x_0}^{x} s(y)\,dy \quad$ where $\quad s(y) := \exp\left(-\int_{x_0}^{y} \frac{2b}{\sigma^2}(v)\,dv\right)$

maps I onto $\mathbb{R}$, then X is recurrent in the sense of Harris, and the invariant measure m on $(I, \mathcal{B}(I))$ is given by

$$m(dx) = \frac{1}{\sigma^2(x)}\frac{1}{s(x)}\,dx = \frac{1}{\sigma^2(x)}\exp\left(\int_{x_0}^{x}\frac{2b}{\sigma^2}(v)\,dv\right)dx, \quad x \in I.$$

Proof. This is a modification of part (3) of the proof of Proposition 9.12 since I is open. For the I-valued semi-martingale X, the transformed process $\widetilde{X} = S \circ X$ is an $\mathbb{R}$-valued diffusion without drift as in step (2) of the proof of Proposition 9.12, thus Harris, with invariant measure $\widetilde{m}$ on $(\mathbb{R}, \mathcal{B}(\mathbb{R}))$. Defining m from $\widetilde{m}$ by

$$m(f) := \widetilde{m}\left(f \circ S^{-1}\right) \quad \text{for } f : I \to [0, \infty) \,\, \mathcal{B}(I)\text{-measurable}$$

we obtain a σ-finite measure on $(I, \mathcal{B}(I))$ such that sets $A \in \mathcal{B}(I)$ with $m(A) > 0$ are visited infinitely often by the process X, almost surely, under every choice of a starting value in I. Thus X is Harris, and the ratio limit theorem identifies m as the invariant measure of X. $\square$

9.13' Example. For every choice of a starting point in $I := (0, \infty)$, the Cox–Ingersoll–Ross process $dX_t = (\alpha - \beta X_t)\,dt + \gamma\sqrt{X_t}\,dW_t$ with parameters $\beta > 0$, $\gamma > 0$ and $\frac{2\alpha}{\gamma^2} \geq 1$ almost surely never hits 0 (e.g. Ikeda and Watanabe [61, p. 235–237]). Proposition 9.13 shows that X is positive Harris recurrent. The invariant probability m is a Gamma law $\Gamma(\frac{2\alpha}{\gamma^2}, \frac{2\beta}{\gamma^2})$. $\square$

9.13" Example. On some $(\Omega, \mathcal{A}, P)$, for some parameter $\vartheta \geq 0$ and for deterministic starting point in $I := (0, 1)$, consider the diffusion

$$dX_t = \vartheta\left(\frac{1}{2} - X_t\right)dt + \sqrt{X_t(1 - X_t)}\,dW_t$$

Section 9.4 *One-dimensional Diffusions

up to the stopping time $\zeta := \inf\{t > 0 : X_t \notin (0, 1)\}$. Then the following holds:

$$P(\zeta = \infty) = 1 \text{ in the case where } \vartheta \geq 1,$$
$$P(\zeta < \infty) = 1 \text{ in the case where } 0 \leq \vartheta < 1.$$

Only in the case where $\vartheta \geq 1$ is the process X a diffusion taking values in I in the sense of Proposition 9.13. For all $\vartheta \geq 1$, X is positive Harris, and the invariant probability on $(0, 1)$ is the Beta law $B(\vartheta, \vartheta)$.

Proof. (1) We fix $\vartheta \geq 0$. Writing $b(x) = \vartheta(\frac{1}{2} - x)$ and $\sigma^2(x) = x(1 - x)$ for $0 < x < 1$, we have

$$\int_{x_0}^{y} \frac{2b}{\sigma^2}(v)\,dv = \vartheta \int_{x_0}^{y} \frac{1 - 2v}{v(1 - v)}\,dv = \vartheta \ln\left(\frac{y(1 - y)}{x_0(1 - x_0)}\right), \quad 0 < y < 1$$

with respect to any fixed $x_0 \in (0, 1)$. The function $s(\cdot)$ in $(*)$ of Proposition 9.13, given by

$$s(y) = \left(\frac{y(1 - y)}{x_0(1 - x_0)}\right)^{-\vartheta}, \quad 0 < y < 1,$$

behaves as $c\, y^{-\vartheta}$ for $y \downarrow 0$ and as $c\,(1 - y)^{-\vartheta}$ for $y \uparrow 1$, with some constant c which depends on ϑ and on x_0. Hence $S(\cdot)$ of $(*)$ of Proposition 9.13 is such that

$$S(0^+) = -\infty, \quad S(1^-) = +\infty \quad \text{in the case where } \vartheta \geq 1,$$
$$S(0^+) > -\infty, \quad S(1^-) < +\infty \quad \text{for } 0 \leq \vartheta < 1.$$

Define a σ-finite measure m on $(I, \mathcal{B}(I))$ by

$$m(dx) := 1_{(0,1)}(x) \frac{1}{s(x)} \frac{1}{\sigma^2(x)}\,dx = 1_{(0,1)}(x)\,[x_0(1 - x_0)]^{-\vartheta}\,[x(1 - x)]^{(\vartheta - 1)}\,dx.$$

Up to a factor 2, this is the 'speed measure' in terms of Karatzas and Shreve [69, p. 343]. In the case where $\vartheta = 0$, this measure has infinite total mass $m(I) = \infty$ such that

$$m((x, x_0]) \sim c\,|\log(x)| \quad \text{as } x \downarrow 0, \qquad m(([x_0, x)) \sim c\,|\log(1 - x)| \quad \text{as } x \uparrow 1$$

for some constant c depending on x_0 and ϑ. In the case where $\vartheta > 0$, m has finite total mass and – arranging for the norming factor – equals the Beta law $B(\vartheta, \vartheta)$ on

$(0, 1)$. We shall also need the function

$$V(x) := \int_{x_0}^x dy\, s(y) \int_{x_0}^y \frac{1}{s(z)\sigma^2(z)}\, dz = \int_{x_0}^x [S(x) - S(z)]\, m(dz), \quad 0 < x < 1$$

as in [69, p. 347]).

(2) In the case where $\vartheta = 0$, X is a local martingale, we have $s(\cdot) \equiv 1$ on $(0, 1)$ and thus

$$V(x) = \begin{cases} \int_{x_0}^x dy\, m([x_0, y)) & \text{if } x_0 < x < 1 \\ \int_x^{x_0} dy\, m((y, x_0]) & \text{if } 0 < x < x_0 \end{cases}$$

by definition. Since $|\log(\cdot)|$ is integrable on small neighbourhoods of 0^+, we have for $\vartheta = 0$

(+) $$V(0^+) > -\infty, \quad V(1^-) < +\infty.$$

According to [69, p. 350)], (+) implies $P(\zeta < \infty) = 1$. Thus in the case where $\vartheta = 0$, almost surely, the martingale $X^\zeta = (X_{t \wedge \zeta})_{t \geq 0}$ is absorbed at the boundary of I in finite time.

(3) Consider the case $0 < \vartheta < 1$ where both $m(I)$ and $S(1^-) - S(0^+)$ are finite. By definition of $V(\cdot)$, this again leads to (+) and thus to $P(\zeta < \infty) = 1$: X hits the boundary of I in finite time.

(4) Consider the case $\vartheta \geq 1$ where $S(\cdot)$ maps I onto $\mathbb{R}$. In this case, [69, p. 345] shows that X has infinite life time in I. Then Proposition 9.13 applies and yields Harris recurrence of X with invariant measure m. Since up to constants m is a probability law, X is positive recurrent. □

The following result on null recurrent diffusions without drift provides exact constants for weak convergence of additive functionals, both for the invariant measure and for the norming functions in Theorem 9.8. It was proved in Khasminskii [73]. We quote part (a) from [73, 2nd ed., pp. 129–130] and part (b) from [73, 2nd ed. pp. 134–136]), embedding it into the Harris setting of Proposition 9.12.

9.14 Proposition. In dimension $d = 1$, for $\sigma(\cdot) > 0$ locally Lipschitz on $\mathbb{R}$ and satisfying a linear growth condition, consider the Harris process $\widetilde{X}$ solution to

$$d\widetilde{X}_t = \widetilde{\sigma}(\widetilde{X}_t)\, dW_t, \quad t \geq 0$$

taking values in $(-\infty, \infty)$. Normed as in Proposition 9.12(b), consider the invariant measure $\widetilde{m}(dx) = \frac{1}{\sigma^2(x)}\, dx$ of $\widetilde{X}$ together with the sequence of stopping times $\tau(k) \uparrow \infty$ as $k \to \infty$:

$$\tau(k) := \inf\{t > \rho(k) : \widetilde{X}_t < 0\}, \quad \rho(k) := \inf\{t > \tau(k-1) : \widetilde{X}_t > 1\},$$
$$k \geq 1, \quad \tau(0) \equiv 0.$$

Section 9.4 *One-dimensional Diffusions

(a) For $k \geq 1$, the i.i.d excursions $\widetilde{X}1_{[[\tau(k),\tau(k+1)]]}$ away from 0 are such that for $f \geq 0$ measurable

$$E\left(\int_{\tau(k)}^{\tau(k+1)} f(\widetilde{X}_s)\, ds\right) = 2\widetilde{m}(f).$$

(b) Assume that there is $\beta > -1$ and constants $A_+ + A_- > 0$ such that the limits

$$\lim_{x \to \pm\infty} \frac{1}{x} \int_0^x |v|^{-\beta} \frac{2}{\widetilde{\sigma}^2(v)}\, dv =: A_\pm \quad \in [0, \infty)$$

exist. Then $\widetilde{X}$ is null recurrent. Defining $\alpha := \frac{1}{\beta+2}$ we have $0 < \alpha < 1$ and regular variation

$$P(\tau(2)-\tau(1) > t) \sim \frac{\alpha^{2\alpha}\,([A_+]^\alpha + [A_-]^\alpha)}{\Gamma(1+\alpha)} \cdot t^{-\alpha} \quad \text{as } t \to \infty.$$

(On the last right-hand side, we have corrected a typing error in formula (4.101) of [73, 2nd ed.], in accordance with (4.110), (4.111), (4.84) and Lemma 4.19 in [73, 2nd ed.]; see also Bingham, Goldie and Teugels [12, p. 349].

Bibliography

[1] T. Anderson, The integral of a symmetric unimodal function over a symmetric convex set and some probability inequalities, *Proc. Amer. Math. Soc.* **6** (1955), 170–176.

[2] J. Azéma, M. Duflo, D. Revuz, *Mesures invariantes des processus de Markov récurrents*, Séminaire de Probabilités III, Lecture Notes in Mathematics **88**, Springer, 1969.

[3] O. Barndorff-Nielsen, *Parametric Statistical Models and Likelihood.* Lecture Notes in Statistics **50**, Springer, 1988.

[4] J. Barra, *Mathematical Basis of Statistics.* Academic Press 1981. French ed. Dunod, 1971.

[5] I. Basawa, J. Scott, *Asymptotic Optimal Inference for Non-ergodic Models*, Lecture Notes in Statistics **17**, Springer, 1983.

[6] R. Bass, *Diffusions and Elliptic Operators,* Springer, 1998.

[7] H. Bauer, *Wahrscheinlichkeitstheorie und Grundzüge der Masstheorie,* 3rd ed., De Gruyter, 1978.

[8] M. Ben Alaya, A. Kebaier, Parameter estimation for the square root diffusions: ergodic and nonergodic cases, *Stochastic Models* **28** (2012), 609–634.

[9] R. Beran, Estimating a distribution function, *Ann. Statist.* **5**, 400-404 (1977).

[10] J. Bertoin, *Lévy Processes,* Cambridge University Press, 1996.

[11] P. Billingsley, *Convergence of Probability Measures,* Wiley 1968.

[12] N. Bingham, C. Goldie, J. Teugels, *Regular Variation,* Cambridge University Press 1987.

[13] C. Boll, *Comparison of Experiments in the Infinite Case,* PhD Thesis, Stanford University, 1955.

[14] P. Brémaud, *Point Processes and Queues*, Springer, 1981.

[15] K. Chung, R. Williams, *Introduction to Stochastic Integration,* 2nd ed., Birkhäuser, 1990.

[16] H. Cremers, D. Kadelka, On weak convergence of integral functionals of stochastic processes with application to processes taking paths in L_p^E, *Stoch. Processes Appl.* **21** (1986), 305–317.

[17] S. Dachian, On limiting likelihood ratio processes of some change-point type statistical models, *Journal of Statistical Planning and Inference* **140** (2010), 2682–2692.

[18] D. Darling, M. Kac, On occupation times for Markov processes. *Trans. Amer. Math. Soc.* **84** (1957), 444–458.

[19] R. Davies, Asymptotic inference when the amount of information is random, in: L. Le Cam, R. Olshen, (Eds): *Proc. of the Berkeley Symp. in Honour of J. Neyman and J. Kiefer,* Vol. II. Wadsworth, 1985.

[20] C. Dellacherie, P. Meyer, *Probabilités et potentiel*, Chap. I–IV, Hermann, 1975; Chap. V–VIII, Hermann, 1980.

[21] M. Diether, Wavelet estimation in diffusions with periodicity, *Statist. Inference Stoch. Processes* **15** (2012), 257–284.

[22] G. Dohnal, On estimating the diffusion coefficient. *J. Appl. Probab.* **24** (1987), 105–114.

[23] P. Feigin, Some comments concerning a curious singularity. *J. Appl. Probab.* **16** (1979), 440–444.

[24] W. Feller, *An Introduction to Probability Theory and its Applications*, Vol. 2, Wiley, 1971.

[25] V. Genon-Catalot, J. Jacod, On the estimation of the diffusion coefficient for multidimensional diffusion processes, *Ann. Inst. H. Poincaré Probab. Stat.* **29**, (1993), 119–151.

[26] V. Genon-Catalot, D. Picard, *Elements de statistique asymptotique,* Springer, 1993.

[27] O. Georgii, *Stochastik*, De Gruyter, 2002.

[28] I. Gihman, A. Skorohod, *The Theory of Stochastic Processes,* Vol. I+II, Springer, 1974 (Reprint 2004).

[29] R. Gill, B. Levit, Applications of the van Trees inequality: a Bayesian Cramér-Rao bound, *Bernoulli* **1** (1995), 59–79.

[30] E. Gobet, Local asymptotic mixed normality property for elliptic diffusion, *Bernoulli* **7** (2001), 899–912.

[31] E. Gobet, LAN property for ergodic diffusions with discrete observations, *Ann. Inst. H. Poincaré PR* **38** (2002), 711–737.

[32] Y. Golubev, Computation of efficiency of maximum likelihood estimate when observing a discontinuous signal in white noise, *Problems Inform. Transm.* **15** (1979), 38–52.

[33] P. Greenwood, A. N. Shiryaev, Asymptotic minimaxity of a sequential estimator for a first order autoregressive system, *Stochastics and Stochastics Reports* **38** (1992), 49-65.

[34] P. Greenwood, W. Wefelmeyer, Efficiency of empirical estimators for Markov chains, *Ann. Statist.* **23** (1995), 132–143.

[35] P. Greenwood, W. Ward, W. Wefelmeyer, Statistical analysis of stochastic resonance in a simple setting, *Phys. Rev. E* **60** (1999), 4687–4695.

[36] Greenwood, P., Wefelmeyer, W.: Asymptotic minimax results for stochastic process families with critical points, *Stoch. Proc. Appl.* **44**, (1993), 107–116.

[37] L. Grinblat, A limit theorem for measurable stochastic processes and its applications, *Proc. Amer. Math. Soc.* **61** (1976), 371–376.

[38] A. Gushchin, On asymptotic optimality of estimators under the LAQ condition, *Theory Probab. Appl.* **40** (1995), 261–272.

[39] A. Gushchin, U. Küchler, Asymptotic inference for a linear stochastic differential equation with time delay, *Bernoulli* **5**, (1999), 1059–1098.

[40] J. Hájek, A characterization of limiting distributions for regular estimators, *Z. Wahrscheinlichkeitsth. Verw. Geb.* **14** (1970), 323–330.

[41] J. Hájek, Z. Sidák, *Theory of Rank Tests*, Academic Press, 1967.

[42] J. Hájek, Z. Sidák, P. Sen, *Theory of Rank Tests* (2nd ed.), Academic Press, 1999.

[43] M. Hammer, *Parameterschätzung in zeitdiskreten ergodischen Markov-Prozessen am Beispiel des Cox-Ingersoll-Ross Modells*, Diplomarbeit, Institut für Mathematik, Universität Mainz 2005.
http://ubm.opus.hbz-nrw.de/volltexte/2006/1154/pdf/diss.pdf

[44] M. Hammer, *Ergodicity and regularity of invariant measure for branching Markov processes with immigration*, PhD Thesis, Institute of Mathematics, University of Mainz, 2012.
http://ubm.opus.hbz-nrw.de/volltexte/2012/3306/pdf/doc.pdf

[45] T. Harris, The existence of stationary measures for certain Markov processes.,*Proc. 3rd Berkeley Symp.,* Vol. II, pp. 113–124, University of California Press, 1956.

[46] R. Höpfner, Asymptotic inference for continuous-time Markov chains, *Probab. Theory Rel. Fields* **77** (1988), 537–550.

[47] R. Höpfner, Null recurrent birth-and-death processes, limits of certain martingales, and local asymptotic mixed normality, *Scand. J. Statist.* **17** (1990), 201–215.

[48] R. Höpfner, On statistics of Markov step processes: representation of log-likelihood ratio processes in filtered local models, *Probab. Theory Rel. Fields* **94** (1993), 375–398.

[49] R. Höpfner, Asymptotic inference for Markov step processes: observation up to a random time, *Stoch. Proc. Appl.* **48** (1993), 295–310.

[50] R. Höpfner, J. Jacod, Some remarks on the joint estimation of the index and the scale parameter for stable prozesses, in: Mandl and Huskova (Eds.), *Asymptotic Statistics*, Proc. Prague 1993, pp. 273–284, Physica Verlag, 1994.

[51] R. Höpfner, J. Jacod, L. Ladelli, Local asymptotic normality and mixed normality for Markov statistical models, *Probab. Theory Rel. Fields* **86** (1990), 105–129.

[52] R. Höpfner, E. Löcherbach, Remarks on ergodicity and invariant occupation measure in branching diffusions with immigration, *Ann. Inst. Henri Poincaré* **41** (2005), 1025–1047.

[53] R. Höpfner, E. Löcherbach, *Limit Theorems for Null Recurrent Markov Processes*, Memoirs AMS **161**, American Mathematical Society, 2003.

[54] R. Höpfner, Y. Kutoyants, On a problem of statistical inference in null recurrent diffusions, *Statist. Inference Stoch. Process.* **6** (2003), 25–42.

[55] R. Höpfner, Y. Kutoyants, On LAN for parametrized continuous periodic signals in a time inhomogeneous diffusion, *Statistics & Decisions* **27** (2009), 309–326.

[56] R. Höpfner, Y. Kutoyants, Estimating discontinuous periodic signals in a time inhomogeneous diffusion, *Statist. Inference Stoch. Process.* **13** (2010), 193–230.

[57] R. Höpfner, Y. Kutoyants, Estimating a periodicity parameter in the drift of a time inhomogeneous diffusion, *Math. Meth. Statist.* **20** (2011), 58–74.

[58] R. Höpfner, L. Rüschendorf, Comparison of estimators in stable models, *Mathematical and Computer Modelling* **29** (1999), 145–160.

[59] R. Iasnogorodski, H. Lhéritier, *Théorie de l'estimation ponctuelle paramétrique,* EDP Sciences, 2003.

[60] I. Ibragimov, R. Khasminskii, *Statistical Estimation,* Springer, 1981.

[61] N. Ikeda, N. Watanabe, *Stochastic Differential Equations and Diffusion Processes,* North-Holland / Kodansha, 2nd ed., 1989.

[62] K. Ito, H. McKean, *Diffusion Processes and their Sample Paths,* Springer, 1965.

[63] J. Jacod, Multivariate point processes: predictable projection, Radon-Nikodym derivatives, representation of martingales, *Z. Wahrscheinlichkeitsth. Verw. Geb.* **31** (1975), 2435–253.

[64] J. Jacod, A. Shiryaev, *Limit Theorems for Stochastic Processes,* Springer 1987, 2nd ed., 2003.

[65] A. Janssen, *Zur Asymptotik nichtparametrischer Tests,* Vorlesungsskript. Düsseldorf, 1998.

[66] P. Jeganathan, On the asymptotic theory of estimation when the limit of the log-likelihood ratios is mixed normal, *Sankhya Ser. A* **44** (1982), 173–212.

[67] P. Jeganathan, Some aspects of asymptotic theory with applications to time series models. Preprint version, 1988. *Econometric Theory* **11**, 818–887 (1995).

[68] Y. Kabanov, R. Liptser, A. Shiryaev, Criteria for absolute continuity of measures corresponding to multivariate point processes, in: J. Prokhorov (Ed.), *Proc. Third Japan-USSR Symposium,* Lecture Notes in Math. **550**, pp. 232–252, Springer, 1976.

[69] I. Karatzas, S. Shreve, *Brownian Motion and Stochastic Calculus,* 2nd ed. Springer, 1991.

[70] M. Kessler, Estimation of an ergodic diffusion from discrete observations, *Scand. J. Statist.* **24** (1997), 211–229.

[71] M. Kessler, A. Lindner, M. Sørensen (Eds.), *Statistical Methods for Stochastic Differential Equations,* CRC Press, 2012.

[72] M. Kessler, A. Schick, W. Wefelmeyer, The information in the marginal law of a Markov chain, *Bernoulli* **7** (2001), 342–266.

[73] R. Khasminskii, *Stochastic Stability of Differential Equations,* 1st ed., Sijthoff und Noordhoff, 1980; 2nd ed., Springer, 2012.

[74] R. Khasminskii, G. Yin, Asymptotic behavior of parabolic equations arising from one-dimensional null recurrent diffusions, *J. Diff. Eqns.* **161** (2000), 154–173.

[75] A. Klenke, *Wahrscheinlichkeitstheorie,* 3rd ed., Springer, 2013.

[76] U. Küchler, Y. Kutoyants, Delay estimation for some stationary diffusion-type processes, *Scand. J. Statist.* **27** (2000), 405–414.

[77] U. Küchler, M. Sørensen, *Exponential Families of Stochastic Processes*, Springer, 1997.

[78] Y. Kutoyants, *Identification of Dynamical Systems with Small Noise*, Kluwer, 1994.

[79] Y. Kutoyants, *Statistical Inference for Spatial Poisson Processes*, Springer, 1998.

[80] Y. Kutoyants, *Statistical Inference for Ergodic Diffusion Processes*, Springer, 2004.

[81] L. Le Cam, *Théorie asymptotique de la décision statistique*, Montréal, 1969.

[82] L. Le Cam, Limits of experiments, in: *Proc. 6th Berkeley Symposium Math. Statist. Probab.* Vol. I, pp. 245–261, University of California Press, 1972.

[83] L. Le Cam, Maximum likelihood: an introduction, *Intern. Statist. Rev.* **58** (1990), 153–171.

[84] L. Le Cam, G. Yang, *Asymptotics in Statistics: Some Basic Concepts,* Springer 1990, 2nd ed., 2002.

[85] D. Lepingle, Sur le comportement asymptotique des martingales locales, in: *Séminaire de Probabilités XII*, Lecture Notes in Mathematics **649**, pp. 148–161, Springer, 1978.

[86] F. Liese, K. Miescke, *Statistical Decision Theory*, Springer, 2008.

[87] F. Liese, I. Vajda, *Convex Statistical Distances*, Teubner, 1987.

[88] R. Liptser, A. Shiryaev, *Statistics of Random Processes* Vol. I+II, Springer, 1981, 2nd ed., 2001.

[89] M. Loève, *Probability Theory*, 3rd ed. van Nostrand 1963; 4th ed. Vol. I+II Springer 1978.

[90] E. Löcherbach, Likelihood ratio processes for Markovian particle systems with killing and jumps, *Statist. Inference Stoch. Process.* **5** (2002), 153–177.

[91] E. Löcherbach, LAN and LAMN for systems of interacting diffusions with branching and immigration, *Ann. Inst. Henri Poincare (B) Probab. Stat.* **38** (2002), 59–90.

[92] E. Löcherbach, Smoothness of the intensity measure density for interacting branching diffusions with immigration, *J. Funct. Analysis* **215** (2004), 130–177.

[93] D. Loukianova, E. Löcherbach, On Nummelin splitting for continuous time Harris recurrent Markov processes and application to kernel estimation for multi-dimensional diffusions, *Stoch. Proc. Appl.* **118** (2008), 1301–1321.

[94] D. Loukianova, O. Loukianov, On deterministic equivalents of additive functionals of recurrent diffusions and drift estimation, *Statist. Inference Stoch. Process.* **11** (2008), 107–121.

[95] H. Luschgy, Local asymptotic mixed normality for semimartingale experiments, *Probab. Theory Rel. Fields* **92** (1992), 151–176.

[96] H. Luschgy, Asymptotic inference for semimartingale models with singular parameter points, *J. Statist. Plann. Inference* **39** (1994), 155–186.

[97] H. Luschgy, Local asymptotic quadraticity of stochastic process models based on stopping times, *Stoch. Proc. Appl.* **57** (1995), 305–317.

[98] M. Métivier, *Semimartingales*, De Gruyter, 1982.

[99] P. Millar, The minimax principle in asymptotic statistical theory, in: P. Hennequin (Ed), *Ecole d' été de probabilités de St. Flour XI 1981*, pp. 75–265, Springer, 1983.

[100] P. Millar, A general approach to the optimality of minimum distance estimators, *Trans. Amer. Math. Soc.* **286** (1984), 377–418.

[101] E. Nummelin, *General Irreducible Markov Chains and Non-negative Operators*, Cambridge University Press, 1985.

[102] E. Nummelin, A splitting technique for Harris recurrent Markov chains, *Z. Wahrscheinlichkeitsth. Verw. Geb.* **43** (1978), 309–318.

[103] L. Overbeck, Estimation for continuous branching processes, *Scand. J. Statist.* **25** (1998), 111–126.

[104] L. Overbeck, T. Ryden, Estimation in the Cox-Ingersoll-Ross model, *Econometric Theory* **13** (1997), 430–461.

[105] J. Pfanzagl, *Parametric Statistical Theory*, De Gruyter, 1994.

[106] G. Pflug, The limiting log-likelihood process for discontinuous density families, *Z. Wahrschein. Verw. Gebiete* **64** (1983), 15–35.

[107] B. Prakasa Rao, *Asymptotic theory of statistical inference*, Wiley, 1987.

[108] B. Prakasa Rao, *Statistical Inference for Diffusion Type Processes*, Arnold, 1999.

[109] B. Prakasa Rao, *Semimartingales and their Statistical Inference*, Chapman and Hall CRC, 1999.

[110] S. Resnick, P. Greenwood, A bivariate stable characterization and domains of attraction, *J. Multivar. Analysis* **9** (1979), 206–221.

[111] D. Revuz, *Markov Chains*, Rev. ed. North-Holland, 1984.

[112] D. Revuz, M. Yor, *Continuous Martingales and Brownian Motion*, Springer, 1991.

[113] G. Roussas, *Contiguity of Probability Measures*, Cambridge University Press, 1972.

[114] H. Rubin, K. Song, Exact computation of the asymptotic effiency of maximum likelihood estimators of a discontinuous signal in Gaussian white noise, *Ann. Statist.* **23** (1995), 732–739.

[115] L. Rüschendorf, *Asymptotische Statistik,* Teubner, 1988.

[116] A. Schick, W. Wefelmeyer, Estimating joint distributions of Markov chains, *Statist. Inference Stoch. Proc.* **5** (2002), 1–22.

[117] R. Schilling, *Measures, Integrals and Martingales*, Cambridge University Press, 2005.

[118] Y. Shimizu, N. Yoshida, Estimation of parameters for diffusion processes with jumps from discrete observations, *Statist. Inference Stoch. Proc.* **9** (2006), 227–277.

[119] M. Sørensen, Likelihood methods for diffusion with jumps, in: Prabhu, Basawa (Eds.), *Statistical Inference in Stochastic Processes*, pp. 67–105, Marcel Dekker, 1991.

[120] H. Strasser, *Einführung in die lokale asymptotische Theorie der Statistik*, Bayreuther Mathematische Schriften, 1985.

[121] H. Strasser, *Mathematical Theory of Statistics*, De Gruyter, 1985.

[122] A. Shiryaev, V. Spokoiny, *Statistical Experiments and Decisions*, World Scientific, 2001.

[123] A. Touati, Théorèmes limites pour les processus de Markov récurrents. Unpublished paper 1988. See also *C. R. Acad. Sci. Paris Série I* **305** (1987), 841–844.

[124] A. Tsybakov, *Introduction à l'estimation non-parametrique*, Springer SMAI, 2004.

[125] A. van der Vaart, An asymptotic representation theorem, *Int. Statist. Rev.* **59** (1991), 97–121.

[126] A. van der Vaart, *Asymptotic Statistics*, Cambridge University Press, 1998.

[127] H. Witting, *Mathematische Statistik I*, Teubner, 1985.

[128] H. Witting, U. Müller-Funk, *Mathematische Statistik II*, Teubner, 1995.

[129] N. Yoshida, Estimation for diffusion processes from discrete observations, *J. Multivar. Anal.* **41** (1992), 220–242.

[130] H. van Zanten, On the rate of convergence of the maximum likelihood estimator in Brownian semimartingale models, *Bernoulli* **11** (2005), 643–664.

[131] V. Zolotarev, *One-dimensional Stable Distributions*, Transl. of Mathematical Monographs **65**, Amer. Math. Soc., 1986.

Index

additive functionals, martingale additive functionals
  definition of, 248, 251
  tightness rates, 249
  regular variation condition, 250, 252, 253, 265
  weak convergence of, 250, 252, 253
approximately equivariant, approximately strongly equivariant, 137, 157

Bayes estimators, 30, 131, 132, 134, 192, 210, 211, 242
Brownian bridges, 69, 70, 79
Brownian motion with unknown drift
  statistical model for, 143, 145
  time change, stopping times, 142. 168, 169
  independent time change, 170, 172, 223, 237

canonical path spaces (C, C, G) or (D, D, G), canonical process, 143, 173, 247
central sequence
  definition of, 181
  results on, 190–192, 194, 196, 205
  else, 200, 224, 237, 242
central statistic
  definition of, 129, 149
  results on, 136, 139, 156, 159
  else, 130–133, 147, 149, 153, 157, 172, 174, 176, 181, 211, 242
contiguity, 87, 92
convergence of martingales, 244, 246, 247
convolution theorem, 133, 154, 157, 193
coupling property, 190, 191, 194, 196, 198, 200, 205

density process (likelihood ratio process)
  definition of, 142
  else, 106, 107, 144, 146, 162, 165, 166, 173, 234

efficient estimator sequences, 194, 210
empirical distribution function, 44, 55, 78, 80
equivariant estimators, strongly equivariant estimators
  definition of, 131, 152
  results on, 133, 136, 154, 156
  else, 131, 132, 147, 192, 242
estimator sequences, 31, 58, 189, 190, 196, 205

filtered statistical experiment, 141, 142

Gaussian martingale, conditionally Gaussian martingale, 244, 245
Gaussian processes, μ-Gaussian processes, 68–70, 77, 79
Gaussian shift models
  definition of, 129, 181
  results on, 131–133, 136, 137, 139, 206
  else, 119, 121, 144, 146, 147, 219, 226

Harris recurrent Markov processes (positive or null)
  setting, definition of, 247, 248
  Harris conditions, 253, 255, 258, 262
  invariant measure, 247, 248, 258, 262, 264
  ratio limit theorem, 248
  else, 213, 214, 228, 261–263
Hellinger distance, 33, 37, 118

information, 117, 118, 129, 152
information process, 217, 234

information with estimated parameter, 203
integrals along the paths of a process, 44, 48, 73

Kullback divergence, 32

L^r-differentiable statistical models, 110, 111, 114, 116–118, 121, 206
LAMN (local asymptotic mixed normality), 180, 192–194, 196, 205, 223, 237
LAN (local asymptotic normality), 121, 181, 191–194, 196, 205–208, 219
LAQ (local asymptotic quadraticity), 180, 183, 186, 189–191, 200, 205, 220, 222
Le Cam's first lemma, 88, 96, 99
Le Cam's second lemma, 121, 206
Le Cam's third lemma, 90, 92, 100
Lebesgue decomposition, 86
likelihood ratio, 86
limit experiment at ϑ, 181
local model at ϑ, 181
local parameter at ϑ, 181
local scale at ϑ, 181
location models, 39, 81, 111, 208
log-likelihood ratio, 86

Markov extension of a statistical model, 186
maximum likelihood (ML) estimators
 definition of, 31
 results on, 37, 39, 131, 149, 191, 205
 else, 25, 29, 132, 200, 209, 216, 217, 221, 223, 225, 233, 237
measurable selection, 59
minimax theorem, local asymptotic minimax theorem, 139, 159, 196
minimum distance (MD) estimators, 58, 60, 63, 64, 77

Mittag–Leffler process V^α, 175, 176, 229, 235, 237, 250, 252
mixed normal experiment
 definition of, 150
 results on, 150, 152–154, 156, 157, 159
 else, 152, 157, 168, 170, 173, 176, 180, 223, 224, 238

one-parametric paths, 23, 112, 198, 207
one-sided stable (stable increasing) process S^α, 175, 176
one-step modification, 200, 205, 208, 239

quadratic experiment
 definition of, 149, 152
 results on, 149, 191, 205
 else, 160, 162, 165–167, 169, 180, 189, 200, 220, 222
quadratic variation of a continuous semimartingale, 167

regular and efficient estimator sequences, 194, 198, 205, 208, 219, 224, 227, 237
regular estimator sequences, 191–194, 210

score, 117, 129, 152
score martingale, 217, 234
score with estimated parameter, 204
stochastic differential equations (SDE)
 setting for, 161
 laws of solutions to, 161, 162, 165–167, 216, 233
stochastic processes with paths in L^p, 45, 47, 56, 69, 70, 73, 77
subconvex loss functions, 134, 135

total variation distance, 137

weak convergence in L^p-path spaces, 45, 47, 79

www.ingramcontent.com/pod-product-compliance
Lightning Source LLC
Chambersburg PA
CBHW081544300426
44116CB00015B/2748